THE SURVIVAL OF SOAP OPERA

THE SURVIVAL OF SOAP OPERA
TRANSFORMATIONS FOR A NEW MEDIA ERA

EDITED BY
SAM FORD, ABIGAIL DE KOSNIK, AND C. LEE HARRINGTON

UNIVERSITY PRESS OF MISSISSIPPI / JACKSON

www.upress.state.ms.us

The University Press of Mississippi is a member
of the Association of American University Presses.

Copyright © 2011 by University Press of Mississippi
All rights reserved
Manufactured in the United States of America

First printing 2011
∞
Library of Congress Cataloging-in-Publication Data

The survival of soap opera : transformations for a new
media era / edited by Sam Ford, Abigail De Kosnik, and
C. Lee Harrington.
    p. cm.
  Includes interviews that provide various perspectives of
the soap opera television.
  Includes bibliographical references and index.
  ISBN 978-1-60473-716-5 (cloth : alk. paper) — ISBN 978-
1-60473-717-2 (ebook) 1. Television soap operas—United
States—History and criticism. 2. Television soap operas—
Social aspects—United States. 3. Television viewers—
United States. I. Ford, Sam. II. Kosnik, Abigail De. III.
Harrington, C. Lee, 1964–
  PN1992.8.S4S87 2010
  791.45'6—dc22   2010020366

British Library Cataloging-in-Publication Data available

CONTENTS

ix Acknowledgments

SECTION ONE
CHALLENGES TO THE FUTURE OF SOAPS

3   Introduction: The Crisis of Daytime Drama and What It Means for the Future of Television
    —SAM FORD, ABIGAIL DE KOSNIK, AND C. LEE HARRINGTON

22 Perspective: Scholars Barbara Irwin and Mary Cassata on the State of U.S. Soap Operas
    (BASED ON AN INTERVIEW BY C. LEE HARRINGTON)

29 Perspective: Historian William J. Reynolds on Memories of *The Edge of Night*
    (BASED ON AN INTERVIEW BY SAM FORD)

31 Perspective: Writer Patrick Mulcahey on Changes in Soap Opera Writing Contracts
    (BASED ON AN INTERVIEW BY GIADA DA ROS)

34 Perspective: Actor Tristan Rogers on Changes in Soaps' Industry, Audiences, and Texts
    (BASED ON AN INTERVIEW BY ABIGAIL DE KOSNIK)

38 Daytime Budget Cuts
    —SARA A. BIBEL

44 Agnes Nixon and Soap Opera "Chemistry Tests"
    —CAROL TRAYNOR WILLIAMS

49 Giving Soaps a Good Scrub: ABC's *Ugly Betty* and the Ethnicity of Television Formats
    —JAIME J. NASSER

58 The Way We Were: The Institutional Logics of Professionals and Fans in the Soap Opera Industry
—MELISSA C. SCARDAVILLE

SECTION TWO
CAPITALIZING ON HISTORY

81 Perspective: Scholar Horace Newcomb on the Pleasures and Influence of Soaps
(BASED ON AN INTERVIEW BY SAM FORD)

83 Perspective: Scholar Robert C. Allen on Studying Soap Operas
(BASED ON AN INTERVIEW BY C. LEE HARRINGTON)

86 Growing Old Together: Following *As the World Turns*' Tom Hughes through the Years
—SAM FORD

101 Perspective: Writer Kay Alden on What Makes Soaps Unique
(BASED ON AN INTERVIEW BY SAM FORD)

104 Perspective: Scholar Nancy Baym on Soaps after the O. J. Simpson Trial
(BASED ON AN INTERVIEW BY ABIGAIL DE KOSNIK)

106 Of Soap Operas, Space Operas, and Television's Rocky Romance with the Feminine Form
—CHRISTINE SCODARI

119 The Ironic and Convoluted Relationship between Daytime and Primetime Soap Operas
—LYNN LICCARDO

130 Perspective: Scholar Louise Spence on Comparing the Soap Opera to Other Forms
(BASED ON AN INTERVIEW BY ABIGAIL DE KOSNIK)

133 Perspective: Scholar Jason Mittell on the Ties between Daytime and Primetime Serials
(BASED ON AN INTERVIEW BY SAM FORD)

140 Preserving Soap History: What Will It Mean for the Future of Soaps?
—MARY JEANNE WILSON

154 Did the 2007 Writers Strike Save Daytime's Highest-Rated Drama?
—J. A. METZLER

SECTION THREE
EXPERIMENTING WITH PRODUCTION AND DISTRIBUTION

163 "The Rhetoric of the Camera in Television Soap Opera" Revisited: The Case of *General Hospital*
—BERNARD M. TIMBERG AND ERNEST ALBA

175 It's Not All Talk: Editing and Storytelling in *As the World Turns*
—DEBORAH L. JARAMILLO

180 *Guiding Light*: Relevance and Renewal in a Changing Genre
—PATRICK ERWIN

187 The Evolution of the Production Process of Soap Operas Today
—ERICK YATES GREEN

191 From Daytime to *Night Shift*: Examining the ABC Daytime/SOAPnet Primetime Spin-off Experiment
—RACQUEL GONZALES

201 "What the hell does TIIC mean?" Online Content and the Struggle to Save the Soaps
—ELANA LEVINE

219 The Evolution of the Fan Video and the Influence of YouTube on the Creative Decision-Making Process for Fans
—EMMA F. WEBB

SECTION FOUR
LEARNING FROM DIVERSE AUDIENCES

233 Soaps for Tomorrow: Media Fans Making Online Drama from Celebrity Gossip
—ABIGAIL DE KOSNIK

250 Soap Opera Critics and Criticism: Industry and Audience in an Era of Transformation
—DENISE D. BIELBY

265 Hanging on by a Common Thread
—JULIE PORTER

272 Perspective: Fan Site Moderator QueenEve on Fan Activity around and against Soaps
(BASED ON AN INTERVIEW BY ABIGAIL DE KOSNIK)

275 The Role of "The Audience" in the Writing Process
—TOM CASIELLO

279 The "Missing Years": How Local Programming Ruptured *Days of Our Lives* in Australia
—RADHA O'MEARA

293 *As the World Turns*' Luke and Noah and Fan Activism
—ROGER NEWCOMB

300 Constructing the Older Audience: Age and Aging in Soaps
—C. LEE HARRINGTON AND DENISE BROTHERS

315 References

331 Index

ACKNOWLEDGMENTS

As is true with many large-scale projects, the path to this book's completion was met with challenges both expected and unexpected: multiple job transitions, first-time parenthood, global time zone coordination, and a rapidly changing world for U.S. daytime television.

We are grateful to the following persons for their ongoing support and enthusiasm for this project: Leila Salisbury, Director of the University Press of Mississippi, who understood our vision of this collection and its potential to intervene at a crucial moment in the history of a beloved television genre; the anonymous reviewers, whose comments on the project proposal proved extremely helpful as the project developed; and our contributors in the worlds of academia, the television industry, and soap fan communities, whose unique perspectives and areas of expertise have shed new insight into the past and future potential of soap opera.

In addition, we thank the other scholars, soap viewers, and industry executives we talked with during this project who provided us with invaluable ideas. We also thank Eric Zinner for his enthusiasm; Henry Jenkins for his support and for helping to create the germ of this idea through his "Gender and Fan Studies/Culture" series on his Confessions of an Aca/Fan blog (http://www.henryjenkins.org/); Amanda Ford and Benjamin De Kosnik for their patience and support; and Kimberly Chen, Christopher Goetz, Caitlin Marshall, and Jennifer Lowe, who were all outstanding research assistants. This project could not have succeeded without all these colleagues and friends.

Institutionally, we would like to thank the Cultural Studies program at Columbia College Chicago, the Berkeley Center for New Media, and the Department of Theater, Dance & Performance Studies at the University of California, Berkeley, the Department of Sociology and Gerontology at Miami University, the Program in Comparative Media Studies at MIT, and Peppercom for the resources and time granted to the editors for the completion of this project. We also thank the Convergence Culture Consortium for bringing many of those involved in this collection together and for providing a framework for a collection aimed at fan, industry, and academic audiences alike.

Finally, we would like to thank all of the industry professionals, trade journalists, Web site moderators, and fans who have, for decades, collaboratively produced the fascinating world of soap operas. Our passion and enthusiasm for this genre provided us with ample motivation to bring this book into being; it has truly been a labor of love.

SECTION ONE
CHALLENGES TO THE FUTURE OF SOAPS

INTRODUCTION
THE CRISIS OF DAYTIME DRAMA AND WHAT IT MEANS FOR THE FUTURE OF TELEVISION

—SAM FORD, ABIGAIL DE KOSNIK, AND C. LEE HARRINGTON

WORLDS WITHOUT END?

This collection aims to intervene at a critical moment in the history of daytime soap operas, as one of television's longest-running genres faces increasingly alarming questions about its economic viability. Scholars have argued that soaps are the most uniquely televisual of all TV genres, designed to capitalize on the intimacy and constant access television provides with a daily, serialized narrative that never ends. From an industry perspective, serials have long been seen as the perfect genre to deliver audiences to advertisers. However, soap opera ratings have been declining for more than a decade, a decline commonly attributed to radical work/lifestyle changes among its core audience (adult women), a proliferation of daytime entertainment choices, and failed creative and promotional strategies. Today, all soaps are struggling to figure out who is actually watching, how to attract and retain new viewers while not losing long-term fans, and how best to capitalize on the deep histories of narratives that have been airing for decades.

Indeed, soaps are currently facing many questions that trouble *all* television production today as we transition from broadcast to digital media. For example: How can TV attract new (younger) viewers who are growing up in an online era rather than a broadcast era? How can the traditional television model interact with the new possibilities offered by cable, time-shifting technologies, archiving technologies, video-on-demand, and Internet distribution models? How can television best communicate with its audiences when many people watching TV programs are not, in fact, watching them on television? Traditional U.S. soap storytelling, particularly its never-ending format, presents what may be television's most difficult challenge in adapting to our new media era. As such, soap operas might be the best test case—as they have often been—for experimentation within the television industry as a whole.

THE FLICKERING LIGHT

When news broke on April 1, 2009, that CBS would cease broadcasting *Guiding Light* (*GL*), bloggers and commentators on numerous soap opera-related Web sites expressed their hope that the story was an April Fool's joke.[1] *GL* debuted on CBS in 1952 and was the oldest U.S. television serial and, after the news program *Meet the Press* (NBC), the second longest-airing television show in history (Hibberd 2009). *GL* originated as a radio show in 1937, and its narrative had continued for seventy-two consecutive years by the time of its final broadcast on CBS in September 2009. *GL* was routinely described as the longest story ever told, and its fans were astonished at the prospect of the story ending.

The bad news gave rise to a question: Could a property like *GL* carry on via an alternative form of distribution, either a cable network or a new distribution format? The passionate reactions of not only the more than two million daily viewers Nielsen estimated (a number that does not include online viewers or establish how many unique weekly viewers the soap opera had), but also those of generations of "lapsed viewers" who have watched *GL* at some point in their lives, demonstrate the power that these narratives have and their prominent place in American popular culture for the past sixty years. The *GL* cancellation was upsetting as well for the many millions who watch the daytime soaps that still aired daily on CBS, ABC, and NBC, a reminder that a set of shows that once seemed permanent was becoming endangered. And there were other danger signals for the genre: the global financial crisis and its impact on the automobile industry (historically a major sponsor of daytime soaps) (Blair 2009); the refusal of all three major networks to air the Daytime Emmys in 2009 for the first time in thirty-six years (cable network CW eventually agreed to broadcast the ceremony) (Associated Press); the budget-induced firing of highly popular soap actors (such as Deidre Hall and Drake Hogestyn from NBC's *Days of Our Lives* [*DOOL*]) (Kroll 2009); the budget-induced pay cuts suffered by other soap stars (such as Eric Braeden of CBS' *The Young and the Restless* [*Y&R*])[2] (Rice 2009a, Rice 2009b); the cost-saving shift of the production for ABC's *All My Children* (*AMC*) from New York City to Los Angeles; and the planned 2012 cancellation of cable network SOAPnet.

Most distressing of all was the news at the end of 2009 that *GL*'s sister show on CBS, *As the World Turns* (*ATWT*), would end its fifty-four year run on the network in September 2010, marking the departure of the only soap company still airing dramas on broadcast television: Procter & Gamble. Fans were particularly upset by CBS chairman Leslie Moonves's statement to *The New York Times* upon the show's cancellation: "Only the special soaps are going to survive. It's certainly the end of the client-owned soap. [. . .] All good things come to an end, whether it's after 72 years or 54 years or 10 years. It's a different time and a different business" (Carter and Stelter 2009). While both Procter & Gamble and *ATWT/GL*'s production company, TeleNext, had indicated continued interest in finding a home for new original episodes of both series (Kiesewetter 2009), no serious prospects materialized. ABC's *One Life to Live* (*OLTL*) has also faced cancellation rumors (Branco 2009), and, if *OLTL* were to go off the air, only five soaps would remain in production, down from eleven in 2000 (and nineteen in 1970).

MULTIGENERATIONAL VIEWING

One reason that the soap genre warrants consideration at present is that, despite their diminishing ratings, soaps have played a role in millions of American and global television viewers' lives far longer, and in many cases much more intensely, than most television series. It is not unusual for people to tell stories of their mothers or grandmothers watching (or listening to) their favorite soap operas in the 1940s, 1950s, 1960s, or 1970s. In private conversation, one hears of favorite rituals associated with soap viewing. In some cases, family or friends come together at regular times of day to follow the trials of beloved characters; in other cases, the soap fan banishes members of the household from the living room to minimize distractions from the TV broadcast. The ritualistic aspect of consuming soaps persists today in new forms: Some soap viewers set aside specific times of night to watch their preferred daytime dramas via time-shifting technologies, such as DVRs, or to catch "their" show's daily re-run on the soon-to-be-cancelled SOAPnet; other viewers watch live broadcasts while participating in online chats or visit their favorite Internet soap opera communities after watching the latest episode. It is rare for news programs or game shows (some of which have aired as long as soap operas) to engender these kinds of ritualistic, intergenerational, and decades-long consumption. The longevity of soaps, as well as certain characteristics of soaps' narrative structure, has given rise to forms of committed viewing that few other television genres can rival.

While *GL*'s seventy-two year run as a daily narrative is unprecedented, (reaching 15,762 episodes at the end of the series), the six soaps currently being broadcast have been airing for decades. The U.S. daytime industry's "youngest" soap, CBS's *The Bold and the Beautiful* (*B&B*), has been on the air since March 1987 and will have broadcast close to 6,000 episodes by the end of 2010. The next youngest show, *Y&R*, first aired in 1973. On these six dramas, younger faces often belong to members of soap families that have been followed by viewers since the shows were launched decades ago. Meanwhile, some soap actors have been portraying the same characters for 40 or even 50 years. *ATWT* featured the most striking examples of such longevity. For example, Helen Wagner was an original cast member on *ATWT* as Nancy Hughes and played the role she originated in 1956 until her 2010 death, only months before the show's cancellation by CBS, while 2010 marked fifty years on *ATWT* for Don Hastings and Eileen Fulton with their respective portrayals of *ATWT*'s Bob Hughes and Lisa Miller Grimaldi. Frances Reid, an original cast member on *DOOL*, played Alice Horton from the show's inception in 1965 through occasional appearances until her 2010 death. Perhaps the most recognizable U.S. soap opera star, Susan Lucci, has played *AMC*'s Erica Kane since the show's 1970 debut, making her role currently the longest-running regular portrayal of a fictional character on broadcast television.[3]

Development of a narrative over decades brought with it shows that built a multigenerational audience. In soap opera fan community discussions, most viewers report they were introduced to daytime dramas by another member of their family or by a close friend who watched soaps regularly. With their ongoing multigenerational casts, daytime soap operas seemed well poised to retain a corresponding ongoing viewership, much the same way that sports franchises consistently recruit new generations of fans. However, cultural shifts in daytime television viewership, coupled with the conflict between this multigenerational model and a demographic focus on eighteen to forty-nine-year-old females, have challenged the long-term vitality of the soap opera. One aim of this volume is to investigate, from a variety of perspectives, how and why soaps have lost their longtime audiences, threatening the viability of the genre as a whole.

THE CULTURAL SIGNIFICANCE OF SOAPS

In addition to the importance that they have held for millions of individuals and families over the last several decades, soaps have also attained a larger significance in American and global society, extending beyond their fan bases. Whether and how the current precariousness of the daytime drama format is affecting this widespread significance is another recurring theme of the essays in this collection. Since they debuted on radio in the 1930s, soap operas have developed into a cornerstone of American popular culture. The genre has adapted radically over the decades, both in terms of technological and production shifts and narrative priorities, but it has remained central to American culture and contributed to storytelling styles around the world. In short, soap operas have adopted the serialized nature of thousands of years of oral storytelling and the serialized narratives of popular writers such as Charles Dickens and transformed it into a unique, daily consecutive narrative, with each soap opera airing more than 250 new episodes each year. Having traveled U.S. airwaves since the dawn of mass electronic media, the term "soap opera" has penetrated mainstream culture. Even those who do not watch daytime dramas know enough about the genre (or think that they do) to use "soap opera" as a common cultural reference. As Spence notes (2005, 7), "One does not have to have watched soaps to know what *The New York Times* means when it says that Claus von Bülow's life is like a soap opera."

Many authors of soap studies—including Brunsdon (1997), Nochimson (1992), and Spence (2005)—have written about the shame that soap opera viewers often feel because they either fear or actually experience disapprobation for their enjoyment of serial melodrama. The soap opera genre has long occupied a lowly rung on the cultural ladder. To many, "soap opera" is a synonym for bad acting, ridiculous plotting, overwrought dialogue, and low production values, while the phrase "soap fans" connotes dull-witted, lazy, unproductive housewives or else delusional viewers who cannot distinguish between reality and the entertainment provided by daytime dramas five days per week all year long. But what will the arbiters of cultural sophistication do if the day comes when soap operas no longer pollute American households and corrupt the minds of TV viewers? The importance and utility of soap opera, as a style of storytelling and televisual production—to both those who adore the format and those who despise it—is such that we can now predict that the concept of the "soap opera" will endure even if soap operas as we know them cease to exist.

Another legacy of daytime drama that may last longer than the dramas themselves is the genre's influence on primetime television. Over the past several decades, primetime television has increasingly incorporated elements of serialized storytelling popularized on television by the soap opera, a move that has revolutionized the concept of "quality television." Many popular primetime series, from action/adventure to science fiction to comedy, share soap operas' focus on ongoing characters, storylines that carry over from one episode to the next, and a predominant focus on an ensemble cast of characters and the development of those characters and their relationships over time.

In a discussion of "the increasing serialization of television," Ross (2008, 182) points out that, "[m]any of the cult shows that have captured media and scholarly attention across the years have relied on heavily serialized plots, connecting them to the soap opera genre and also to the world of comic books." Ross regards the engagement and interactivity—both on- and off-line—that television producers today actively attempt to cultivate among their viewers as having been pioneered, tried, and tested by soap operas (and by other often-degraded genres such as comics and video games). In Ross's analysis, soap operas helped to blaze the trails on which much of television production and reception now travel. Following the soaps' lead, TV networks are increasingly presenting serialized stories, and viewers are actively engaging with these texts by contributing spoilers, speculations, summaries, and interpretations of particular programs to pop culture Web sites such as Television Without Pity and to online fan communities dedicated to these shows.

The pleasures audiences derive from immersing themselves in the continuous story worlds that have begun to populate primetime in recent years certainly tie into the pleasures that soap opera fans have enjoyed for decades. According to Ross (2008, 212), "'[T]hings soap operatic' may be at work with the fandom surrounding what Mittell (2006) describes as narratively complex texts. I would argue that, if fans of narratively complex texts who enjoy considering the mechanisms at work in a show do not have an *equally* strong investment in characters and relationships, at some point the operational aesthetics of that show will become all 'bells and whistles' [...] and will inevitably lead to viewer *disengagement*." The incorporation of elements of soap opera narrative and soap viewer protocols by many other media formats speaks to the lasting legacy of the genre and to the substantial impact daytime dramas have had, not just on popular culture, but on the nature of storytelling and media audience engagement. The relationship between soap opera and primetime "complexity" is an area of study throughout this collection.

MASCULINE, FEMININE, AND HYBRID TELEVISION

One marked difference between the modes of viewer reception and interaction stimulated by serial primetime shows and those engendered by daytime dramas is that very little stigma accompanies the enjoyment of primetime "puzzle" narratives. While soap fans, as well as fans of "cult" series (usually in the sci-fi/fantasy category), have tended to be dismissed as overzealous and obsessive about their favorite TV shows, fan activities related to newer primetime serials are "comparatively stigma-free" (Ross [2008, 182]). Some of this de-stigmatization of fandom associated with TV serials might be attributed to the fact that the media industry, seeking to take advantage of the Internet's potential to generate "buzz" that boosts shows' ratings and DVD sales, now welcomes and sometimes cultivates fannish levels of enthusiasm and interactivity with their programs. In comparison, in earlier decades, devotees of shows such as *Star Trek* (NBC) and the soap opera *Search for Tomorrow* (CBS/NBC) were derided for launching "save-our-show" letter-writing campaigns and creating edited tapes or fanzines filled with discussions, stories, and artwork related to their favorite programs.[4]

However, another likely reason for the large-scale acceptance of present-day media fandom relating to serial TV is that, whereas soap opera has always been considered a primarily "feminine" format, most primetime serials are coded as what Fiske (1987) calls "masculine" programming, or what we might regard as a "hybrid" of masculine and feminine television.[5] Followers of soap opera have often been depicted as unproductive and powerless, as housewives enraptured and captured by the weepy melodrama supplied by daytime dramas during the same hours of the day that more powerful members of American society (adult men and women employed outside the home) are at their most productive. Followers of primetime serials, however, are widely seen as productive in their fascination with their preferred programs, approaching them as (to borrow Ross's and Mittell's terms) "puzzles" or "games" to be played, worked out, and won. Thus, speculating on the intricacies of primetime serials today is seen as similar to the voluntary tinkering or problem-solving of hobbyists: an active, intellectual, masculine, and mature form of participation. In contrast, speculating on the intricacies of soap opera plots still connotes addiction, surrendering one's will to the television in a passive, mindless, feminine, and child-like form of consumption. The difference in widespread perceptions of what women (particularly women at home during the day) do when they watch daytime dramas, and what men and working women do when they watch primetime serials partially accounts for the stigma that attaches to one group and not the other.

Primetime serial television's utilization of hybrid narrative styles, rather than predominantly feminine styles of storytelling, may also account for women's declining interest in soap opera and their rising interest in serialized primetime programs. Since *Hill Street Blues* launched on NBC in 1981, increasing numbers of television shows have been hybrids of the formerly distinct masculine and feminine genres. Cop shows, legal dramas, hospital dramas, action and military and espionage shows, and other male-oriented formats now also display traits characteristic of feminine formats. Not only serial plot lines, but a focus on interpersonal relationships, intimate conversations, romantic and familial conflicts, domestic settings, emotionally aware and sensitive male characters, and professional and powerful female characters—all cited by Brown (1987) as key elements of soap opera—are integral features of apparently masculine primetime television series such as FX's *The Shield* and HBO's *The Sopranos,* as well as of more female-oriented primetime shows such as CW's *Gossip Girl* or ABC's *Desperate Housewives*.

In recent years, many sitcoms have similarly combined male humor about sports, work, and sex with serial narratives focusing on relationships and the domestic (NBC's *Friends*, CBS' *How I Met Your Mother*), addressing both male and female audiences. Reality shows, whether more male-focused (CBS' *Survivor* and *Big Brother* and NBC's *The Apprentice*) or female-focused (CW's *America's Next Top Model*, Bravo's *Top Chef*, Bravo/Lifetime's *Project Runway*) are inherently fusions of masculine and feminine formats. They alternate between personal conflicts in domestic settings (many reality shows feature contestants who live together for the duration of the competition) and time-limited challenges that often require professional or physical prowess, and consist of an overarching "plot" ("who will win?") that only concludes at the end of each season. Collins (2008, 89–90) describes some reality series, such as MTV's *The Real World*, as "docusoap:" "[The 'docusoap'] places [individuals] in 'natural' settings, and uses documentary style production values to focus on their everyday lives but with the intervention of soap opera structuring techniques." One might conclude that television series blending feminine and masculine elements succeed because they appeal to both female and male audiences, while both female-oriented soap operas and male-oriented, closed-narrative shows have more limited or niche appeal. Perhaps would-be viewers of soap opera are finding their needs for the pleasures of daytime drama satisfied by hybrid primetime programs.

SOAPS, "STRIPPED" TELEVISION, AND RAPID-SUCCESSION VIEWING

Soaps' influence may be seen in the structure and reception of not only first-run primetime television series, but of rerun (via broadcast and online) and repackaged (on DVD and Blu-Ray) television as well. Laura Stempel Mumford (1994) notes that the practice of "daily stripping" older TV shows—that is, airing one episode of a series every week day, five days a week—effectively converts programs that had originally been broadcast once per week into soap operas. Mumford argues that viewers interpret shows quite differently when they consume television texts in quick succession. Amanda Lotz (2006) applies this insight to another form of rapid-succession viewing: watching an entire season or multiple seasons of a program on DVD in a compressed period of time, an activity she calls "consecutive viewing."

At the time of Mumford's and Lotz's writing, consuming TV online was not yet standard practice, but contemporary media audiences have myriad options for streaming or downloading multiple episodes of a given program and watching them one after another. This makes the comparison of "stripped" TV shows to soap operas more relevant today than it was even five years ago. The benefits audiences derive from rapid-succession viewing online, or on

DVD/Blu-Ray, are similar to the benefits that soap opera audiences derive from watching soaps daily. To paraphrase Mumford, one of these benefits is intimacy, a felt personal connection between the show and the viewer, who has a sense that a fictional program is a part of what John Tulloch (1990, 228) calls "viewers' daily rituals."

Through stripping, David Marc (1984, 5) writes, "[s]imple identification and suspense yield to the subtler nuances of cohabitation." This is a kind of intimacy that a viewer can experience with any genre of television show merely by dedicating a certain amount of time to a given program and is not specifically tied to the soap opera format's frequent use of closeups or focus on domestic content, which have both been cited as sources of women's intimacy with the soaps they watch (Mumford 1994, 170–1). In the same vein, shows that are viewed in compressed time periods offer, according to Mumford, "a particularly intense version of the quintessential viewing process: the (comforting and pleasurable) repetition of the same experience, using the same technology, in much the same viewing relationship, and even, in some cases, watching the very same televised material" (173–4).

In addition to providing specific forms of intimacy and comfort, rapid-succession viewing allows a program's narrative threads to tie together differently than they do when the show is consumed in weekly installments. Mumford notes that when watching *thirtysomething*, a show that originally aired on ABC in the late 1980s, stripped daily on Lifetime Television in the mid-1990s, "a number of *thirtysomething*'s themes resonate more immediately with each other when the episodes are viewed several days in a row, instead of being separated by a week.... [S]triking connections *between* the episodes ... emerge, not only at a broad thematic level ... but equally at the level of individual scenes, which acquire new significance when viewed on consecutive days" (181). Mumford implies that *thirtysomething* became more of a soap opera when it was stripped on Lifetime because the viewer tended to make sense of each episode by linking it to moments and themes from previous episodes seen in rapid succession This kind of "interpretive work" (177) depends on long-term narrative memory, which the soap opera has always required of its audiences. Time-compressed viewing encourages the audience to remember and forge their own narrative linkages to an extent that once-weekly viewing cannot match.

These types of consumption are on the rise, with narrowcast cable networks, DVD/Blu-Ray packaging of already-aired series, and Internet streaming and downloading all facilitating viewing and re-viewing a television program's episodes in quick succession. As modes of rapid-succession consumption become more popular, content producers and distributors have much to learn from the daytime serial genre about encouraging and satisfying consumer demand for expansive and ritually viewed television texts.

THE UNIQUENESS OF SOAPS

Despite the incorporation of soap opera-esque devices in a variety of other content and distribution forms, this collection argues that there remains a distinctive set of principles that sets daytime soaps apart from other television genres and storytelling forms. In his description of "immersive story worlds," Ford (2007) identifies six elements of this category of narratives. First, immersive story worlds have expansive backstories that cannot be neatly summarized or explained. Second, these narratives employ a vast set of ensemble characters, including both those that are currently involved in the narrative (frontburner) and a much larger group of characters that are part of the fictional universe (backburner). Third, the narratives of these worlds tie their contemporary stories to their sprawling histories and sets of characters. Much like communities in the "real" world, immersive story worlds possess both history and memory that contextualize current plot and character development. Fourth, immersive story worlds are built by multiple creative forces, usually consisting of generations of teams who have expanded inherited stories. In addition, these narrative worlds are centered on serialized stories that build on one another and continue to expand over time. Finally, these five elements combine to create a sense of permanence for these worlds, which constitute a universe that will continue to exist throughout time.

Ford argues that current and past U.S. soap operas fall into the "immersive story world" category, as do professional wrestling narratives and the superhero universes of Marvel and DC Comics. We argue here that the elements that set this category apart are particularly pronounced in soap opera, a fact which explains why soaps have long been called "worlds without end" and why, even as primetime dramas incorporate many elements of the soap opera genre, daytime serials still offer a unique narrative style and an unparalleled experience for viewers that cannot be replaced by other forms of entertainment. Therefore, if soap opera production ceases in the U.S., one cannot predict confidently that the lost genre will "live on" in other formats or modes of reception that it has influenced. If soap operas disappear, no other genre will wholly replace them. They are unique in the history of television and in the history of narrative. One of the goals of this book is not only to put forward contemporary analyses of a long-running genre and investigate why daytime drama seems on the verge of collapse, but also to ask what might be done to save this format, for which there can never be an adequate substitute.

DECADES OF SPEAKING ABOUT SOAPS

The Survival of Soap Opera is the latest entry in the ongoing critical conversation about U.S. daytime drama that began when the genre launched. The massive popularity of soap opera narratives garnered the attention of academics and cultural critics from the start. For the first few decades of the soap opera's existence, studies of the genre adopted a social science approach, with researchers focused on the audience's fascination with "their stories" and the (often presumably negative) social effects that were believed to ensue from devoted viewership. In the 1970s, with the rise of more serious criticism of television aesthetics, academics such as Horace Newcomb (1974, 161–82) began to turn their attention to soap opera narratives and their unique artistic contributions. Within a few years, this focus expanded to include production and televisual qualities, driven by work such as Timberg (1981), Intintoli (1984), and Butler (1986). This work continued alongside more sociological approaches to soaps, such as Cantor and Pingree (1983), a division explicitly addressed in Cassata and Skill (1983).

By the beginning of the 1980s, soap operas were also becoming an increasingly crucial text for a community of scholars who applied feminist methodologies to the study of television. These scholars took up the soap opera, long perceived as having low cultural status, not only because of its direct reference to the commercialism of the art form (with "soap opera" stemming from the soap companies that initially sponsored soaps), but also because of the lower cultural status of soaps' primary demographic: women at home. For instance, Modleski (1979) examines the soap opera specifically as it relates to the female audience's reception, arguing that keeping that audience in mind shapes how one understands and evaluates the genre, and Seiter (1982) looks at how soap opera narratives reflect the contradictions in women's lives, as do a variety of other feminist analyses of U.S. and British soaps.[6] The study of female audience perspectives on U.S. soaps has continued throughout the past three decades (e.g., Mumford [1995], Blumenthal [1997]).

One indication of this renewed interest in the study of the soap opera is Robert Allen's (1985) seminal book *Speaking of Soap Operas*, which focuses on the elements of the soap opera that set it apart as a narrative form. Allen suggests that U.S. soap operas perhaps constitute a unique form of television art, one in which fictional narratives are "over-coded"—that is, "characters, events, situations, and relationships are invested with signifying possibilities greatly in excess of those necessary to their narrative functions" (1985, 84). Allen's focus on excess meaning and the feminist reception studies of the late 1970s and 1980s generated increased scholarly interest in the active nature of soap opera fan communities, with soap fans cited as active participants in the meaning-making of serialized texts. As part of the emerging literature on media fan studies, Harrington and Bielby (1995) examined the longstanding practices of soap opera fans, from the formation of fan clubs and gatherings, to the audience's support of the soap opera press. Subsequent research by Baym (2000) brought soap studies into the Internet age, exploring the formation and practices of online communities through early discussion forums devoted to soap operas. And substantial studies on soap opera viewer practices by Hayward (1997), Scodari (2004), Spence (2005), and others have given us a nuanced understanding of the many ways soap fans relate to soap opera texts and to one another.

These studies are only a sampling of the extensive academic work conducted on U.S. soap operas over the past four decades. For instance, the attention granted soap opera research at the Popular Culture Association's annual national gathering has generated a variety of publications on the subject, including by Williams (1992) and Frentz (1992), and the PCA's tradition of soap opera research is carried on today by area chair Barbara Irwin of Project Daytime. Further, studies positioning U.S. soaps in an international context (Allen 1995; Matelski 1999) and focusing on ethnicity in U.S. daytime serials (Jenrette, McIntosh, and Winterberger 1999)[7] have provided essential context for understanding these shows. The work reviewed here demonstrates the overall trajectory of work on soaps, from studying the effects soap operas have on viewers, to taking daytime serial dramas, the female viewing perspective, and soap fans themselves seriously (to borrow a phrase from Intintoli [1984]).

THE SOAP OPERA WORLD KEEPS TURNING

Groundbreaking academic research on soaps continues; however, scholars and journalists increasingly refer to soaps as a "historical" genre, a relic of the past. In classrooms, students are often assigned to read scholarship that speaks to the soap opera landscape of decades ago, not that of 2010. We assert that it is crucial to understand changes in production and reception that define the soap opera viewing experience today, and this collection includes diverse scholarship that reinforces our assertion.

Soap opera ratings have dropped consistently over the past decade, and the once solid position of the genre in the U.S. network broadcasting schedule is in serious jeopardy. Since the early 1990s, ten daytime dramas have been cancelled. Seven of the programs were quite "young": ABC's *Loving* (1983–1995) and spin-off *The City* (1995–1997), as well as *Port Charles* (1997–2003); and NBC's *Santa Barbara* (1984–1993), *Generations* (1989–1991), *Sunset Beach*

(1997–1999), and *Passions* (1999–2008). However, three of the cancelled soaps were considered grande dames of the genre: NBC's *Another World* (1964–1999) and CBS' *GL* (1952–2009) and *ATWT* (1956–2010), all produced by Procter & Gamble. As of this writing, continuously declining ratings, ongoing budget cuts, and repeated moratoriums both in print and online have sapped industry and viewer confidence alike. Soap operas have reacted with financial restructuring, experimentation in storytelling, and dramatic changes to how those stories appear onscreen, as several of the essays in this book will illustrate. Alongside these budget and production changes have come major shifts in how soap operas are viewed and discussed by fans and increasing experimentation with transmedia storytelling.

Further, few dispute that soap stories themselves have changed—and not for the better. Fans and industry veterans agree that the quality of soap opera storytelling has declined along with ratings and budgets. Stories began focusing less on the overall community of characters and more on granular, isolated plot elements that seem in direct opposition to what originally defined the soap opera form. Such changes suggest that soaps are at a crossroads, not only in terms of the genre's economic viability, but also in terms of the very meaning of the daytime serial drama. Thus, we believe this to be a particularly important juncture to discuss the soap opera genre's past, present, and potential future.

The first section of this book looks at the underlying tensions challenging the future of the soap opera form. Irwin and Cassata are interviewed about many of the challenges covered in this book and about their work studying health, aging, and the family through Project Daytime. Reynolds is interviewed about his view that today's soaps pale in comparison to the shows of yesteryear, such as CBS/ABC's *The Edge of Night* (1956–1984). Mulcahey (interviewed by Giada Da Ros) discusses the development of shows and networks contracting writers directly and the impact this change has had on the daytime industry over the past three decades. Rogers is interviewed about the many changes in soap operas during his stints on today's longest-running U.S. network soap opera, ABC's *General Hospital* (*GH*), in a role that has spanned almost three decades. Bibel analyzes the many distressing industry changes in the U.S. soap opera landscape, concentrating on 2008 and examining that banner year in relation to the realities of soap opera budgets. Williams provides insight from the perspective of soaps pioneer Agnes Nixon, as she watches the current struggles of her two long-running creations, ABC's *AMC* and *OLTL*. Nasser looks at developments in two industries related to the U.S. daytime soap opera, Latin American telenovelas and U.S. primetime drama, with special attention to two biases that played out in critical reception of U.S. primetime series adapted from telenovelas: gendered bias against the soap opera format and ethnic/cultural biases against telenovela content. Finally, Scardaville examines fan and industry perceptions of the decline in quality of soap opera narratives and posits that two distinct and competing organizing principles are at play: aesthetic logic and economic logic.

THREE AREAS OF OPPORTUNITY

Despite the grave tone of industry reports and soap opera fan discussions, this collection is not intended to be an obituary for a genre whose time has passed; rather, we aim to offer a range of strategies for its long-term survival. We argue that, despite the industry's decline, our media landscape would suffer a major loss if daytime serialized drama stopped airing on network television. No other media property can boast the detailed history, the transgenerational development, and the intergenerational complexity of the immersive story worlds of daytime dramas. The best solution for all parties concerned—for the networks that have spent decades building their brands, for the faithful audiences who love them, for the advertisers that have invested heavily in the genre, and for television as a whole—is for soap operas to experience a revival.

Soap opera producers are certainly seeking ways to rejuvenate their shows. For instance, *GH* struck a deal with famed film actor James Franco to join the soap for a two-month stint, starting in late 2009. According to *GH* Executive Producer Jill Farren Phelps, Franco joined the *GH* cast just "for the experience of it" (Kung 2009), and the significant amount of press coverage about Franco's appearances highlighted the skill of soap opera acting. In one story, Franco commented on having to memorize a single day's shooting script "that was as thick as a film script," without ad-libbing or using a TelePrompTer (Stein 2009).

The remainder of this book examines current experimentation and areas of potential exploration for the genre. In particular, the second, third, and fourth sections of this collection focus on *three key challenges* for the soap industry: capitalizing on history, experimenting with production/distribution, and learning from diverse audiences. If the industry can execute change in these areas, the soap genre's most innovative and exciting years may prove to be the future, rather than decades in the past.

CAPITALIZING ON HISTORY

The most prominent factor differentiating soap operas from other narrative forms is their rich narrative history, which is both an asset and a potential burden. While soaps can successfully utilize this history to develop the communities and characters depicted onscreen, it can also present seemingly insurmountable challenges to ever-shifting teams of writers and producers who may not know this story as well as the viewers know it—or respect it as they do. We assert that these deep archives contain many of the resources soap operas need to mine in order to retain their relevance. For lapsed viewers and potential new viewers alike, it is this rich and vast backstory that stands the best chance of attracting viewers away from higher-budget, less immersive alternatives. With the increased serialization of primetime narratives, this history is the most significant and unique factor of soap opera storytelling. Embracing the past requires embracing the model of multigenerational fandom that has long been the genre's strength, but which has been ignored by target-marketing efforts since the 1970s. Furthermore, with new distribution outlets for ancillary products, the deep archives these shows provide are an untapped resource for creating alternative revenue streams and attracting fans who remain deeply engaged with these complex narratives. Soap operas have long drawn on the power of their history by bringing characters/actors back for new stories or cameo appearances. For instance, to celebrate its fifty-fourth and final broadcast anniversary, *ATWT* brought back Julianne Moore's Frannie Hughes in April 2010 for the twenty-fifth wedding anniversary of her father and aunt/stepmother, Bob and Kim Hughes. However, we contend in this collection that there are a variety of ways soaps can more deeply and thoroughly make use of this unparalleled history.

Our collection offers a variety of perspectives about this deep history from experts writing from industrial, academic, and/or fan-based perspectives. Newcomb is interviewed about the influences and pleasures of soap opera narratives. Allen is interviewed about his work on soaps and the rewards of studying a text as prolific as U.S. daytime serial dramas. Through a case study of *ATWT*'s Tom Hughes, Ford examines the storytelling abilities of the soap opera, which allow audiences to watch characters develop in "real time," ultimately arguing that these unique attributes have to be emphasized and supported for the soap opera genre to remain viable. Alden is interviewed about the elements that make soap operas unique: duration, frequency, character, and "writing for the ear." Baym discusses the changes in her personal relationship with soaps after the mid-1990s O. J. Simpson trial. Scodari looks at the migration of feminine forms of storytelling to primetime and calls for soap operas to return to the style of storytelling that appealed to millions of women during the genre's most successful periods. Liccardo focuses on the soap opera genre's increasing incorporation of over-the-top and campy elements alongside the development of many classic storytelling traits in primetime, arguing that the decline of quality in soap opera narratives has been driven in part by the industry's self-loathing. Spence is interviewed about public comparisons of soap operas with other texts and forms. Mittell discusses his work on complex television narratives and contends that serialized primetime shows today did not evolve from daytime soaps. Wilson looks at the role of public archives in preserving U.S. soap opera history and maintains that institutional practices have been crucial in shaping how the genre's history is being recorded and recalled. Finally, Metzler examines the recent writers strike as a formative moment in the development of *Y&R* and asserts that the show's creative renaissance was driven by a return to respect for the show's history.

EXPERIMENTING WITH PRODUCTION AND DISTRIBUTION

Faced with increased economic pressures, soap operas have to find ways to innovate with lower budgets. Soaps have a rich history of experimentation, from encouraging active fan involvement in narrative development to demonstrating bold innovation in balancing art and commerce. All media industries can learn from the successes and failures of the genre as it once again reinvents itself in the face of shifting technologies, viewer practices, and revenue streams. In short, we argue that "as go soaps, so goes television." Soap operas are the testcase for many problems that not just all television production, but all media industries will encounter. Because of the sheer volume of the soap opera text, the genre has been a particularly active site for industrial and artistic experimentation. For the television industry, the current state of soap operas provides crucial lessons about the transition from broadcasting to cable, about rapidly changing audiences as baby boomers who grew up with daytime serials "age out" of the eighteen to forty-nine-year-old demographic, and about the potential continued existence of television content in an online, globalized, "post-racial" world. Further, our focus on the future of soap operas poses questions about the promise of time-shifting technologies and new distribution platforms, the potential of archival practices for preserving and presenting television history, and new ways of communicating between industries and audiences.

This section includes a variety of essays detailing specific examples of soap opera narrative and visual experimentation. Timberg and Alba compare the changing use of the camera on *GH* from the show's debut in 1963 until today. Jaramillo examines a particular episode of *ATWT* to demonstrate how, even with limited budgets, soap operas innovate and create new meanings for stories through editing. Erwin looks at the revamp of *GL* in 2008 and the many production changes the show experimented with during its final run on CBS, while Green focuses on production changes in soaps' use of cameras. Gonzales examines the development of *GH*'s spin-off *General Hospital: Night Shift* on SOAPnet and the poor reception of the show's first season in comparison with much stronger feedback for its second season. Through an examination of a blog by *GH* character Robin Scorpio, Levine details the development of online ancillary content for soap operas and fan reception to ways in which that content conflicts with character continuity and longstanding fan viewing practices. Finally, Webb looks at the development of video production and editing practices among *GH* fans in a study of how fans use YouTube to distribute historic soap opera content.

LEARNING FROM DIVERSE AUDIENCES

Soap opera audiences are the *raison d'être* of daytime dramas. While online fan communities for these shows constitute only a subset of soap opera viewers and an echo chamber for viewer opinion substantially influenced by groupthink, these groups still afford producers unparalleled insight into the consciousness of soap opera fans. Media industries in general, together with advertisers, have still not completely understood the best way to tap into this wealth of publicly available input. Nevertheless, the opportunity remains for shows not only to learn from viewers but to develop direct ways of communicating with them, potentially providing soaps with avenues for testing ideas with viewers ahead of time and fosterng feelings of collaboration amongst soap opera fans and the industry. Such an opportunity is especially welcome at a time when the audience and the industry must come together if the genre is going to survive. In addition, listening to soap opera fans provides the industry with a chance to learn about the practices of "surplus audiences" that are outside the target demographic, but nonetheless might provide significant advertising and other revenue opportunities. Most importantly, embracing a multigenerational audience model would allow soap opera producers to find new value in older audience members, who can act as the proselytizers to help bring new viewers into the world of daytime soaps (Ford 2007).

In this final section, De Kosnik proposes that media fans have begun to construct their own soap-like narratives out of non-fiction material—specifically celebrity gossip—in order to satisfy their desire for serialized melodramatic narrative. Bielby looks at the role of traditional and fan critics in relation to the soap opera industry today. Porter details her own experiences with online fandom through the shifting nature of the soap opera industry. QueenEve is interviewed about the role of soap fan communities in resisting the direction of their shows and providing greater pleasure for fans through community online interaction. Casiello looks at the role of "The Audience" in the process of soap opera writers. O'Meara focuses on international viewing contexts for U.S. soaps through *DOOL* in Australia, exploring viewers' reactions to the Nine Network's decision to skip four years of episodes. Roger Newcomb focuses on fans of the popular *ATWT* gay couple Luke and Noah as a surplus audience. Finally, Harrington and Brothers look at an aging U.S. audience base and the subsequent rise in purchasing power among elders, arguing that soap operas should be focusing on narratives—and advertising—that target this underserved market.

A NEW KIND OF ANTHOLOGY

While this collection adopts an anthology format, a central argument runs through every component. The claim at the core of this volume is that soap operas remain crucial texts, and that scholars, fans, advertisers, and media industries should remain invested in them—and their future incarnations—despite declining ratings.

The diversity of our contributors is an important component of this book's interventionist strategy. Included throughout the collection are excerpts, presented as short "Perspectives" pieces, of interviews with industry veterans, longtime fans, and pioneering scholars who have researched the genre. (The "Perspectives," based on interviews that the editors [and, in one case, a television critic] conducted, were written by the co-editors of this collection. Each essay consists of a selection of quotes from a given interview presented in first-person format, and all were approved by the interview subject before going to press.) Many of our contributions are written by what might be called "second-generation" soap scholars who became interested in new questions about industry/audience/fan relations that arose in the 1990s, and by rising figures in media studies who are keenly interested in how soaps might best adapt to the new media era. This latter group of scholars is perhaps most interested in pioneering new ways of thinking about soap opera. The industry's current crisis requires fresh and innovative approaches, and these emerging scholars represent a new generation of experts committed to the genre's adaptation and survival. Finally, we have also included essays from a variety of soap opera critics, journalists, and writers who not only add to the range of voices in the collection but open soap opera scholarship to the industry as well as soap opera viewers.

This collection does not, of course, provide an exhaustive study of the soap opera genre. Much remains to be said, for instance, about close textual analysis of specific stories themselves, audiences that remain under-studied (particularly audiences of color), as well as changes in the nature of soap opera storytelling that are mentioned but not covered exhaustively in this collection. Our hope is not to produce the definitive collection of work on the soap opera, but rather to provide a catalyst for renewed interest in this genre at a time when deep thinking about the past, present, and future of soaps is not only particularly appropriate but needed.

For scholars, longtime soap opera viewers, and television industry professionals—the collective intended audience for this book—the current moment provides a chance to reflect on one of the most important formats in television history at a time of significant transition. This collection aims to provide a variety of perspectives on the challenges and opportunities the genre faces in this critical period and serves as a testament to our dedication and passion for a genre which we hope to see not only survive, but thrive once again.

NOTES

1. See, for instance, comments at Adalian (2009) and James (2009).

2. Braeden quit *Y&R* in early October 2009 after being asked to take a mid-contract pay cut. However, when offered a three-year contract two weeks later—still for lower pay—he accepted the deal.

3. Fellow *AMC* original castmate Ray MacDonnell likewise debuted in 1970 as Dr. Joe Martin but left the show as a regular once *AMC* moved its production to Los Angeles.

4. For more on the "offline" precursors to online soap opera fan activity, see Ford (2008a).

5. For Fiske's writing on soap operas and feminine television, see Fiske (1987, 179–97).

6. For a review of feminist studies on the soap opera, see Brunsdon (2000) and Geraghty (2005).

7. Also, see Johnson's (1982) brief essay on African Americans in soaps.

PERSPECTIVE
SCHOLARS BARBARA IRWIN AND MARY CASSATA ON THE STATE OF U.S. SOAP OPERAS

BASED ON AN INTERVIEW BY C. LEE HARRINGTON

Barbara Irwin is professor of communication studies at Canisius College and chair of the soap opera area for the Popular Culture Association's annual national conference. She focused her dissertation, entitled "An Oral History of a Piece of Americana: The Soap Opera Experience," on soaps and has researched and written on soaps for more than twenty years. Irwin began watching soaps casually in high school, and began watching them seriously as a graduate student at the University at Buffalo, studying, first, the audience and, later, the historical significance of the genre.

Mary Cassata is associate professor of mass communication at University at Buffalo, the State University of New York. She is co-editor of *Daytime Television: Tuning in American Serial Drama* with Thomas Skill and has written and presented research on soap operas throughout the past thirty years. Cassata began watching soaps after teaching a course on soap opera research in the late 1970s.

Together, Cassata and Irwin are the directors of Project Daytime, a comprehensive research endeavor Cassata founded, which focuses on the study and monitoring of the daytime television industry. At its inception, Project Daytime was one of few scholarly endeavors that championed the importance of soaps in the everyday lives of its audience. In addition to conducting multiple research projects together, Cassata and Irwin have published two books on *The Young and the Restless*.

In this piece, Irwin and Cassata provide their perspective on the three main themes of this book, previewing many of the issues contributors will tackle in the three sections to come: capitalizing on history, experimenting with production and distribution, and learning from diverse audiences. The piece ends with an overview of several aspects of soap operas explored in depth by Project Daytime.

CAPITALIZING ON HISTORY

U.S. soap operas today are struggling. Ratings are down, longtime viewers have left, and younger viewers are not replacing them. These shows still attract millions of viewers every day, but—with fewer soaps on the air and changes in our society, our lifestyles, and our media habits—soaps have lost more than three-quarters of their audience over the last twenty years (seventy-eight million viewers in 1990 as compared with seventeen million in 2009). Further evidence of the struggle came with the September 2010 cancellation of CBS' *As the World Turns* (*ATWT*) on the heels of the September 2009 cancellation of CBS' *Guiding Light*, thus ending the era of Procter and Gamble soap operas that had spanned seven decades.

Some credit the O. J. Simpson trial for the dramatic decline in soap opera ratings. In 1995, the trial was broadcast for thirty-seven consecutive weeks. The trial broadcast resulted in regular preemptions and interruptions of the daily soap opera line-up. The trial also received wall-to-wall coverage on cable's Court TV, now truTV. It could be argued that the real-life drama unfolding before viewers' eyes was more dramatic than what the soaps had to offer. Many viewers did not return to their soaps after the trial ended, having discovered that the reality played out on Court TV and other cable networks was more worthy of their viewing time.

The Simpson trial cannot be blamed for single-handedly causing a crisis in the soap opera industry, but during the time it aired, loyal soap opera viewers became aware of the vast array of viewing options available to them. And broadcast and cable programmers noted the types of programming viewers responded to. Reality-based programming began to flourish, and the sordid lives of real people were played out on myriad talk shows, court shows, magazine programs, and tabloid shows, all competing for—and often winning over—the soap opera audience.

Primetime drama has "borrowed" from the soaps for years in a relationship that multiple essays in this book examine. In the 1980s, the "primetime soap opera" became a ratings winner with *Dallas* (CBS), *Dynasty* (ABC), *Knots Landing* (CBS), *Falcon Crest* (CBS), and many others. Clearly, the primetime drama industry was capitalizing on what the soap opera did best: creating a continuing story with enough interest in each episode to bring viewers back for more. Other groundbreaking primetime dramas such as *Cagney and Lacey* (CBS) and *Hill Street Blues* (NBC) borrowed from the soaps and presented stories with a "cumulative narrative" structure, in which one story has a distinct start and end point within an episode, while other stories continue from week to week and season to season. Many of today's primetime series follow this same model.

For a time, the soap opera industry seemed to respond to primetime's encroachment by attempting to be more like primetime. Soap operas picked up the pace of their storytelling, and while many would hold that the focus on minutia and the extremely slow pace of the early soaps needed to change with the times, some critics claim that soap operas today are moving too quickly. With the pressure to bring in new and younger viewers, the industry perhaps has a false perception that viewers will not wait for a story's denouement. By speeding up the pace, it may be that the ramifications of characters' actions are not fully explored, and the impact of stories may be compromised as a result.

The soap industry has responded with various attempts to increase viewership while at the same time trying to reduce costs. Some of these efforts seem misguided. For example, soap opera creators have long recognized the value of bringing in younger viewers to build an audience for the future. In the 1980s and 1990s, this goal resulted in increased attention on younger characters during the summer months when high school and college students were home to watch. In the last five to ten years, in their overzealous attempt to win this younger audience, many shows' focus on the younger set has become unbalanced, while they often abandon the core families and older characters that have endeared the majority of their viewers to their shows. Research suggests that viewers of all ages appreciate and relate to characters across the age spectrum, so the better solution might be to present a collection of intertwined stories that give younger characters a reason to interact with their older counterparts.

EXPERIMENTING WITH PRODUCTION AND DISTRIBUTION

In January 2000, with the launch of SOAPnet, new opportunities developed for the distribution and reception of soap operas. Initially, the Disney/ABC-owned cable network rebroadcast same-day episodes of ABC soaps during the early morning and evening hours, and also aired weekend "marathons" in an attempt to reach working adults when they were at home. And, although offerings expanded to include rebroadcasts of CBS' *The Young & the Restless* and NBC's *Days of Our Lives,* SOAPnet's program schedule included a heavy dose of non-soap content when news of its 2012 cancellation broke, featuring reruns of Fox's *Beverly Hills 90210* and even theatrical release films.

While SOAPnet provides the opportunity for some soap viewers to watch their shows at convenient times, DVRs offer yet another option for time-shifting. With the introduction of these new technologies comes the end of habitual, ritualistic viewing.

Of course, the Internet also provides another avenue for delivering soap opera content. Network soap opera sites and other platforms such as YouTube and Joost offer full episodes, clips, and features for soap fans, a phenomenon several essays in this collection examine. In 2005, The National Academy of Television Arts and Sciences created a category for the Daytime Emmy awards that recognized original programming created specifically for computers and mobile devices, thus acknowledging the variety of Webisodes and podcasts being created for soaps.

Webisodes and online soap operas represent an innovation in soap opera storytelling. By generally focusing on one particular storyline with a limited story arc and a definite start and end point, they diverge from the traditional soap opera. Although still in its infancy, this means of delivering soap opera content does hold out the possibility of developing into something significant. If original online content can successfully dovetail with the stories being told on the soaps in their traditional broadcast delivery, they have the potential to drive increased viewership of the network soaps among lost viewers, and also to build a new and different audience. The new model may offer soaps in a downloadable form that viewers can take with them on their mobile phones and PDAs. However, the success of such efforts has yet to materialize. *Passions* (NBC) hoped that a ten-week campaign of two-minute Webisodes would build audience for their ailing show, but despite a reported ten million hits, the show could not be saved.

The advancement of technology has the potential to help the struggling soap industry. However, technology also has created an environment in which the average household has access to well over 100 channels of television programming (vastly more, in fact, as digital channels are launched). This explosive growth in viewing options, in combination with the unlimited online world, has enormously increased competition for viewers.

LEARNING FROM (DIVERSE?) AUDIENCES

Fan feedback has always been an important barometer for soap opera writers and producers. Until recently, this feedback came in the form of letters (and occasional telephone calls) to the production office. Today, feedback comes in many forms, including online communities, message boards, discussions and surveys on show Web sites, and blogs. Two key ingredients set today's feedback apart from that of the past: It comes instantaneously, and it allows fans to communicate readily with other fans. To some degree, this feedback may provide the soap industry with a sense of what viewers are responding to with regard to characters and storylines. However, because of their interaction, fans may be responding more to what other fans are saying than directly to what they are experiencing through their own viewing of the soaps. They say misery loves company, and often, online chats become "gripe sessions" for fans. The threads of discussions sometimes veer far afield from the characters and stories presented. As a result, such feedback serves little purpose for the soap industry and may in fact fuel a general dissatisfaction among viewers who participate in these discussions. Ultimately, participation in these online forums may displace actual viewing for some. If fans can connect with other fans to find out what's happening on their soaps, the need to view regularly diminishes.

PROJECT DAYTIME

Since 1985, the attention of Project Daytime has been focused on analysis of audience, industry, and message systems. Through our oral history project, we have explored the pressures, limitations, and constraints faced by the myriad individuals and entities involved in the production of soap operas. Interviews with creators, writers, directors, actors, technical crew, and network and sponsor executives connected with all of the soap operas allowed these players to tell their own stories about making the dramas millions view every day. While it is regrettably too late to capture the perspectives of the originators of the form (e.g., Irna Phillips), we were fortunate to connect with the second generation of soap opera creators, William J. Bell and Agnes Nixon, both protégés of Irna Phillips in the 1940s.

Starting in the early 1990s, Project Daytime has, through its Cultural Indicators project, examined various aspects of soap opera content with the aim of identifying trends in message systems. We'll focus our discussion here on three of the areas we've explored: health, aging, and families.

HEALTH
In the early days of soap operas, characters tended to be afflicted with strange, oftentimes fictitious illnesses. Soap opera writers purposefully avoided dealing with health issues that actually affected our society. This trend continued into the 1960s and 1970s, when the soaps typically did not identify specific diseases and seldom connected disease and death. Far more characters died from external causes (car accidents, murders, drug overdoses, plane crashes) than from illness. Psychiatric disorders were more frequent than any other health-related occurrences on soaps, but portrayals were not realistic. And it's worth noting that more than two-thirds of all professionals on soap operas were medical professionals—a gross misrepresentation of real world statistics.

Our recent study of soap opera health revealed that soaps today are populated by characters who, for the most part, are not afflicted with physical or mental health problems. However, mental illness was the most prevalent health problem faced by soap opera characters. In a departure from earlier days when those affected by mental illness were traditionally women, our study shows that mental illness today equally affects male and female characters—though, again, the number of those afflicted is small. We've also found that the health of older characters on soap operas is and has been generally portrayed as good. Advancing age does not typically affect soap characters' health.

More recently, soap operas have included realistic portrayals of health issues, often as part of a concerted effort to provide factually accurate information to audience members. Soap operas have launched successful campaigns to address a number of health issues, including breast cancer, AIDS, strokes, heart attacks, kidney failure, diabetes, and obesity, to name just a few.

AGING

Daytime serials, in contrast to primetime drama, were for many years a safe haven for aging actors. In our early content analysis studies in the late 1970s, older characters—who were portrayed as being both physically and emotionally healthy—were overrepresented as compared with their actual numbers in the real world. They played important parts in the stories being told. This trend of presenting positive images of older individuals appeared to hold steady up until the mid-1990s, when our observations led us to the inescapable conclusion that subtle changes were indeed taking place. While the number of aging characters was at first not blatantly diminishing, it was becoming clear that many of them could disappear from the canvas without affecting the storyline at all. More of them were being recast in minor roles, and these once prominent tentpole characters were now making cameo appearances or being used for comic relief as their stories dried up. These once powerful matriarchs and patriarchs were sitting on the sidelines, until the turn of the twenty-first century, when—although the actors were physically healthy—their characters began to be killed off as cost saving measures swept through the networks.

FAMILIES

Soaps have always been about family. One of Project Daytime's first studies on television's families, conducted in 1982, reached the conclusion that daytime families, which aired all of their dirty linen publicly, were more realistic than the idealistic families of primetime. While daytime stories were centered around multiple families, primetime generally concentrated on one.

In a second Project Daytime study on families a decade later, we examined a sample of three soaps: *Another World* (NBC), *ATWT* (CBS), and *One Life to Live* (ABC). The data revealed that a number of different family types exist in the soap opera world, with single parent families with female heads of households representing the largest proportion (nearly 35 percent). This observation further suggested that soap operas are reflective of the increase in single parent families in the real world. The traditional family—a married couple with biological children—was still alive and well as the next most-represented family type. The vast majority of the families were middle-class-to-wealthy. Families featured almost equal numbers of female and male breadwinners, and power was equally distributed, suggesting family unity.

However, today, almost a decade-and-a-half later, we note with disappointment that soaps seem to have lost respect for the family unit and that hearth, home, and family have been torn apart. The history of soap operas is important to loyal viewers who always enjoyed the soaps' bringing families together for holidays, weddings, and family gatherings of all kinds—even funerals. Though times change, family, to our minds, is and always will be the heart of the soaps.

PERSPECTIVE
HISTORIAN WILLIAM J. REYNOLDS ON MEMORIES OF *THE EDGE OF NIGHT*

BASED ON AN INTERVIEW BY SAM FORD

> William J. Reynolds is a published historian, focusing on U.S. presidencies and the Ossining, New York, area. He also researches and writes on the history of U.S. soap operas and participates in both online and offline soap opera fan community events. *The Edge of Night* and *As the World Turns* have been among his personal passions since he first made his own "debut" shortly after their 1956 premieres.

I literally grew up with both *The Edge of Night* (*EON*) and *As the World Turns* (*ATWT*), as I was born only three weeks after their CBS debuts in April 1956. When I was in college some twenty-five years ago, I was enrolled in a course entitled "The History of American Popular Culture." One of my class projects was "The History of Radio and Television Soaps." It was while I was doing this project that I was fortunate to meet my now dear friend, Dagne Crane, who played Sandy, the second Mrs. Bob Hughes, on *ATWT* from 1965 to 1971.

The way I first met Dagne was most unusual. I guess it is every soap fan's dream to wake up in a hospital setting and have your favorite soap star hovering over you and caring for you. Well, that's exactly what happened to me. Dagne was working as a paramedic for our local volunteer ambulance corps. Even though she had been off the show for a dozen years, I recognized her immediately. A short while later, I got up enough nerve to interview Dagne for my "History of Soaps" project. To this day, I never tire of hearing her stories of how it was, back then, to do a *live* thirty-minute drama, five days a week, fifty-two weeks a year.

One of the things I enjoyed most about *EON* was the fact that, for its entire run on CBS (1956–1975), the show was done live. [*EON* was subsequently picked up by ABC from 1975 to 1984.] To be able to successfully pull off a half-hour of mystery every day, with all the pitfalls of a live show, was masterful. The scholars of today can also look back on a show that was not afraid to take risks. For instance, the monologue of Adam Drake in his closing summation during the

Julie Jamison murder trial, which took up an entire episode, was soap opera at its very best, and has never been duplicated. Similarly, a decade later, when the two leading characters Mike and Nancy were facing marital separation, the only two actors used on that pivotal day were Mike and Nancy's portrayers, Forrest Compton and Ann Flood. Nothing like that has ever been remotely duplicated.

In particular, *EON*'s format stuck out from the rest of the pack. The primary focus of its soap opera stories was based on crime and mystery. The standard soap themes of jilted lovers, extramarital affairs, and teen pregnancies took a back seat on a show that was mystery-oriented. One of my all-time favorite *EON* storylines involved the Whitneys, whom head writer Henry Slesar, who held the position from 1968 to 1983, fashioned after America's real-life political royalty, the Kennedys. The revelation that the younger son Keith was not dead but actually disguised as hippie Jonah Lockwood was masterfully told. In order to keep his secret, Keith/Jonah sent many a character to an early grave. The storyline was riveting and truly kept you on the edge of your seat every afternoon. And, if I recall, the storyline played out over an eighteen-month period, culminating with Keith's fatal fall off a tower in one of those early on-location shoots.

It does my heart good to see polls that indicate that not only was *EON* one of the top soaps of all time, but it is still warmly regarded by its fan base a quarter-century after its cancellation. In more recent years, Web sites that offer vintage clips and episodes of old TV shows have been offering *EON* and opening the doors for a new generation of fans to enjoy. The soaps of yesterday, which were only thirty minutes in length, in my opinion, told more in-depth stories than today's hour-long shows. Yes, I know that boggles one's mind and does not make mathematical sense, but it's true. Further, today's soap producers feel compelled to outdo themselves and their competition with large-scale special effects and exotic remote location shoots. Soaps feel compelled to give us tornadoes, floods, and explosions to draw the audience in. However, sets do not have to be elaborate, nor do special effects have to be over-the-top. Soaps in their radio days did more to capture the attention of their audiences, only through the use of a script, than any of the shows today. I swear they try to outdo each other with Emmy reels of nominated shows full of car blow-ups, murder, kidnappings, rapes, etc.

The American soap opera is, sad to say, on life support. I grew up in America's golden age of television, both primetime and daytime. I saw the soap opera genre at its best and lost one of my lifelong favorite shows 25 years ago. To see it like it is today is like seeing a dearly loved family member with a long terminal illness. In 2010, with great reluctance, we had to say goodbye to *ATWT* on CBS after more than fifty-four years on the air.

It's a crying shame, and I am not optimistic.

PERSPECTIVE
WRITER PATRICK MULCAHEY ON CHANGES IN SOAP OPERA WRITING CONTRACTS

BASED ON AN INTERVIEW BY GIADA DA ROS

> Patrick Mulcahey is currently on the writing team of *The Bold and the Beautiful*. He has won four Daytime Emmys and three Writers Guild of America awards for nearly three decades in the soap opera industry. He has worked on *General Hospital*, *Guiding Light*, *Loving*, *Santa Barbara*, *Search for Tomorrow*, and *Texas*.
>
> Giada Da Ros has been a television critic for an Italian weekly newspaper for the past eighteen years. She has published essays on a variety of primetime serials, including *Buffy the Vampire Slayer* (WB/UPN), *Gilmore Girls* (WB/CW), *The L Word* (Showtime), *Lost* (ABC), and *Queer as Folk* (Showtime). She started watching U.S. soaps as a teen in the 1980s when they were a booming phenomenon in Italy.

I started out in daytime working for Bill and Joyce Corrington on *Search for Tomorrow* (CBS; later NBC), and then for Douglas Marland on *Guiding Light* (CBS). I didn't work *for* the shows. The Corringtons hand-wrote the paycheck I picked up every week. Douglas's agent wrote the writers' paychecks. My bosses worked for the shows, and the shows contracted with them for all writing services. Douglas and the Corringtons were free to hire whatever writers they liked. We were a team. We were also a package deal. If Proctor & Gamble was unhappy with the way a story was developing, they had to decide whether they were unhappy enough to have a showdown with Douglas—who, if he became disgruntled enough, could walk and take the entire writing team with him. A writer with a team thus structured had tremendous clout, unthinkable today. (Series in primetime, however, and on cable channels like HBO, still work the same way.) The ability of the network or producing company to control the writing of the shows was drastically curtailed. The writer generally got his or her way, or else.

Over the course of the 1980s, networks grabbed not just ownership of most soaps but control of the writing process, too, by contracting for each writer individually. We had little to no say over with whom we'd be working. Members of a writing team were presumed to be interchangeable, the assumption being that any writer can and will work with any other.

For example, on *Santa Barbara* (NBC), the writing was novel and eccentric from the start because that's who creators Bridget and Jerry Dobson were. We whom they hired thought we worked for them, but, technically, we didn't; we were paid by the production company they had partnered with financially. When that company came to a final impasse with the Dobsons, it had either the clout or the cheek to oust them and lock them out—and, suddenly, we who'd been working with Bridget and Jerry were subjected to a stream of head writers we didn't want and who didn't know the show the way the rest of us did. It was a new experience for us at the time. What we did was to throw caution to the wind and write the show like the team we were trained and accustomed to being—Jerry's and Bridget's team—with little regard for who was nominally our new boss. We did it so well that nobody stopped us. Finally, we ran out of steam; network executive Jackie Smith and executive producer John Conboy finished us off. The team scattered, and the show died.

I am consistently tough, I know, on Smith and Conboy. I do blame them for the demise of that wonderful show. But the truth is that, by the early 1990s, all the shows (with the singular exception of the Bell shows) were being written the same way. And they were sinking fast.

Also, when I started in soaps in 1979, we were instructed that "everything must be spoken." The presumption about our viewers was that they were busy doing housewifely things, more listening than watching. A scene without dialogue was out of the question. Even what we still quaintly call "subtext" had to pass muster to the ear: Will a woman ironing in the next room understand it?

Being expected to write for people who weren't paying attention never sat well with me. Douglas Marland attacked the very notion by developing a dense narrative texture that *demanded* the audience watch closely, remember details, match them to new information, learn to compare them to the characters' understanding of the same facts. Even the advent of the VCR didn't dislodge that ironing housewife. She was presumed to work during the day, then come home and iron while she watched her soaps on tape. One technique initially considered edgy on *Santa Barbara* in the mid-1980s was that we dared to do silent scenes, purely visual business; if your eyes weren't on the screen, you had no idea what was going on. Brad Bell [executive producer of CBS' *The Bold & the Beautiful*] tells me that *B&B*, which kicked off in 1987, has always done scenes without dialogue.

Today, I, for one, am convinced that the lives we present in daytime serials resemble less and less the lives of anyone watching them. Surely the prevalence of no-holds-barred talk and "reality" shows—*Judge Judy* [syndicated], *The Montel Williams Show* [syndicated], and the gang—demonstrates that the mores on which our shows are traditionally predicated aren't broadly shared anymore. We're no longer "racy"; we're quaint. There's no such thing anymore as a cutting-edge soap. The proliferation of daytime viewing options—even as fewer people are watching, day or night—means the shows' revenues have shrunk, and so have their budgets. We're under pressure to stretch each fictional "day" over more episodes to get more mileage out of the actors' clothes. We use fewer sets, create simpler shows that can be produced in shorter studio days, and it's harder to get a remote shoot or even a raise approved. It's all led to a ratings panic at the networks, which has spelled the end for the risk-taking soaps were famous for in their heyday. *Stick with what always worked before!*

Now that's something that makes my job tougher—less satisfying too—but can I blame DVR viewing, streaming video, mobisodes, etc.? No. The biggest creative hit daytime took during my career had nothing to do with cable or the Internet or technology at all: It was the radical change in the way writers and writing services are contracted.

PERSPECTIVE
ACTOR TRISTAN ROGERS ON CHANGES IN SOAPS' INDUSTRY, AUDIENCES, AND TEXTS

BASED ON AN INTERVIEW BY ABIGAIL DE KOSNIK

> Tristan Rogers is best known for portraying Robert Scorpio on *General Hospital*, a role he played from 1980 until 1992 and in subsequent stints in 1995, 2006, and 2008. He also played his character on the second season of *General Hospital: Night Shift* in 2008. Rogers writes regularly about the soap opera genre on his official Web site, http://www.tristanrogers.com/.

The current economic crisis has led to soaps basically being talked about, from the networks' point-of-view, as being cancelled. Cancellation was always in the works, but now it's out in the open, and there is a plan to do it. In five or six years—or it could be six months—that's going to be it, unless something drastic happens to change it, but even that would only be a delay. The investment from the network standpoint today is strictly dollars and cents.

This crisis was a great opportunity that the soaps completely blew. They had a chance to bring in storylines that reflected the times, that showed how people were getting through, that generated a feeling of hope. That chance slipped by all of the soaps, and a great opportunity was lost. When Gloria Monty was hired by ABC to run *General Hospital* (*GH*) in the late 1970s, she was told to get the show ready for cancellation. Instead, she threw away the rulebook and reinvented the show. Because of that, we got the golden age of soaps in the 1980s, with *GH* as the vanguard. I never got a chance to talk to Gloria about what it meant that she was brought in to prepare *GH* for cancellation. Was she going to put all of the characters onto a plane and fly them into a mountain? Put them on a boat and sink them? What do you do to wrap up seventy years of history?

NBC Universal President and CEO Jeff Zucker gave an address at the 2008 gathering of the National Association of Television Program Executives about the state of network television. He was very frank. He said that everyone at the networks was operating with a business plan that was sixty years old, and, if they didn't change the plan, they wouldn't be in business very long (Albiniak 2008). In particular, the advent of the Web has changed everything. I still believe to this second that the networks don't get what the Web is. There are some clued-in people, but, as you move higher up the ladder, you see their thinking is that the Web is just another channel. However, the Web is far from being just another channel—it's bigger than that. It's far more complex, but, in another sense, it's much simpler. Everything you pitch now has to be multiplatform: You have to pitch ideas for a TV show, a Web show, and a show for 3G mobile devices. They've all got to work together, but they can't all be the same. The future of television will be all the platforms mingled together to enhance each other.

Alongside the second season of SOAPnet's *General Hospital: Night Shift* (*GH:NS*), I did something for the Web called *Backstage Pass*, which was me taking a camera into dressing rooms and just having a chat with whoever was there. That got a huge amount of traffic; no one had seen that before. That's part of what we've got to do in the future. Soaps have to let people in more, give them access, and show them what's happening on the other side of the cameras. Reality is obviously going to play a much bigger role in TV—I don't think it will be reality for reality's sake but rather reality with structure. I'm impressed with what I've seen with some of the reality shows out there, but I believe they can go a bit further out. Getting the networks to believe in that is a struggle, though.

The Web has also changed how fans relate to soaps. Back in the 1980s, all we had was mail, and it was carefully monitored. All of it was read and catalogued. Criticisms were watched, and the more erudite comments were specially marked. The mail count could radically influence a storyline and the speed of development given a character. A high mail count could do wonders for you.

With the appearance of the Web, this has changed. Overall, fewer people now comment on their favorite soap, even though the means to do it online has made the process quite easy. Therein lies the problem. With the Web, it is much easier to "manipulate" a subject. When it was "snail mail," you had to go to some trouble to write a letter and physically post it. Today, you go to a computer and just do it. What this has brought forth is the rise of the "bitch post." These types of rants are, for the most part, comments posted online on a minute-by-minute basis by, in some cases, rabidly psychotic fans who feel compelled to get issues of no real importance off their minds. While there isn't anything wrong with this, this level of comment rarely serves anything. In the early stages, the networks probably read these comments and may have been horrified. Now, they know better.

However, I also have a far greater appreciation of the fans now than I ever did back in the 1980s. There are a lot of smart people out there who don't watch any more because they realize their soaps have changed too much. But they still like to talk about them. I love hearing what they have to say. Many of their comments are common sense and have direct implications with regard to what is happening today. It's a shame that common sense has such little street value in the soaps industry these days.

One of the great frustrations today is wondering why TPTB ["the powers that be"] don't look back for some inspiration. The reasons for this are based more in ego than anything else. Last year, there was a vocalized desire to attract lapsed viewers. What happened? So far, there has not been any departure from the current agenda, and the ratings are sliding. Without question, lapsed viewers are probably the only audience this genre has left. My guess is that this is a hard sell to the marketing department, which worships at the altar of the eighteen-to-thirty-four-year-old demographic.

And here lies the big problem. In these uncertain times, all of the available information amassed over the last fifty years on viewing habits and demographics is probably useless. The audience shifts on a daily basis, and that's hard to keep up with. But daytime drama has an audience, so why not try accommodating it? Here, we get into a complex mix of indecision, fear, lack of imagination, and the overall desires of the marketing department, where programming statistics trump common sense.

What is effectively the strength and the weakness of the daytime drama genre is its age. No other genre even comes close to the age soaps have. This age, although impressive, is a problem because it makes change difficult and, at times, not well received. CBS' *Guiding Light* (*GL*) is a good example of this. With all the changes *GL* implemented, it was ultimately rejected by the fans and cancelled. Further, the story structure of soaps is just not something that younger people really respond to. Storylines move at a much slower pace, and this isn't something that keeps the attention of the younger crowd. Their idea of a soap is *The Hills* (MTV), which started life as a reality show but has now become much more structured.

The difference between this and a show like ABC's *All My Children*, for instance, is stark, so any introduction of this style of writing to daytime soaps is generally rejected by the core group of viewers. And this is the group that really anchors daytime. The genre has tried to attract a new and younger group of viewers, but so far this has failed. The daytime soaps as they stand will always belong to the baby boomers.

Ultimately, though, soap operas have a meter that is all their own. Change that, and it really isn't daytime anymore. We had a glimpse of how it could be done back in 2008 with the second season of *GH:NS*. On *GH:NS* season two, we had a mix of daytime and primetime formats seamlessly working together, embodying many of the more endearing elements of daytime with the timing of primetime. When I went back to *GH* for a brief reappearance in November 2008, the differences were stark. I thought at first it was me, but, simply put, I'd gone back twenty years.

These shows had their golden period in the 1980s, and that is what they will always be remembered for. This is not a genre that will be around in another fifty years. And, in many respects, it isn't necessary. They have made their mark, and almost every type of medium owes something to the way the soaps have been put together, whether they want to admit it or not.

DAYTIME BUDGET CUTS

—SARA A. BIBEL

> Sara A. Bibel is a soap opera writer and the author, since May 2008, of the "Deep Soap" column on Fancast (http://www.fancast.com/). She was a member of the 2006 Emmy award-winning writing team for *The Young and the Restless*, where she wrote from 2003 to 2007, and she has also written for *All My Children*. She started watching soaps as a child during the 1980s.

2008 will surely be remembered as the year a lot of American institutions collapsed, leaving Wall Street decimated and the automotive industry circling the drain. Though it's generated a lot less publicity and has a cultural rather than economic impact, the American daytime drama also found itself in danger of dying an ignoble death by year's end. As someone who has been part of the daytime television industry as a fan, a writer, and a blogger, I have felt like I was watching a beloved family member slowly succumb to a terminal illness.

I began the year marching in endless circles with my fellow striking writers. A month before our strike started, I was cut from ABC's *All My Children* (*AMC*) after my trial deal ended. It was humiliating and disappointing to lose my dream job before I even had a chance to prove myself. Earlier that year, I was let go from CBS' *The Young and the Restless* (*Y&R*), the show that had become a second family to me. At first, striking perversely seemed like an improvement on unemployment. I had some place to be every morning. I got to know the writers of ABC's *General Hospital* (*GH*), as well as primetime and film writers. The *Grey's Anatomy* (ABC) cast bought everyone pizza. But, rather than end after a couple of weeks, the strike kept going and going. Daytime shows continued to be produced, written by a combination of scabs and writers who resigned the union and went fi-core. Writers who go fi-core effectively resign from the union, though they continue to pay dues and receive union benefits. This strategy allows them to continue working during strikes. An increasing number of writers decided that a paycheck and saving their jobs were more important than the union.[1]

Unfortunately, by the time the strike ended, the shows had learned an unfortunate lesson: They could be created by smaller writing staffs without significant ratings declines. When it finally ended, shows used the strike as a way to get rid of writers who had fallen out of favor. Worse, CBS' two Procter and Gamble Productions/TeleNext shows, *As the World Turns* (*ATWT*) and *Guiding Light* (*GL*), and *AMC* opted to eliminate the breakdown writing position, saving hundreds of thousands of dollars a year. As someone who has worked primarily as a breakdown writer, I know how much of the actual story and show structure comes from this position, whose main role is to structure the show, creating a detailed scene by scene outline of each episode. Losing this layer of writers will do long-term damage to the genre; indeed, it may have contributed to both *GL*'s and *ATWT*'s eventual cancellation. As *GL*'s run concluded, news surfaced that some of its writers would become breakdown writers for *ATWT*, but the final fate of *ATWT* may have already been all but decided by that point.

Meanwhile, NBC's *Days of Our Lives* (*DOOL*) was on the verge of cancellation. In January 2007, NBC Entertainment president Jeffrey Zucker announced that he planned to discontinue the show when its contract expired in 2009. *DOOL* struggled with budget problems for years, in part due to an ironclad contract with Jim Reilly that forced the show to keep paying him after he was fired—and even after he died. The show languished in the ratings, as new head writer Hogan Sheffer's storylines failed to engage the audience. During the strike, Executive Producer Ken Corday opted to replace Sheffer with former head writer Dena Higley. After the strike, he was stuck paying off Sheffer and all the other writers who were fired. This put a further strain on the bottom line. The show's internal problems were exacerbated by alleged conflict between Higley and Executive Producer Ed Scott. Corday opted to replace Scott with Gary Tomlin. Finally, after months of negotiations, the show was renewed for eighteen months in November. The catch was that the budget had to be cut by 40 percent. The first casualties were prominent stars Mary Beth Evans, Stephen Nichols, and Thaao Penghlis, and the iconic Diedre Hall and Drake Hogestyn, who played the roles, respectively, of Marlena Evans and John Black. Numerous other actors on the show agreed to large pay cuts.

NBC also cancelled *Passions* in 2008, after moving it to an exclusive channel owned by satellite provider DirecTV. Some hoped that this experiment would prove that a small cadre of loyal fans could keep a show afloat. Unfortunately, NBC and DirecTV determined that it was simply too expensive to keep producing *Passions* for such a small audience, although the experiment paved the way for the critically acclaimed primetime show *Friday Night Lights* to split its production costs between the two companies by airing first on DirecTV and then on NBC.

The network's daytime problems were the result of decades of mismanagement, leading to consistently bad decisions. One prominent example was NBC's 1970s expansion of *Another World (AW)* to ninety minutes, making it impossible to maintain the show's quality while producing a greater volume of material. Then, in the 1980s, NBC allowed New World to fire *Santa Barbara*'s creators just as the show's ratings were starting to rise. Subsequently, *Santa Barbara*'s ratings fell and never recovered. The network then cancelled this critically acclaimed soap (my all-time favorite) in 1993 and replaced it with game shows. Several years later, the network launched *Sunset Beach*, another soap set in Southern California. The show was consistently the lowest rated soap; however, when the network decided to cancel a soap to make room for *Passions*, they chose to keep *Sunset Beach* and cancel *AW*, a show that had been on their lineup for decades. One problem is that NBC's daytime line-up performs better in smaller media markets than in the big urban centers where NBC owns stations. ABC has the opposite viewer profile, doing well in places where it owns stations. As a result, NBC has less of a financial incentive to promote and invest in its soaps. However, *DOOL*'s ratings have risen, and the show was consistently ranked second in key demos by mid-2009. As of this writing, it appears that *DOOL* has improbably saved itself.

The outlook might be particularly bleak for the future of the genre on NBC, but the situation hasn't been much brighter for the other network line-ups. Elsewhere in this book, Patrick Erwin looks at the changes that were put in place at *GL* in light of the budget crunch and *GL*'s eventual cancellation on CBS, while the network's other shows have been tightening their taping schedules, eliminating longtime actors, and casting an increasing number of parts as non-contracted roles.

Meanwhile, things are no better for the Disney/ABC Daytime soaps, which should theoretically be most secure because the network owns its three soap operas directly. Even SOAPnet, the Disney-owned cable network dedicated to the soap opera genre, turned its back on daytime in 2008. I was thrilled when the channel launched, envisioning it as a showcase for classic episodes that would honor and uplift the genre. I imagined a line-up of classic episodes, *Entertainment Tonight*–style news about the genre, and original soap operas. Instead, since the head of ABC Daytime took over the channel, it has veered away from soap operas and toward becoming a lowbrow, general interest women's channel. For instance, the second season of the network's *GH* spin-off, *General Hospital: Night Shift*, received tremendous critical acclaim on SOAPnet, yet it seems unlikely the network will eventually pick up a third season. Also, in 2009, when CBS declined to air the Daytime Emmy telecast,

SOAPnet seemed like the obvious home for the show. The channel was not asked to air the telecast itself but declined to air a red carpet show or a rerun. The awards instead eventually landed on the CW Network, which has no daytime television line-up. SOAPnet's new programming instead included movie telecasts and reality shows. Although these shows didn't do well in the ratings, the network was trying to prepare for the post-daytime television universe (but will leave the air in 2012).

On top of NBC's struggles and SOAPnet's move further away from traditional daytime soaps, a *Portfolio* magazine article caused major waves at the beginning of 2009 by implying that the soap opera was an endangered species and breaking the news that daytime's biggest star, *AMC*'s Susan Lucci, had agreed to a substantial pay cut (Lidz 2009). Soaps have been quietly cutting actors' salaries during contract renewals for several years. When it happened to Lucci, it generated headlines in the mainstream entertainment press. *GL* went through similar across-the-board salary cuts several years ago, and it seems that every show will experience significant cutbacks during the time of this book's publication.

In 2009, *Y&R* started demanding that its highest paid actors take pay cuts before their contracts expired. Jess Walton and Melody Thomas Scott briefly left the show before agreeing to work for lower salaries. *Y&R* also took a hard line with Eric Braeden, allowing him to walk away from the show for weeks until he agreed to a substantial reduction in his salary. Some speculate that Sony, which is involved in producing both *Y&R* and *DOOL*, took note of *DOOL*'s ratings growth after the departure of Hogestyn and Hall and concluded that veteran actors are expendable.

Though talent may be the most publicized focus of budget cuts, these measures affect every aspect of a daytime drama, from the writing team (as I mentioned earlier) to the sets. These budgets have been slowly declining since the 1990s. Falling ratings transformed the shows from extraordinarily profitable entities that helped keep the networks in the black to loss leaders. With significantly lower ratings, daytime ad revenues dropped significantly. The networks responded by cutting the shows' budgets. In the late 1970s, *GH* allegedly generated one quarter of ABC's profit. Now ABC is but one piece of the Disney empire, and *GH* is a show with fewer revenue streams than many of its other properties. Gone are the lavish sets and location shoots. Popular actors can no longer count on the near seven figure salaries standard for almost any actor in a successful primetime series.

What do these budget cuts actually mean for the production of these shows? When I worked in the production office at *Y&R*, we went through a less extreme belt tightening. Like any long standing business, *Y&R* was subject to some genuine inefficiencies. People were accustomed to doing things in a certain way and didn't adopt newer, better technologies when they became available. I was shocked to learn that *Y&R*, and most other soaps, employed script typists well into the 1990s even though affordable, user-friendly screenwriting software was available. During "my" budget cut, logical improvements included switching from Polaroids to digital cameras for continuity photos (used to make sure actors' hair, make-up and wardrobe match throughout multi-episode soap days), saving thousands of dollars a year. Less pleasant was the end of annual cost-of-living salary increases for much of the staff.

In daytime, actors, writers, directors, and producers are paid very well relative to the average American. Most of the crew is unionized but hourly, which guarantees them a solid middle class salary as long as they work enough hours. Those who work in the production office (assistants, production coordinators, receptionists, etc.) do vital work. Their salaries are low, especially for Los Angeles or New York. Some of the staffers are recent college graduates who are hoping to move up the ladder and understand that entry-level positions in the entertainment industry have terrible salaries. Nevertheless, plenty of people have chosen to stay with shows that they love for many years for the equivalent of junior secretarial salaries. Actors who make four hundred grand a year can afford to take a 25 percent salary cut without significantly altering their lifestyle. They also have the ability to pursue additional income opportunities, including guest starring on primetime and public appearances. People who work sixty hours a week for $30,000 a year can't afford a pay cut and don't have the time to work a second job. The result is that talented and ambitious young people who could be the future of daytime are going to flee to greener pastures.

Soap wardrobe departments make good use of Loehmann's and department store sales, but it wouldn't be plausible to outfit characters who are supposed to be millionaires in clothes from Target. The same holds true for the sets themselves. I used to fantasize about living in Cruz and Eden's *Santa Barbara* beach house. I can't think of a single newer daytime set that is similarly visually enticing. Nor can I recall the last memorable soap couple's love song—music licensing fees are expensive, as is hiring a composer to write something original that the show owns. Television is a visual medium. Production values matter. Good storytelling does not require a lavish location shoot in Rome, but a bedroom is often necessary.

Budget cuts also mean that shows have less time to tape each episode. This is bad for the actors, who no longer have time to rehearse scenes. Further, the condensed schedule means viewers can forget about seeing fun party scenes and big weddings. They take all night to shoot. Shooting an episode in eight hours instead of ten means no overtime for the crew. That's great for the bottom line but lousy for the staff who counted on that income.

One of our mandates on *Y&R* was to minimize the number of set moves—the sets that changed from one episode to the next—because it takes significant time and manpower to move them. If we could go a whole week without any moves, no stagehands were needed, and they didn't get paid. This trend is why viewers are driven insane watching characters have private conversations in the middle of crowded restaurants and sex scenes on living room couches. It is also why *GL* moved to standing sets.

I realize that, in order to survive, daytime must find a way to deliver high quality entertainment for far less money, especially since talk and court shows are less expensive forms of programming that have proven popular with the daytime audience. In 2009, I visited *GL* and watched the non-union crews tape actors at lightning speed, both on tiny, standing sets and on location. After an incredibly rocky start for its new production model, the show improved dramatically in its final few months on CBS. I'm beginning to think this model just may be the future of daytime. Another strategy might be to persuade actors to take large pay cuts in exchange for the freedom to pursue other projects. *DOOL*'s ratings have held steady despite the firing two of its biggest stars, indicating that shows may not need all of their veteran actors to maintain their audience. These ideas are the opposite of what most fans would say they want, but daytime shows are facing a tough choice: change or die.

With SOAPnet out of the picture for bringing new life to soaps, the Internet is another possible delivery system for daytime dramas. CBS streams its remaining two soaps on multiple Web sites hours after their broadcast airing, and *Y&R* does particularly well, ranking in the top ten shows for time spent online as of December 2008 (Nielsen Online 2009). Episodes are available of all three ABC soap operas on ABC.com, and *DOOL* episodes are available at NBC.com and for purchase on iTunes. Given that soap opera fans are incredibly active online, this approach seems strategic, but it's been difficult to monetize Internet television viewing, especially with the impact of the recession on advertising revenues. Meanwhile, both fans and P&G/TeleNext have questioned whether an alternative distribution scenario might give *GL* or *ATWT* a future beyond network television. If the genre has any hope of surviving, it must find a way to become profitable through a combination of broadcast airings and on-line streaming.

NOTE

1. For more on the writers strike, see Metzler's essay in this collection.

AGNES NIXON AND SOAP OPERA "CHEMISTRY TESTS"

—CAROL TRAYNOR WILLIAMS

> Carol Traynor Williams is a retired professor of humanities at Roosevelt University and a freelance writer in Southwest Harbor, Maine. She is author of *"It's Time for My Story": Soap Opera Sources, Structure, and Response*, as well as other published work on the soap opera genre. Williams first became a fan of soaps on sick days away from school, listening to radio soaps like *Portia Faces Life* and returned to the genre in 1980 when recuperation from surgery led to her following *Ryan's Hope* and then other ABC shows: *All My Children*, *General Hospital*, and *One Life to Live*.
>
> Agnes Nixon is the creator of ABC's *One Life to Live* and *All My Children*, a show for which she continues to serve as a creative consultant. She also was co-creator of ABC's *Loving* (1983–1995). Prior to launching her own shows, Nixon worked under soap opera pioneer Irna Phillips as a writer for CBS' *As the World Turns* and as head writer for CBS' *Guiding Light* and *Search for Tomorrow* and NBC's *Another World*. She has won five Writers Guild of America Awards and five Daytime Emmys.

My conversations with soap opera pioneer Agnes Nixon took place between November 2008 and April 2009. This was the time the country's economy was failing, and the same can be said of soaps. CBS' *Guiding Light* was cancelled. Classic stars were fired, incuding Deidre Hall and Drake Hogestyn (Marlena and John) and Stephen Nichols and Mary Beth Evans ("Patch" and Kayla) from NBC's *Days of Our Lives*. Other major stars became ancillary, such as Erika Slezak (Viki) from ABC's Nixon creation *One Life to Live* (*OLTL*), and even Susan Lucci (Erica) of Nixon's "baby," ABC's *All My Children* (*AMC*) was marginalized. Soap writers were scattershooting new pairings all over their yards, or testing for "chemistry," which is how *AMC* head writer Charles Pratt,

Jr. labeled moving the character Kendall from her fan favorite pairing with Zach to Ryan instead (*Soap Opera Digest* 2009a).

Fans, soap critics, casts, and—most tellingly—behind-the-scenes creative staff people, including Agnes Nixon, consultant to *AMC*, spoke out for more stories for the long-term (and better) actors, for stories true to character, and for "money couples," like Kendall and Zach. For fans, perhaps the most unbelievable aspect about moving Kendall from Zach to Ryan was Pratt's statement that Kendall had "subconscious[ly]" loved Ryan throughout her four years with Zach (*Soap Opera Digest* 2009a).

Shocked by the Kendall-Zach split in mid-February, Nixon said, "I have no control over story anymore." And for the first time in our relationship, which dated from 1987, she allowed me to quote her saying something negative about her co-workers. Meanwhile, Alicia Minshaw (Kendall) spoke out furiously: "Even if it's written that it's over for us . . . I am not going to play that! I refuse. I have invested seven years into this character, four years with him, and . . . [a]ny chance I get to play that she still loves this man, I'm going to show it" (Levinsky 2009).

Meanwhile, this creative decision also endangered the groundbreaking story of *AMC*'s lesbian lovers Bianca and Reese. In November 2008, Nixon had spoken lovingly of Bianca and Reese as the latest in her long string of "issue" stories, adding that "[C]redit is due to Chuck Pratt," and also to the network higher-ups—because "[E]ven [they] have said, 'Keep it in your tradition.'" "But," she added, "they [also] want to move fast and attract the young eighteen-to-forty-nine audience. Sometimes it's difficult to do both at the same time." This was a prescient summary indeed, as in early 2009 Bianca and Reese were broken up. This couple "in Nixon's tradition" appeared to be separated to aid and abet the Kendall-Zach breakup, as Zach and Reese were said to have "chemistry," to give Zach a new pairing now that Kendall had supposedly always been in love with Ryan.

Nixon's forte has always been the social issue story. In 1983, Nixon had taken on the then rarely addressed subject of women's alcoholism, giving the story to a character who was a sweet, young stay-at-home mother. This plot led to Nixon's first try at writing a lesbian storyline for a soap opera, making the young mother attracted to a lesbian child psychologist. But the attempt was aborted. "It was too early," Nixon said. Not so with Bianca and Reese. The audience had watched Bianca being born to the iconic Erica, had seen Bianca grow up, and had lived through Bianca's and Erica's pain at Bianca's coming out. Helped by the acting of portrayers Eden Riegel and Tamara Braun and the slow and complex pace of Bianca and Reese's love story, the audience invested deeply in their relationship, making *AMC*'s "chemistry" experiment all the more difficult for fans, for Nixon—and for Riegel, Bianca's portrayer, who said of Reese: She was "absolutely distraught when the bad stuff went down" (*Soap Opera Digest* 2009b).

Another recent social issue storyline for Nixon deals with war. *AMC* is well known for taking on the Vietnam War. Mary Fickett, playing Ruth Martin, mother of a missing U.S. soldier, was awarded the first Emmy given to a daytime actor. Nixon humanized the Vietnamese people by writing a plot in which a Vietnamese veteran rescued Ruth and Joe Martin's adopted son, Phillip Brent. As Nixon said, "[Y]ou saw the fuselage coming down and Phillip lying on it, and Binh saving him, and then when Binh got out of the water, he had no legs, and it was very, very affecting." Nixon laughs that she and the actors in the story (including Ruth Warrick and Hugh Franklin playing "hawks" Phoebe and Charles Tyler) were all "doves to the teeth." But, as in all of her social issue storylines, Nixon showed both sides of the conflict, aiming to teach both sides empathy.

Her reasoning grew from what the noted social organizer Saul Alinsky said to her in the early 1970s. As Nixon recounted (Williams 1992, 100):

> [He said,] "When I go to a meeting, . . . I make a point of not bringing out the baby pictures because once you bring out the baby pictures and pass them around, you can't negotiate anymore. You've become friendly. . . You [television people] can do that, you can bring out the baby pictures, and explain the hard hats to the liberals, and so on.". . . I thought that charge applied particularly to daytime, and I did it in a lot of ways.[1]

In 2008, *AMC* tackled a new war through J. R. Martinez, not a seasoned actor but an Iraq War veteran badly burned and scarred when his Humvee hit a land mine in 2003. Martinez played Brot Monroe, a battle-scarred vet in what was called "a landmark story" (Logan 2008). Brot was brought on as the presumed-dead fiancé of Taylor. Yet in April, Pratt confirmed that pairing Taylor with longtime *AMC* character Tad was a move "definitely in the experimental stage, the chemistry test stage," because Taylor fans wanted "her story [to] heat up." Meanwhile, his opinion on the Brot Monroe story? "I'll be honest, it's kind of gotten back-burner and boring" (*Soap Opera Digest* 2009c).

In our first conversation for this piece, Nixon laughed that, in spite of her wish to retire and enjoy her family (she is the mother of four and grandmother of eleven), she was still deep in the fray, fighting for the creative health of her "baby," *AMC*. At that time ("Black November," as it was being called because of the economy), I asked Nixon about soaps' declining viewership. She answered like any business executive: "It's the financial situation that's very important. We're all fighting for ratings because of cable, and I would say only because of cable." But other culprits quickly came up: The technical difficulties of charting time-shifted viewings, and the competition posed by ubiquitous (and cheap) reality shows. In particular, she dwelled on the networks' focus on youth, what she called "the demographic battle":

> That's what [the networks] can charge [advertisers] on. It's a mistake to ignore the old. I learned that years ago [on *AMC*] with the triangle of Dr. Charles Tyler, his wife Phoebe, and Mona Kane. That was a *romance*. Do you remember Mona following Charles and Phoebe to Italy, and hiding in doorways? . . . Of course that was romantic comedy. No one was more surprised than I was at how popular that story was.

This led us to discuss the popularity of older "grandparent" characters, who fit into soap canvases as supporting characters, rather than plot-drivers. Nixon knew how much the audience loved *AMC*'s Grandma Kate Martin, the show's bulwark. Nixon wrote, and still loved, the "quiet moments" between characters of different generations, for example, when Kate comforted—and gently wised up—her then-roguish grandson, Tad. The genre's names for these wise elders are "tentpole" characters or the "Greek Chorus." During the time of my interviews with Nixon, *AMC* lost one of those characters, Myrtle Fargate, when portrayer Eileen Herlie died in late 2008. Nixon even appeared in a December 19, 2008, episode dedicated to Myrtle, blowing a kiss at Myrtle's portrait during her memorial on the show.

When I brought up the common viewer criticism that soaps today recycle "tired old repeated plots," Nixon linked that, too, to the "demographic battle." "Of course, I believe there are no new plots. There are just new characters to live the old [folk and fairy tale-based] stories. But, today, the plots don't seem fresh because they move too fast. We can't get hooked on the characters." She cited a focus group she had been to in Philadelphia the week before, where the sample viewers complained that "[T]he pace of events is too fast for us to get to care about the characters." Nixon added ruefully, "Today, [viewers] do not have the empathetic experience."

Maybe there's hope, though. Bianca and Reese were reconciled just prior to both characters leaving *AMC* in April 2009—a bittersweet reunion, for sure, but a chance to keep a groundbreaking story intact. Meanwhile, just a week after announcing Kendall's "subconscious" love for Ryan and the new pairing, Pratt said that, in May, Ryan would decide that the "real love of his life" was Greenlee (*Soap Opera Digest* 2009c), and events would bring Zach and Kendall back together again. Through all the stress-laden ups and downs of the "perfect storm" of winter 2008–2009, Nixon and the *AMC* producers and cast—and fans and reporters—may have made their influence felt. Writing that stays true to established characters may prevail—at least for this moment.

NOTE

1. This quote was from an interview conducted with Agnes Nixon on March 17, 1987.

GIVING SOAPS A GOOD SCRUB
ABC'S *UGLY BETTY* AND THE ETHNICITY OF TELEVISION FORMATS

—JAIME J. NASSER

> Jaime J. Nasser is a lecturer at the Film and Digital Media Department at the University of California Santa Cruz. He recently completed a postdoctoral fellowship at the School of Cinematic Arts at the University of Southern California after finishing his doctoral dissertation, "Exporting Tears and Fantasies of (under) Development," which focuses in part on the emergence of the telenovela alongside local and global politics of modernization. He watched his first telenovela, *Rosa Salvaje*, with his grandmother in the midst of El Salvador's civil war when he was eight years old.

INTRODUCTION: "MAKE WAY FOR THE SUPERSOAPS"

The above headline (Dominguez 2006) is attached to just one of the many popular and trade press articles appearing in the Summer of 2006 that positioned Latin American telenovelas as completely different from, and perhaps "better" than, U.S. daytime serials.[1] Around this time, the television industry in the U.S. started generating significant buzz regarding the possibility of adapting telenovelas for the American market. In Fall 2006, two attempts at translating telenovelas into American television programs took place. One of these attempts occurred as the result of the merger of UPN with the WB to form the CW Network. News Corporation, the parent company of Fox, acquired the remaining UPN stations and formed MyNetworkTV (MNTV). As a new network, MNTV had to find an identity quickly. To achieve this end, network executives took a step that was considered very bold—or, as one reporter wrote, as "breaking new ground" (Lisotta 2006a)—when they decided that the new network's primetime schedule would be dominated by adaptations of Cuban and Colombian telenovelas, each airing six nights a week during the

8 p.m. and 9 p.m. EST slots: *Fashion House* and *Desire*. During that same season, ABC premiered its adaptation of the Colombian telenovela *Yo Soy Betty La Fea* ("I am Betty the Ugly," hereafter *YSBLF*) on Thursday nights at 8 p.m. EST. While ABC's adaptation of the telenovela format for U.S. primetime was considered a Nielsen ratings success, MNTV's adaptations failed to maintain a significant audience.[2]

This essay will focus on how Anglo-Saxon media industry trade press and critics praised ABC's *Ugly Betty* (*UB*) for its difference from its Latin American telenovela forebears and framed *UB*'s success as a type of sensational transformation rather than a continuation and adaptation of a format that is itself an adaptation of the American daytime soap opera. For two reasons, I contrast the media's treatment of *UB* with their more negative treatment of MNTV's telenovela experiment. First, the critical dismissal of primetime telenovelas reinforces the notion that there is a cultural bias against daily melodramatic serials. Second, the importance of the daytime soap opera to the development of international television formats and genres is completely erased by representing the telenovela as a foreign format, an impulse that is informed by xenophobic tendencies dismissive of Latin American culture.

THE CULTURAL BIASES AGAINST THE TELENOVELA AND DAYTIME SOAP OPERA

A longstanding cultural bias against daily melodramatic serials such as the soap opera and the telenovela became evident in the press coverage of the telenovela experiments that took place in the fall of 2006. First, the media treated MNTV's efforts to maintain the original telenovela format as an experiment, despite the telenovela's obvious ties to and emergence from U.S. daytime soap operas. Journalists expressing doubts about the MNTV telenovelas' prospects for ratings success explicitly highlighted the foreignness of the telenovela format and also employed adjectives commonly regarded as ethnic stereotypes, as exemplified by a headline in *Variety* in July 2006 which read, "MyNetwork turns up heat on spicy telenovelas" (Larmonth 2006). Press characterization of the telenovela as foreign went hand-in-hand with erasure of any similarities (of which there are plenty) between the telenovela and the soap opera, one of America's oldest broadcasting formats. The telenovela format in fact originated as an adaptation of the American soap format in Latin American countries such as Cuba, Mexico, and Venezuela when television was introduced in the late 1940s and 1950s. For example, a *TelevisionWeek* article discussed how TV executives believed it would be necessary to educate American audiences about the telenovela format as if it were something entirely foreign and different (Greppi 2006). Bill Lamb, the president and general manager of the Louisville, Kentucky, Fox affiliate, was quoted as saying: "We're kind of introducing a new way of watching TV in the United States for the Anglo public or the non-Hispanic public ... There's going to be a little education required." The *TelevisionWeek* reporter reinforced Lamb's perspective by stating, "[T]he network's format is far different from the primetime norm, and therefore a gamble for not only its owners but also the local stations, like WMYO, who sign on. That's especially true in markets like Louisville, where the MyNetworkTV format is largely unfamiliar to most viewers. The format borrows heavily from Spanish-language telenovelas, which are popular throughout the world but familiar in this country only to viewers of the Spanish-language networks" (Greppi 2006). This statement minimizes—to the point of erasing—any acknowledgement of the impact soap opera conventions have had on U.S. primetime. In other words, these media stories from 2006 expressing doubts that American primetime audiences will understand telenovelas are a continuation of the press's disregard for daytime soaps' influence in general.[3]

Telenovelas were born out of American government involvement in Latin American broadcasting after World War II. The U.S. took an interest in Latin American radio and television in order to discourage government-owned and operated broadcasting networks because the U.S. government feared the rise of socialist forms of national mass media industries after the fall of Nazi Germany and the rise of communism. Therefore, the U.S. government sought to promote a commercial model of broadcasting in Latin America. The telenovela and the U.S. daytime soap share a common ancestry and thus share narrative and aesthetic conventions.[4] The American trade press's avoidance of this link between telenovelas and American television, as indicated by the examples above, and the suspicion that Latino television styles would not appeal to American viewers both demonstrate bias. One bias is that American tastes cannot accept "degraded," or culturally less valuable foreign programming—even though it is a historical fact that this type of foreign programming originated with American tastes. A second preconception is that daytime formats cannot possibly translate successfully to primetime. The examples discussed above show how both Latino television and daytime soaps are regarded as low culture and degraded genres by the American media trade press.

U.S. PRIMETIME AWAKES FROM A TELENOVELA NIGHTMARE

Episode twenty-four of ABC's *UB*, entitled "How Betty got her Grieve Back," begins with a fantasy that attempts to imitate a low-budget telenovela. In this scene, Betty is a maid brushing the portrait of her boss as she cries because she has a crush on him. Betty's love interest, Henry, comes in wearing a cowboy hat. Henry tells Betty that he loves her, and they embrace and kiss. At that moment, Henry's ex-girlfriend Charlie enters, pregnant with Henry's baby, and tries to shoot Betty and Henry. The telenovela ends as soon as Betty wakes up and realizes that it was all a bad dream, and the opening credits for the show begin. This dream sequence is stylistically reminiscent of the telenovela that the Suarez family watches on their television set. The vast majority of these brief telenovela scenes can be read as tongue-in-cheek references to the original form, paying a type of homage through parody to the telenovela format. What is significant about episode twenty-four's opening is that it explicitly represents these fading "Latino telenovela moments" as a bad dream from which Betty must awake. By having short scenes embedded within *UB* that are stylistically different from the polished look of the series, the primetime drama distances itself from the original low-budget telenovela *YSBLF*. Therefore, these telenovela scenes serve a double function: On the one hand, they pay an homage to the telenovela format; on the other, they distance the low-quality aesthetics of the Latin American, Spanish-language telenovela from the American re-interpretation of the telenovela, with the latter portrayed as a high-budget, primetime, award-winning, "quality television" show, *UB*.

U.S. primetime wishes to differentiate itself from daily serials such as the telenovela and U.S. daytime soaps could be said to represent a larger cultural bias against forms of melodrama that are primarily aimed at female and Latino audiences. As early as December 2006, just a few months after *UB*'s premiere, *New York Times* writer Bill Carter (2006b) observed that U.S. television networks' desire to experiment with telenovela adaptations which "was sizzling a year [before]" had vanished. Carter's article argued that *UB* is not a telenovela in its original form. In order to explain the transformation, he quoted Ben Silverman, who originally planned for *UB* to follow the telenovela format. Silverman changed his mind when, he said, it became clear that the show wouldn't have been able to hire a star of America Ferrera's caliber or to shoot on location in New York if they adhered to a budget typical of the telenovela genre. Carter paraphrases Silverman as saying that, "[T]he expensive look of the 'Betty' series, which is set in the glamorous world of New York couture magazines, could never have been fashioned on the budget of a real telenovela." Adding to this, Carter discusses how network executives thought that they would only alienate viewers if they provided shows of a "distinctly lesser quality than their regular [primetime] shows."

Carter's focus on aesthetics and budgets shows an effort to separate *UB* from its original form, effectively foregrounding the tendency to frame ABC's adaptation of *UB* as a type of "transformation" from a lower cultural form, the telenovela, into a more sophisticated type of program. Implicit here is a narrative of progress informed by a hierarchy of taste that regards daily melodramatic serials as inferior. This narrative of change frames the adaptation of UB as a form of cultural transformation wherein translation for American audiences required an obvious improvement on the Colombian original. This change from low-budget to high-budget is rendered "natural" in episode twenty-four, by deploying telenovela aesthetics to represent the nightmare from which Betty must awaken. Similarly, the regular appearance of telenovela scenes on the Suarezes' television set naturalizes the distance between the telenovela format and American primetime by relegating this type of melodrama to a Latino neighborhood in Queens. Simply put, UB has been denatured of its telenovela ancestry in order to appeal to a more "discerning" American audience that presumably would not accept cheap melodrama—especially if it comes from Latin America.

Even though it is common practice for successful program formats (such as sitcoms and reality shows) to be adapted for U.S. television, these imports are mostly from Europe and Australia. Far less frequently, a television program from a Third World nation achieves notoriety in nations in the "developed" world. That UB has gained a successful reputation— garnering high ratings beyond Latin American countries—as a direct result of the common assumption that, while the U.S. might frequently adapt formats originating in other Anglo nations and U.S. formats will often be adapted by Latin America and other Third World regions, Latin American formats aren't likely to succeed in the American market.

THE GEOPOLITICS OF TASTE: "HOW *UGLY BETTY* CHANGED ON THE FLIGHT FROM BOGOTA"

This section's headline quotes the title of an article from the *New York Times* on *YSBLF*'s transformation into *UB* (Rohter 2007). The headline explicitly maps a sense of geopolitical space onto the aesthetic changes that ABC made from the original *YSBLF*. This reference to travel and immigration is utilized to characterize the adaptation of the original version from Colombia. As argued earlier, some news articles regarded this change as not only geographical, but also as an alteration in quality and taste.

One of the most colorful metaphors used to characterize *UB*'s arrival was Britain's *Telegraph*'s "How *Ugly Betty* turned into a swan" (Wilson 2007).[5] This headline necessarily complicates our discussion of this shift in format as being a simple "adaptation," "change," or "translation," as it implies a different type of narrative of change: from an inferior Colombian original to a superior American version. The *Telegraph*'s headline reflects a hierarchy of taste that is not only biased against women's programming (as argued by feminist television scholars like Ian Eng [1985], Charlotte Brunsdon [1997; 2000], Ellen Seiter [1982], and Tania Modleski [1979; 1982], among others) but also biased against Latin American culture.

Another British article stands out because of the way it blurs the line between global politics and fiction. In the piece, critic Mark Lawson, writing for *The Guardian*, observes: "Although the pilot episode of [*UB*] had the misfortune to share several jokes and one plotline with *The Devil Wears Prada*, it immediately seemed a more mature and surprising version of the Cinderella at Vogue storyline" (Lawson 2006). After describing the basic premise of the plot—that it is precisely Betty's "disadvantages" (ugliness) that lands her the job at Meade fashion magazine—Lawson proceeds to describe *UB*'s success in the U.S. as an adaptation of a 1999 Colombian telenovela, even though "America has an export rather than import mentality." The review flows from descriptions of the fictional character's ugliness, to the "Cinderella at Vogue" plot, to references to global politics, and finally concludes with the uneven exchange of television programs between Latin America and the U.S.:

> [*UB*] is the story of Betty Suarez (America Ferrera), a gauche Latino woman who lands a post as assistant on a top U.S. fashion magazine, *Mode*, despite a coiffure that knows only bad-hair days, *dental braces that resemble the fences at Guantanamo Bay, eyeglasses issued by the Colombian National Health Service, and a wardrobe bought from discount brochures*" (Lawson 2006, emphasis added).

This description of Betty foregrounds the geopolitics embedded in these types of hierarchical articulations of taste, as it clearly maps global politics onto the wardrobe and body of a lower middle class Latino woman. The global violence and human rights violations synonymous with the Guantanamo detention facility are mapped onto Betty's braces, and Betty's big eyeglasses are described in terms of global social and class inequality through a reference to Colombia's health service. Such a description was preceded by a comment that the U.S. series "has some way to go before it belongs in the company of *Dad's Army, Fawlty Towers, Only Fools and Horses* and *The Office*," all BBC series (Lawson 2006). This judgment is significant because not only is Betty's "ugliness" associated with crises in Latin America (that are in great part influenced and worsened by the international policies of the U.S. towards Latin

America, the program is also found inferior to British comedies, which have long been a recurring source of stories for American TV.

Lawson's article describes the success of the adaptation of the Colombian telenovela in the U.S. in terms similar to his "Cinderella at Vogue" approach to *UB*'s plot, pointing out how rare it is for American television to import programs from Latin America. This analysis effectively renders the Colombian adaptation for U.S. primetime a Cinderella story itself, having transformed from a degraded, overlooked, and cheap Latin American format to into a highly praised glossy primetime program on U.S. screens.

Presenting an alternative to the patronizing attitude in the trade press to the telenovela, *UB* producer Silvio Horta was quoted as saying, "[W]e always wanted to have a depiction of a first-generation Latin American family. To me, the appeal was always this girl straddling these two different worlds of working-class Queens and a glamorous Manhattan setting" (Dominguez 2006). Horta's description of the series appeared in an article framing *UB* as possibly paving the way for the introduction into U.S. television of more telenovelas originating in Latin America. A telenovela director in the U.S., Jayme Monjardim, is quoted in the same article as saying, "[T]his is something that we know how to do very well [making successful telenovelas], and when we figure out how to translate our sentiments, well, then just watch out" (Rohter 2007). This quote points to the ways in which a television format is made to embody the experiences of a particular culture yet does not imbue that culture with negative implications. When Mojardim's statement is considered alongside Horta's articulation of the appeal of *UB* as being the story of a Latino woman "straddling [. . .] two different worlds" and Lawson's inscribing of geopolitics onto Betty's body, we can see that *UB* was widely represented as a "first-generation" attempt to adapt a telenovela to U.S. primetime. In this way, the initial appeal of a "Cinderella at Vogue" storyline was conflated with the Cinderella-like transformation of a Latino telenovela into a primetime series.

It may seem ironic that the press's focus is on the transformation of the telenovela format to American primetime when one considers the original storyline of Colombia's *YSBLF*. This plot is centered on Betty's unrequited love for her promiscuous boss, whom she marries in the final episode after undergoing a radical transformation and makeover from an "ugly" lower-middle class assistant to a "beautiful" executive who takes over the magazine. This doesn't mean that the "Cinderella narrative" has lost its appeal for American audiences. Rather, I would argue that the transformation of the American Betty already took place at the highly publicized and widely acknowledged level of television formatting and transnational format reformulation (transformation).

CONCLUSION

While the U.S. version of *UB* found a sizable audience during its first season, not all telenovela experiments were as successful. MNTV tried to import the telenovela format at the same time that ABC launched *UB*, yet none of the telenovelas launched by MNTV was considered a ratings success. Critical reaction to the failure of MNTV as compared with the success of *UB* emphasized MNTV's inability, contrasted with the ability of *UB*'s producers, to change Latin American telenovelas into high-quality, high-budget American shows. For example, five months after the debut of the CW and MNTV, *TelevisionWeek* ran an article summing up attitudes towards both networks held by critics and the television industry. In the article, MNTV's primetime dramas were described as "disposable television," and Doug Elfman from the *Chicago Sun-Times* said that the new network deserved its low ratings for investing so little money on programming, adding "[P]orn is the only thing that can bring up its ratings at this point" (Hibberd 2007). This reaction mirrors cultural critics' longstanding and widespread negative reaction to American soap operas and their pigeonholing of soaps as inferior to many other cultural forms, including primetime television formats.

UB's adaptation and the trade press's reaction to it allow us to see how this bias, previously associated with women's programming, is also being deployed against shows that target ethnic audiences. In an era of increased global exchanges, intercultural collaboration, and format experimentation, scholars and critics bear a responsibility to foreground the way in which televisual pleasures that emerge out of specific program formats from different national cultures are interconnected historically. Television has always been a radically hybrid medium, and successful TV programs and genres have borrowed heavily from radio and cinema conventions. If cultural critics and trade journalists were to stress the similarities, while simultaneously respecting and understanding the differences, between the Latin American telenovela, U.S. daytime dramas, and U.S. primetime melodramas, they would likely produce far more generative and inclusive perspectives on the contemporary television market. By maintaining tunnel vision regarding U.S. importations of other countries' TV formats, focusing exclusively on quality or ratings while failing to discuss issues of nationality or history, today's critics and industry workers fail to understand a key attribute of most successful television: Having an emotional impact on audiences is culturally relevant.

NOTES

1. Just as Dominguez's title suggests that regular soaps become super soaps with American telenovelas, the following title promotes "newness" and difference: "A New Kind of Prime-Time" (Cridlin 2006).

2. News Corporation's incursion into the telenovela adaptation business was originally intended for syndication in local stations through Twentieth Television, but after the acquisition of the affiliates that were left out of the UPN-WB merger, Fox decided to turn these shows into the new network's primetime line-up as a last minute attempt by executives to provide programming for MNTV (Lisotta 2006). *Desire* and *Fashion House* were the first two shows destined for syndication, and the original plans were to produce three telenovelas a year. These two telenovelas were the only ones that aired as originally envisioned: five nights a week with a Saturday recap. Ratings were so low that the network cut back on the number of episodes aired during the week and replaced them with reality programs.

3. Television scholars have been discussing the impact of soap opera conventions in primetime dramatic serials from very early on. For more on this, see Ang (1985), and Seiter and Wilson's (2005) argument regarding Fox's *The O.C.*

4. I explore in greater detail such similarities and differences in Nasser (2008).

5. In his article, Wilson (2007) trashes the show in the manner that soap operas would be dismissed: "The show has preposterous plotlines, few big-name stars, garish sets, and a moral framework so simplistic that even Aesop would have balked at it."

THE WAY WE WERE
THE INSTITUTIONAL LOGICS OF PROFESSIONALS AND FANS IN THE SOAP OPERA INDUSTRY

—MELISSA C. SCARDAVILLE

> Melissa C. Scardaville is a doctoral candidate in sociology at Emory University, where she studies mass media and culture. She is the author of "Accidental Activists: Fan Activism in the Soap Opera Community" for *American Behavioral Scientist*. She also served as the former *Guiding Light* editor for *Soap Opera Digest*. Her love of soaps started with an elementary school fascination with *Dallas* and eventually a pre-teen love of NBC soap operas, particularly *Another World*.

INTRODUCTION

A common complaint among soap opera fans is the belief that no one within the daytime industry is listening. Letters to soap opera magazines and posts on message boards reveal fan anger with "the idiots in charge" (or TIIC, as soap opera fans often refer to them), as well as an abiding frustration with the fact that this formally stalwart genre is dying with no one trying to save it. Industry professionals bemoan rapidly falling ratings and search for the right combination to lure back old viewers and attract new ones, yet find themselves continually stymied because the magic formula seems out of reach. While executives wonder why viewers are not tuning in, fans condemn executives for tuning them out. This project is an exploratory study aimed at understanding how the two groups "talk past" each other and why this relationship often turns contentious. Drawing on in-depth interviews with soap opera professionals and longtime fans, this essay offers a glimpse into industry and fan perspectives on soaps, the job responsibilities within the field, and the dynamics between these two often-contentious parties.

DUAL MEMBERSHIP: SOAP OPERA PROFESSIONAL AND FAN

In the interest of full disclosure, I was once a soap opera professional. I worked as a writer for the magazine *Soap Opera Digest* (*SOD*) for six years in the 2000s. I have also been an avid fan of the genre since my middle school days, faithfully following five or six daytime dramas at any given time. Once I arrived at *SOD*, my membership in the professional group, now coupled with my longtime fan status, afforded me a unique perspective. I sympathized with and often agreed with fans' criticisms about the shows, yet I sometimes found myself siding with executives who deemed the barrage of fans' emails and calls to the studio to be problematic. I understood why fans contacted the shows, as I myself had a long history of writing letters as a fan, but I also understood why their feedback was sometimes greeted with disdain. What baffled me most, however, was that professionals and fans by-and-large seemed to agree on two points: that something was wrong with soap operas, and, if these shows were to survive well into the twenty-first century, that change was needed. Though both acknowledged a problem, professionals and fans differed sharply over what caused soaps' downfall. Fans often blamed soap opera professionals who, in turn, often blamed fans. This essay acknowledges that tension but takes a different approach. Instead of asking who is at "fault," I want to ask: What are the fundamental assumptions about the soap opera world that guide decision-making for professionals and fans? Do these different assumptions explain the communication gap?

INSTITUTIONAL LOGICS DEFINED

A notable group of organizational scholars describe the types of fundamental assumptions at play here as "institutional logic," a collection of precepts, rationales, and resources participants use to interpret and judge courses of actions within a given field (Jackall 1988; Thornton and Ocasio 1999). These scholars argue that how we ultimately make sense of the world is shaped by the assumptions we have about it. That is, we interpret our world through the lenses of institutional logics. The power of these assumptions becomes especially clear when we look at opposing logics within one industry. In the realm of media industries, the classic example involves the rationale that stresses profit and the bottom line (which people often attribute to giant media corporations) versus a logic that stresses the importance of artistic expression above and beyond profit margins (which people often connect with independent companies and individuals). Sometimes, one way of thinking gives way to another. For example, Thornton and Ocasio (1999) find that the institutional logic in the book publishing industry shifted over time from editorial logic (where publishing was seen as a profession) to market logic (where publishing was seen as a business). In the first model, the quality content of books, the knowledge they contributed, and the publishing houses' relationships with authors are paramount, whereas financial returns for best-selling books and series dominate the latter market-based logic. Thus, in some industries, the product remains the same, but the yardsticks gauging its worth and purpose may drastically change.

Particularly germane for my analysis, however, is research that examines how competing logics coexist within a given setting. Regev (1997) examines the Israeli music video scene and finds that multiple types of institutional logics operate within the field simultaneously and complement, not compete, with one another. While record companies see music videos as a way to promote sales, musicians, critics, and film directors view videos as an emergent art form. The impetus behind making the music videos differs, but industry players see the end result—the video—as being profitable and having artistic content simultaneously. In contrast, Glynn (2000) demonstrates that, in symphony orchestras, certain institutional logics—such as economic and aesthetic—can co-exist until external changes disrupt the balance. For instance, musicians extol the importance of the music itself (aesthetic logic), while the board and administrators voice concerns over ticket sales and donations (economic logic). The concert becomes a contentious site because the aesthetic values of the musicians are seen as antithetical to profit.

Although not framing it in terms of institutional logic, scholars have argued that two types of ideology operate within the daytime television universe: economic (soaps must make money) and aesthetic (soaps have artistic value) (Allen 1985; Cantor and Pingree 1983; Edmondson and Rounds 1976). How these two distinct logics—economic and various incarnations of the aesthetic—developed is beyond the scope of this essay. Here, I want to stress that these two paradigms do exist, and that professionals and fans can potentially draw upon economic and/or aesthetic logics at different times to evaluate or make sense of particular phenomena.

DATA AND METHODS

For this project, I interviewed thirteen individuals, seven longtime soap opera fans and six industry professionals (including writers, producers, network executives, production coordinators, and others) in order to see how competing logics take shape in the soap world. To assure their anonymity, the particular jobs and shows of those interviewed are not given. Each network (ABC, CBS, and NBC) is represented, and at least one professional currently works or has worked on a show that is watched by the fans in this study: ABC's *All My Children (AMC)*, *General Hospital (GH)*, and *One Life to Live (OLTL)*; CBS' *As The World Turns (ATWT)* and *Guiding Light (GL)* (both of which were still airing on the network at the time these interviews were conducted); and NBC's *Days of Our Lives (DOOL)*. Professionals and fans were recruited by using personal and professional contacts and online message forums. At the conclusion of the interview, participants were asked if they could recommend anyone for this project. Given this technique and the small number of participants in the sample, this study cannot be taken as representative of the soap opera industry.

Following Harrington and Bielby's (1995) soap opera fan classification scheme, I recruited "breadth" fans who watched or had watched multiple daytime dramas for several years. The fans in the project—six women and one man—began watching soap operas as early as 1959 and as recently as 1990. Their ages ranged from twenty-two to seventy-five years old. They lived in the South, the Midwest, the Mid-Atlantic, and northeastern and western United States. At the time of the interviews, all were watching at least one soap opera, although all had watched other soaps in the past. Each participant was interviewed individually and asked questions about the genre itself, the job responsibilities of soap opera professionals, and the state of the relationship between industry professionals and fans. I identified participants as drawing on aesthetic logic when they discussed soap operas in relation to art, craft, and/or the importance of emotional resonance and literary or film theory. Discussions about profit, "the bottom line," demographics, advertising, and the importance of ratings were labeled as economic logic. As the chart below shows, some categories (such as the "right" kind of fan or who should be the driving creative force) operate in both logics, but their function and interpretation differ, depending on whether the issue at hand is viewed through an economic or aesthetic logic.

INSTITUTIONAL LOGICS IN THE SOAP OPERA INDUSTRY

According to the professionals and fans interviewed, both the economic and aesthetic paradigms have co-existed in the soap opera industry, albeit uncomfortably, for years. This tension is certainly not unique to this industry; it frequently operates in the arena of mass media. Soap operas traditionally had to make money first and foremost, but the staff had a relative degree of creative freedom and often aired episodes that the participants considered to be art. As long as enough profit was made to satisfy the requirements for economic institutional logic—which was more certain in the 1970s and 1980s—the creative force of the show (the head writer) was able to craft the type of long-term, character-driven aesthetic storytelling emblematic of the genre. The responsibilities of executive producers during this time period were more akin to primetime episodic "show runners," mainly tasked with bringing the writing team's vision to life. In the late 1980s and early 1990s, the producers and network executives' roles expanded, in large part because of the vacuum created during the 1988 writers' strike. Aesthetic logic gave way to economic logic, which professionals and fans agree now dominates the soap industry. Especially to the fans interviewed, the upswing in economic logic caused aesthetic logic to wane. Whatever the cause of the aesthetic logic's descent, ironically, soap operas were most profitable in the era that logic prevailed. Today, however, both logics appear to be failing. Professionals and fans consider today's soap operas to lack storytelling quality and character exploration. Meanwhile, ratings, which determine advertising revenue, continue to drop dramatically.

INSTITUTIONAL LOGICS IN THE SOAP OPERA INDUSTRY		
	ECONOMIC	AESTHETIC
Soaps as a Genre	cheap to produce; high profit returns	unique artistic medium
Organizational Identity	soap opera as business	soap opera as craft
Mission	to earn profit	to tell quality stories long term
Legitimacy	high ratings	critically acclaimed quality product
Profit	resource for networks	resource to reinvest in creative staff
Role of Network	intense involvement in all aspects of production	involvement in business decisions only
Creative Leader	executive producer, head writer, and network executives	head writer
Role of Fans	consumer	connoisseur
Type of Desired Fan	women under 30	longtime viewers

AESTHETIC LOGICS IN THE SOAP OPERA INDUSTRY

Professionals and fans invoked aesthetic logic when discussing past soap operas and considered current daytime dramas to pale in comparison to their previous incarnations. In fact, the participants' only common ground was their shared agreement that soap operas no longer offer high quality storytelling on a consistent basis. This assessment may or may not be qualitatively true; perhaps the tinted glass used to reflect on yesteryear's soaps gives them a rosier-than-deserved glow. The important point for this essay is that professionals and fans both viewed daytime dramas from previous decades as superior to those airing today. Every participant pointed to other time periods as being exemplars of quality daytime television: the rich character development of the 1970s; the multi-generational families and romances of the 1980s; the powerful social issues storylines of the 1990s. In contrast, the 2000s were generally seen as having few highlights and a plethora of low points, with several respondents commenting that soap operas had, in the last decade, transformed into what the general public had always considered them to be: simple-minded, illogical melodramas populated with caricatures instead of characters.[1] Fans specifically stated that they continued to watch out of loyalty (either to the show or to particular actors), habit, or the promise shown by the occasional memorable scene or show, not due to long-term story arcs or personal excitement or satisfaction. Several professionals who worked in the soap opera industry at the time of the interviews admitted that either they or people they knew no longer had the drive and attachment to the genre they once had.

The overwhelming reason cited by both professionals and fans for inferior soaps today was the perspicuous decline in the quality of soap opera storytelling, a component of aesthetic logic. One professional, who was also a longtime fan, explained:

> Now, you have two people: they meet, a week later, they're a couple, then a week later they're married and then they are broken up and people have forgotten about them. There is no pay off. There's no watching them get together, no getting to know them as characters or as a couple. A lot of the fun and humor is gone. There are no characters that are cool and eccentric. They're all the same characters played by different people.

Another professional noted:

> There has been a real loss of community on soaps today. You get everybody so isolated in their own storylines. There's not a whole lot of crossover. I feel like everything is so isolated now. I've encountered it in [meetings]: What are we going to do with the kids this week? What are we doing with the thirty-somethings? Are we going to show the vets this week? It's all done in a very isolated way.

One regular fan of *ATWT*, *GL*, *GH*, and *OLTL* explained her dissatisfaction this way: "There are too many extreme or shock-value storylines and not enough about *real life* family dramas. Every soap doesn't have to have murderous villains, and most of them have very few decent, likable characters left on their shows. There just aren't enough laughs interspersed with all the drama." Another fan, who watches *AMC*, felt similarly: "You can't have everyday life anymore. This is why those of us who have watched soaps for over thirty years watch them—because they are almost like family. Now it's just completely different. Everything has to be the big bang for the buck."

Given that the product created by the professionals and consumed by the fans is widely believed to be inferior in quality, both groups operate within the same paradigm, judging soap operas based on aesthetic logic. When using aesthetic criteria, daytime dramas are found to be either failing or consistently missing the mark. However, the main source of tension between the groups is rooted not in conflicting perceptions of current shows but in the use of different logics to explain what caused this decline and what can be done to remedy it. Fans invoked aesthetic logic to interpret falling ratings. To them, soaps' decline is related to inferior storytelling that began in the 1990s and the industry's loss of respect for the audience. Professionals, however, turn to economic logic to understand the current state of soap operas.

ECONOMIC LOGIC: THE BUSINESS OF SOAP OPERAS

The majority of professionals drew on economic logic to explain soap opera's decline, citing changing audience demographics and increased competition from other entertainment outlets (cable, DVDs, the Internet, etc.). One professional who has worked in soap operas since the 1990s explains:

> Society has changed. If we look at the last thirty years, there are more and more women in the work force, so there are not as many women at home watching. I think that is the first factor [for declining ratings]. I think, number two, there are too many choices. If you think, "Hey, it's 1 p.m."—twenty years ago—"What am I going to do? Well, there's ABC, NBC, CBS, read a book, or go outside." Now, when it's 1 p.m., I can go to ABC, NBC, CBS, read a book, go outside, surf the Internet, or watch 5,000 other channels. We used to be a big giant tree in a very small forest, and now we're a big, giant tree in a very big forest.

Later, he added that, given the expansion in programming choices, audiences' tastes have changed. "One of the things that soap opera has built its whole identity around is its sense of continuity and consistency but also its very methodical storytelling. As compared to everything else, soap opera storytelling is very slow. Audiences right now are used to instant gratification—thirteen episodes in a season is too long for them—so that's another challenge. How audiences have consumed entertainment has changed in terms of their attention span." Another professional, a veteran of the soap opera business, concurs:

> What has been dogging soap operas for years is that our primary audiences are drifting away. Stay-at-home moms aren't stay-at-home moms anymore, and that was our bread and butter for our soap operas in the '50s and '60s. Stay-at-home moms who don't work [no longer] sit at home and watch television shows all day long. Trying to get new viewers has been a real problem for daytime television. It's hard for us to compete with what's going on on cable and on the networks right now. For us to try and get viewership from the same eighteen-to-twenty-five year old audience is tough. The competition is huge. It's bigger than it's ever been. There's a lot of stuff out there, and soap operas are a pretty unique genre, and it requires people to really get involved with it over a long period of time over many weeks. If people really have several hours a week anymore, I don't know if they [have it] to dedicate to one program.

In contrast, fans in this study believed that citing increased competition for viewers as the culprit was merely a way to shift the blame for falling ratings onto the viewer rather than the professional. Almost all fans interviewed felt that audiences, regardless of other entertainment options, would watch daytime soap operas if they were worth watching. One fan, who has watched soap operas for fifty years, noted:

> Soaps need to focus on storytelling but not all that peripheral stuff. It wouldn't bother people if the sound was bad and the camera jerked around if what they were telling was worth watching. It seems like, when ratings go lower, the soaps do something else [outlandish] to get people to watch when good storytelling would do it. No one is willing to admit that their stories aren't any good. The plots—to quote William Faulkner—"they are not about the human heart in conflict with itself." Shows are producing some really good episodes, but when you string them all together, the story is nothing. It has no base. They don't know how to tell a story anymore, and they don't know what the audience wants to see.

Overwhelmingly, the audience members in this study believed that the fall in ratings was directly related to the decline in the quality of the shows, a reaction characteristic of aesthetic logic. Quality deteriorated because the type of storytelling offered was at odds with the multi-dimensional emotional tone intrinsic to the soap opera genre. Most fans reported that their shows had become too dark and violent, with beloved characters transforming into one-note villains, only for their transgressions to be forgotten and glossed over by the show. "There's a general philosophy now that any character can do anything," explained one fan. Another longtime fan described the connection between quality and authentic soap opera storytelling:

> Shows want to be something other than a soap opera. They want to be a dark, edgy cable show. You can't change characters from who they've been for twenty-five years. One of the perks of long-term fandom is that you can see things unfold and when something happens twenty years later that hearkens back to what you watched, for a fan, that is a very fulfilling thing. So, when you create something out of nowhere and it doesn't fit in with what they've watched before, they feel like it's violated the time and energy they put into it, which doesn't seem to be something that the people who run these shows get at all. They think that people forget or that it doesn't matter and people will accept it and move on. If you rip that out, you are again ripping out a lot of what makes soaps unique and enjoyable.

As noted earlier, professionals also bemoaned the lack of good storytelling, but most did not see it as the reason for the falling ratings. Instead, according to their analysis, the shrinking audience, which translates into declining profits, caused the shows to funnel less money to their writers. The tightened economic situation led to smaller writing staffs, increased creative burnout, and an unwillingness for new talent to enter the ranks of soap opera scribes. Thus, shows resorted to a fast burst of plot-driven action in hopes that former and new viewers would be enticed to watch. In actuality, these stories often lacked the emotional resonance desired by soap opera audiences. Most of the professionals interviewed admitted that these quick fixes have failed over the long term, yet the continued decline in viewership (and thus falling ad revenues) perpetuates the cycle. Several professionals cited the ever-shrinking show budget as the reason the position of breakdown writer—one who helps shape the beats of a particular episode—was recently eliminated on many soap operas. While this cost-cutting measure saves the shows money, the professionals acknowledged that this move will likely affect the quality of storytelling in years to come, an instance of economic logic constraining aesthetic logic.

Fans, on the other hand, believe that the consistently poor storytelling, which began anywhere from the early-to-late 1990s according to their timelines, is the cause of falling ratings. Thus, the failure of the aesthetic logic led to the industry's reliance on the economic logic. To fans, past economic success was a happy by-product of well-executed aesthetic logic, not the result of perfectly executed economic logic. Over the years, audience members continued to tune out because the type of purposeful, character-driven, family-centric storytelling that initially enticed and later sustained viewership was replaced by often violent, plot-driven action where consequences and ramifications never materialized. Specific reasons for this switch were unclear to the audiences: Some fans pointed to the writer's strike of 1988, others to the O. J. Simpson trial of 1995, still others to daytime's ever-increasing desire to replicate primetime story techniques. Whatever the turning point, fans said that it was not the events themselves but the reaction of those within the industry that led to a decline in viewership. Regardless of the contributing factors cited, these fans believed that the shows' choice of simplistic and fractured storytelling demonstrated that they did not value viewers' intelligence or their commitment to the genre, giving audience members even less incentive to watch. Thus, the demolition of the aesthetic logic led to current economic problems.

In terms of institutional logics, both the professionals and the fans agreed that the *quality* of storytelling (aesthetic) and the *quantity* of viewership (economic) have significantly declined over the years. The point of contention resides in which logic one uses to interpret the current situation and the question as to which factor was the cause and which was the effect. For the professionals, increased competition in the entertainment field and the decline in the number of women who stay at home led to fewer viewers. Lower ratings over time meant smaller show budgets, which caused the writing staffs to shrink. These changes increased the need for a particular plot- driven type of storytelling to garner quick ratings bumps. In that sense, factors not unique to the daytime industry (such as the 1988 writers' strike) initially affected ratings, which led to the perceived need for economic logic to dominate over aesthetic logic. Because profits faltered, resources had to be redirected from aesthetic concepts to economic ones. Fans, however, believed that the erroneous courses of action espoused by daytime executives generally *led* to the shrinking audience. The caliber of storytelling declined, which caused viewers to tune out, which led to falling ratings that then triggered a downward spiral in the quality of drama.

Professionals' and fans' differing determinations of cause and effect also shaped their recommendations for how to rebuild the daytime audience. The fans interviewed maintained that a return to quality and consistent storytelling would lure back and even expand the audience. Thus, the audience argued for the aesthetic logic to be revived in the industry. Although they felt this option would not be explored by the shows, the fans believed that, if the caliber of the product was restored, fans could resume their recruitment of others to their chosen show or to the genre itself. Regarding people's negative attitudes towards soaps, one fan said:

> We are right where we were in the mid '70s when I started watching them. I think that things improved for a while because there was enough really good stuff out there that you could point people in the direction of soaps, and if they would actually approach it with an open mind and watch it, they would see that they didn't fit the stereotype. But unfortunately, we have gone back to the point where most of what's making it on the air fits every bad stereotype about soap operas that was ever promulgated.

To the fans interviewed, the soap opera genre has been deteriorating for at least a decade. Although salvageable, it would take time to rebuild the audience, time—the fans acknowledge—that the daytime industry might not have.

Conversely, professionals were less certain about which corrective remedies to take. While those interviewed acknowledged that daytime has always faced its fair share of problems, all but one noted that, in the last few years, the erosion somehow felt different. Several professionals mentioned that daytime felt like it was "hemorrhaging" and that "no one can stop the bleeding." One veteran professional gave this account of how the industry changed: "I feel like daytime is at this state where it's every man for himself and that includes the soap press, the actors, the writers. Everyone is just hanging on to whatever lifeline they have, and they're hoping they can survive the next thirteen-week [contract] cycle. It's a really scary time. I don't know anyone who feels secure and confident in their job right now." Later in the interview, this person added, "ABC is petrified of bringing in new blood, but they also don't want to bring in these old war horses who are recycling stories because that is not healthy either, so they are kind of at a standstill. P&G doesn't even have [high-level executives] looking at their shows anymore. We all know about NBC: they wanted out for years, and they might just get their wish. I definitely feel like there has been this erosion that has happened [in 2008] where everybody is either stuck in place or free falling." Another veteran professional relayed the following story:

> The entire industry feels different. I also happen to be part of the negotiating team for my union, and there's no question: the mantra from the industry side of it is that everything is changing. The fact that people fast-forward through commercials now [on a DVR] is freaking out the advertisers to no end [and] advertising dollars are being pulled more and more from the television screen. The entire business is changing, and everybody is freaking out about it. It sounds very gloom and doom, doesn't it? It certainly feels different

to me, and I've done [more than twenty-five] years of this. I don't believe I have ever experienced anything like what we are facing now. And I've been through several times when it's been rumored that [my] show will be going off the air any minute, and all of those times I've kind of laughed and gone, "Yeah, whatever; we'll see." Obviously, it hasn't happened. But, this time around, I really don't know. I don't know what the future is for this show or any of the shows.

This high level of uncertainty leaves those in charge unsure of what course of action to take. Although no professional believed that soap opera storytelling would completely disappear, all did predict that the traditional televised daytime drama afternoon line-up would change substantially over the next five years. As one professional put it, "I don't think soap operas are ever going to die. They've been around too long. To find serialized storytelling, you can go back to Dickens, you can go back to Homer; it's way too embedded in human desire to find out what happens next. I think soap operas are always going to be there, but the wrapper that they are in is going to change." Another professional stated, "I really hope there is a future for soap operas. But, right now, I think the future for soap operas is a complete unknown at this point. They could completely go away and then return again in some other form or other shape, or they could stay on the air but will probably have to change the way they do things." Fans also believed that the future of soaps looked dim. One longtime fan said, "I think the genre will be dead. I don't see daytime soaps surviving because I don't see [the shows] taking the steps for them to survive. They've lost so much audience. Ironically, it seems like soap-oriented storytelling in primetime has never been hotter. Primetime has basically stolen elements from daytime and done it better than daytime trying to steal elements of primetime."[2] In other words, the aesthetic elements that had once worked for daytime have now infused primetime with new life.

PERCEPTIONS OF PROFESSIONALS IN THE SOAP OPERA INDUSTRY

Although all institutions face uncertainty, especially institutions involved in the creative aspects of production (Bielby and Bielby 1994; DiMaggio and Powell 1983), the parameters that define who should and should not be allowed to be creative have particular salience in the soap opera industry. Three positions in the soap opera field elicited the interviewees' strongest opinions: the network executive, the head writer, and the executive producer. In fact, the fans interviewed did not think that some soap opera professionals such as actors and dialogue writers had much creative input at all. The fans' comments were directed almost exclusively toward the top executives. Professionals and fans alike had firm beliefs concerning what duties these jobs should entail, descriptions that were not always congruent with the type of management that occurred during the day-to-day running of shows. How professionals and fans decided who should and should not be permitted to act as the show's creative force correlated to the two institutional logics—economic and aesthetic—and the increasing dominance of the profit model.

Given the current supremacy of economic logic, the demarcation between creative and business jobs is now murky. Aesthetic decisions are now considered in the purview of executives, while the head writer (previously considered a strictly creative position) has to think constantly about how the business model intersects with storytelling. Fans traced many of daytime's current troubles to the growing importance of the executive producer and network executive, coupled with the diminishing power of the head writer. To fans, neither "network suits" nor executive producers should be involved in any creative aspects. These fans believed that, in past years, network executives embraced economic logic, head writers the aesthetic logic, and executive producers acted as a mediator so both paradigms could operate. Now, the fans see the aesthetic paradigm as completely subsumed by the business one. Professionals, on the other hand, thought network executives should be involved in some creative decisions but that their level of involvement had become detrimental to the shows. Input from the network was necessary and sometimes useful, but the degree to which executives had become involved restrained the writing and production staff. While fans felt that many of daytime's current problems could be ameliorated if the network executives left the shows alone, the professionals generally felt that the institutional logic had swung so far toward the economic paradigm that network executive involvement in creative decisions was a permanent fixture in the soap industry. Network executives, head writers, and executive producers now constitute a team where each position is responsible for balancing economic demands with art, with profit being the most salient concern.

NETWORK EXECUTIVES

Although both professionals and fans thought network executives exercised too much control over day-to-day creative decisions, the fans felt this involvement arose due to the network's lack of interest in the soap opera genre. As a veteran *DOOL* fan stated, "The networks are not supporters at all. If they could, they would take off all the soaps and put on talk shows. I don't think any of them care." Many fans believed all three networks—ABC, CBS and NBC—wanted daytime dramas off the air. Thus, the networks' involvement was not viewed as supportive of the genre, but rather as an attempt to dismantle it.

The professionals, however, thought that the networks wanted their soap operas to succeed, but that their attempts to help the soaps were misguided and often backfired, in part because network decision-making occurs in such a high-risk environment. One professional observed, "ABC is the biggest supporter [of daytime] because they own their shows outright. ABC loses a lot if they lose their shows. So ABC has a real devotion to their shows, and I think that then manifests itself as micromanaging. Ken Corday [executive producer and part owner of *DOOL*] is the same way. He is so devoted to saving his parents' show that he sometimes suffocates it." Another professional directly linked the network's intense involvement with the growing economic uncertainty in the entertainment industry:

> There is a big fear of television in general going the way of the music industry, where people just start trading programs online. Uncertainty has always been part of being in this business, but the overall uncertainty of the business now is kind of frightening. The [network executives] don't know where this business is going anymore. [They look at] the music industry a lot. What went wrong with the music industry? How did it lose control over its product? That seems to be the big question now: How do we maintain control over our product? Will people go to the network's Web site to see yesterday's episode, or will they go to twenty other places where they don't have to sit through streaming advertising?

Despite understanding the impetus behind the network executives' motives, many of the professionals were extremely frustrated with constant executive participation. Said one professional:

> I think the network should be more hands-off with us, with every show. Sometimes they come up with ideas where certain people have to be fired just because they don't like them. What about this great story we have planned? All of a sudden someone at the network says, "No, that's not happening" and [the show] has to completely rewrite, and the writers' hearts aren't into it because it's not their idea. I understand why [the network is] doing it, but there are too many people running the show, and they're not all on the same page.

Another professional relayed this story:

> I worked at one show where they literally had the ratings down to the minute. They said to us in a meeting once, "Whatever you're doing at the forty-six-minute mark, stop doing it because we lose viewers then." We all kind

of looked at each other like, "Are you kidding? Is this a joke? *Whatever we are doing at the forty-six-minute mark, stop?*" It was shocking. Or they would be like, "Find something really thrilling to happen on the forty-six-minute mark." What? Why? Never mind that it could just be that it's 2:45 [p.m.], and people are going to pick up their kids from school. Maybe it has nothing to do with the show's storytelling at all, and it's that people have to leave their homes.

Those in the sample who were directly tied to the network interpreted executive actions differently. One such respondent notes:

I was one of those network people who could be accused of micromanaging, so I come at it from a unique perspective in that I was in the thick of things. My experience has always been that any notes, criticism, thoughts were a) given in a collaborative sense and b) never given for the sake of giving it. It was always for the betterment for the show. I have heard too [about network interference], but my experience has always been a very collaborative one. There is nobody who is like, "Do this!" Everybody is trying to work together—head writers, executive producers, network executives—trying to make the show as great as it can be.

EXECUTIVE PRODUCERS AND HEAD WRITERS

While this study cannot say how representative the experience of the network executive cited directly above is, the notion of collaboration—no matter the interpretation—among head writers, executive producers, and network executives was problematic to many fans. Overwhelmingly, the fans in the study expressed concern that network representatives and executive producers were involved in any decisions related to creative endeavors. One fan who has watched most daytime soaps at some point explains:

There are people in the business area who are now in control of creative decisions, and they are more concerned with the bottom line. I don't think they have the passion for it that a creative person does. When you watched a show written by someone like [Douglas] Marland or [Agnes] Nixon, you can see that they love their shows. They love their characters. They enjoy this format. When I watch some of the shows now, there is a disconnect. You don't get that feeling of love being put into it, like a pride of craftsmanship. It's more about putting out a show that sort of follows some of the same lines as before but without the intimacy that used to be in the format.

In a sense, fans in this study echoed the talk of the professionals: Both groups believed the business model now reigns. However, they also dismissed the idea that good soap opera can develop from this approach. To the fans, the executive producer's job should be to facilitate the storytelling and manage the business side of the show. As one fan said, "The executive producer is the person who takes the material generated by the creative staff and makes sure that it gets on the air in a form that is accessible to the viewers. That's what I think it should be and what it once was. But I think now it's become a jack-of-all-trades thing where the executive producer participates in generating story, editing scripts, negotiating with executives above him or her. Basically it's become a creative force." For these fans a good executive producer is invisible. "You shouldn't really notice the way the wheels turn," remarked one fan. "If people are noticing the man behind the curtain, then you are doing something wrong. A performance is supposed to be magic and feel like it's real, and, if we're thinking about the producers, it's because they are doing a bad job. People are noticing them more because they are doing a worse and worse job over time."

Head writers, on the other hand, should be given more room to shine, fans and professionals agreed. "The head writer used to be the *auteur*, which is why they used to be better shows, the way directors are the *auteurs* in [movies]," said one fan. Both professionals and fans were unsure how talented the current crop of daytime head writers were because the overriding business paradigm prevents writers from carefully and personally crafting stories. As one fan noted, "Head writers used to be experts; they understood the genre completely. It's hard for me to say how good some people are because I don't think they have the same kind of power that the others did. Left to their own devices, they might very well produce a good show." One professional explained how the network and executive producers' power affected long-term story: "Nobody wants to take the time to build any kind of vision. Soap writers used to have the next year, year-and-a-half planned out in their heads, and now they have like three months tops. They just can't think any further than that because everything else can change." Another professional observed:

> It's easy for me to look back on someone like Doug Marland or Pam Long and say, "Wow, those people were really extraordinary storytellers," but that would imply that people who are writing soaps today are not, and I don't know that that is necessarily true. Maybe it's because it's become more of a business now than it ever was. Maybe [soaps] are not so much a storytelling thing. Now, you have executive producers taking control, telling writers how

many sets they can use in a week or how many actors they can use, [which has been] really detrimental to storytelling.

Overall, professionals and fans shared the perception that head writers should be the creative force behind soap operas. They also concurred that this role's scope and autonomy have changed dramatically since the 1980s because executive producers and network executives have been added to the creative team. Professionals generally did not view these new alliances as inherently problematic but did concede that the current level of involvement, especially by network executives, hampered good storytelling. To fans, however, the mixing of creativity and finances was extremely problematic for daytime dramas. Given the current dominance of the business model, fans thought that one way to revive the genre was to reinvigorate the logics of aesthetics. Once writers reclaimed creative power and executive producers and network executives handled business decisions exclusively, these two logics would once again be in balance. For the professionals, the preeminence of economic logic was predicted to endure because the entire entertainment industry is in too much uncertainty and flux.

PERCEPTIONS OF SOAP OPERA FANS

The most striking difference between professionals and fans was their perceptions of one another. In general, professionals thought their relationship with fans was relatively amicable, and they were, to some degree, baffled over fan criticism that the shows did not care what viewers thought. One professional explained, "When the fans are happy with us, it makes us that much happier at the studio. It's like, 'We're on the same page! Let's keep it up!' When we are proud of something and it doesn't go over well, a somber mood comes over the studio. 'We tried so hard; why don't you like it?'" In the interviews, professionals routinely discussed the ways in which they try to incorporate fan feedback into their shows, arguing that they listen to fans as best they can.

Fans, on the other hand, firmly believed that, not only do the shows not listen to them, the shows purposefully make their relationships with fans antagonistic. As one longtime fan explained, "When these shows were run by people who really enjoyed what they were doing, they took it seriously, and you could feel that. They had a respect for the intelligence of the audience. Now, 'the powers that be' are there to slap the audience." Another fan described how storytelling on *AMC* insults the audience: "It's all about shock value. They either have to un-abort an aborted fetus or rape the lesbian and get her pregnant. If that isn't insulting enough, then they kidnap her baby and then, if that's not insulting enough, have her forgive the kidnapper and, if all that wasn't insulting enough, then have her fall in love with a woman with a penis."

Every fan interviewed attributed the source of this acrimonious relationship to the belief that those in charge of the soaps assume the audience consists of unintelligent and stereotypical "fanatics." Said one veteran fan:

> I don't think "the powers that be" have a very high opinion of us. It's almost as though we are tolerated as a necessity. They have to have viewers. Judging by what's making it on the screen, they don't think we deserve very much. They don't seem to think that we pay attention or are particularly intelligent. I am sure there is no uniformity across the industry, but things have come out in the press and I think, "God, this person is in charge of producing a soap or network? This person doesn't get it and really has a bad attitude."

For instance, in a 2008 *SOD* interview with *ATWT*'s Executive Producer, Chris Goutman, he stated that he never listens to fan feedback and suggested that viewers not send items such as letters to the show (Lenhart 2008b, 43). Nonetheless, each professional interviewed in this study expressed a high opinion of the audience (and many were soap opera fans before working in the field), although each noted that stereotypes continue to lurk within the industry. As one professional explained, "There's a perception about soap opera fans, and there's a reality of soap opera fans. Culturally, if we look at the larger universe, everyone thinks that the soap opera fan is an overweight woman that eats bon bons and is really lazy and lives in a trailer park. Internally, that's not what [the network] sees their fan as. The stereotyped idea of the soap opera fan as a lazy, ignorant person is definitely not true."

Why does the unflattering portrait of soap opera fans endure? Some professionals argue that, because soaps are such a stigmatized medium, fans of the genre must also be of a low status. Others point to the extreme behavior of some fans, which might color the perception people have of all viewers. One professional, who is also a longtime soap fan, notes:

> I feel like daytime is its own worst enemy because what happens is, there's a psycho group of fans that sends 80,000 boxes of mac 'n cheese to support a couple, which leads the show to look at all their fans and say, "Whatever. These are the people who spend thousands of dollars on mac 'n cheese." There's a contempt for fans, and there's a contempt for "the idiots in control," which they call us on the message boards. It doesn't make any sense to me. It hurts daytime in the long run because what happens is you have fans who watch the show and don't give a crap and you have people who are creating the show who don't give a crap.

Why is the professional/fan relationship antagonistic? One cause may again be the competing institutional logics. Professionals and fans have different responsibilities under economic and aesthetic logics. These different perceptions of roles affect how fan feedback is interpreted. In general, the professionals see fan reaction—letter-writing, calling the studio—as positive because it means viewers are engaged. Fans in this study tended to classify fan communication with the professionals in two ways. First, negative feedback does not equate to active engagement with the show and instead signifies a last-ditch effort before a viewer may stop watching the show regularly. Secondly, the fans, like the professionals, were aware of the sheer number of fan campaigns and felt these drives often misrepresented what a majority of viewers want. Importantly, while under the aesthetic model long-term viewership is rewarded, the economic model sees fans as consumers. These different categories have distinct responsibilities attached to them.

These assertions about fans raise the question: Who is the soap opera viewer? Professionals and fans did not think one could sketch, with any detail, a portrait of the typical viewer. Fans believed that this lack of uniformity was positive because it indicates the true diversity of the soap opera audience. Professionals, however, found such inconsistency distressing because they need to know their market in order to attract advertisers. This dilemma is at the heart of one of the problems for soap operas: No one is quite sure who is watching. Viewers can watch soaps in ways that are not captured in ratings (college campuses, offices, online), so the Nielsen system is far from foolproof (as almost all professionals and fans recognize). Even when demographic information about viewers is known, professionals are not sure what version of their show viewers expect. Due to the rapidly revolving door of head writers and executive producers that has operated since the late 1980s, fans who started watching under different regimes have different ideas about what is appropriate for their show. Here, a professional discusses the fractured audience:

> I almost feel like we can do no right by these fans and these characters. They have been watching for so long. These characters have been through so much; they lived ten different lives, and anything we come up with is not going to make most people happy. Everyone is so attached to these characters and these worlds—I am, too; don't get me wrong. You get people who watched [DOOL] from Jim Reilly on who want the show to be one thing, and then you have people who watched before Jim Reilly [pre-1992] who want an entirely different thing. You get these divided fan bases, and no one is helping the show.

In this sense, the importance of long-term viewers in the aesthetic model conflicts with the needs of new consumers in the economic model, yet the economic model has not been able to replace lapsed viewers with new ones. Therefore, the economic logic struggles (few new consumers), while the aesthetic logic languishes (no incentives for long-term viewers).

CONCLUSION

As it turns out, there is one "truth" which industry professionals and fans can agree on: Soaps used to be so much better. The heyday of superior storytelling and windfall profits will never be seen again in the soap opera industry, according to these professionals and fans. While there have long been two institutional logics—economic, where profit is paramount, and aesthetic, where quality storytelling is the goal—professionals and fans disagree over how the genre's decline occurred. Soap opera professionals, like their counterparts in primetime, have not yet developed a solution to their problems, and the weakening economic logic has led to deterioration in the aesthetic logic. These dual, and often contradictory, frameworks may explain why soap opera professionals and fans often feel like they are speaking different languages. At times, the only common ground they share is that everyone is nostalgic for "the good ole days."

NOTES

1. For more on this perception, see Liccardo's essay in this collection.
2. See essays by Scodari and Liccardo in this volume for more on the daytime/primetime comparison.

SECTION TWO
CAPITALIZING ON HISTORY

PERSPECTIVE
SCHOLAR HORACE NEWCOMB ON THE PLEASURES AND INFLUENCE OF SOAPS

BASED ON AN INTERVIEW BY SAM FORD

> Horace Newcomb is professor of telecommunications in the Grady College of Journalism and Mass Communication at the University of Georgia and director of the George Foster Peabody Awards Programs. He wrote about the soap opera genre in his 1974 book, *TV: The Most Popular Art*. Newcomb and his wife were fans of *Another World*.

I began writing about soap operas because I had noticed that soaps offered more complex narrative strategies than primetime shows. Even if some of the performances and topics were conventional, or even stereotypical, the possibilities for storytelling were much greater than in episodic primetime series.

When *Dallas* (CBS) and *Hill Street Blues* (NBC) hit screens in the late 1970s and early 1980s, it was clear that episodic TV had to change. All the moral complexities of serialized TV made primetime more complex. This opened doors for dealing with socio-cultural, political topics in more powerful ways. It's not that episodic television had ignored issues but that it relied on viewer memory to note nuances. Serialization keeps issues alive in character choices and changes. Both arenas are defined, in my view, by being subcategories of melodrama, the great narrative creation of post-Enlightenment culture (Brooks 1995).

For instance, a major factor in why people become soap opera fans is the power of cognition—the "game" aspect, if you will—in which viewers "know more" than characters—know the frames and schema that are in play—and are always watching for variation, violation of norms, and new strategies. Allen (1985) had much to say about this. His notion that events have a paradigmatic as well as syntagmatic "effect" as information ripples through couples and groups in soap worlds is a major insight that I continue to apply in all forms of television, especially anything serialized.

On television today, there's so much competition for viewer attention. I now compare television to a newsstand or bookstore or library or warehouse. We often think of viewer choice in terms of time shifting, etc., but the real power we now have is the power to browse, to select our preferences from numerous offerings that match or tease or modify those preferences. If soap operas were to cease to exist in particular, though, what would be lost is a great deal of viewer pleasure. Personal investment—emotional, psychological, moral—in specific fictional worlds is a powerful experience. The parasocial relationships are stronger among soap opera audiences than some other genres.

PERSPECTIVE
SCHOLAR ROBERT C. ALLEN ON STUDYING SOAP OPERAS

BASED ON AN INTERVIEW BY C. LEE HARRINGTON

> Robert C. Allen is professor of American Studies, History and Communication Studies at the University of North Carolina at Chapel Hill. He is the author of *Speaking of Soap Operas* and the editor of *To Be Continued . . . : Soap Operas Around the World*, along with several other publications on soaps.

My initial interest in soap operas emerged within the context of graduate course work in cinema studies at the University of Iowa in the mid-1970s. Cinema studies was still very much a discipline in formation at the time, drawing its approaches and methods from a range of other fields of inquiry, including anthropology, political philosophy, and literature. A number of us took a seminar on narrative theory, taught in the English department. We spent a lot of time discussing the basic elements of narrative and the relationship between narrative and narration, using myths, folk tales, short stories, and novels as our case instances. I remember asking how narrative theory might handle a form of narrative whose narration had begun decades in the past and which was predicated upon its ending being indefinitely deferred. I was thinking, of course, of soap operas. The seminar, I think, found this to be a mildly interesting outlier of an example of narration/narrative relationship, but no one seemed particularly interested in pursuing the implications of soap opera's baroquely excessive narrational form or the relationship of that narrational style to the form's enduring popularity. Shortly thereafter, I had an opportunity to offer my own course on broadcast criticism, and I included a unit on soap operas. The more I looked at soaps, the more fascinated I became, and the more I looked for serious critical consideration of soaps—of which there was very little in the mid-1970s—the more convinced I became that there was an opportunity here for a scholarly account of soaps.

However, it quickly became apparent that soaps resisted being treated as "texts" in any normative notion of that term. It seemed to me that it made more sense to ask questions about how soap operas were situated as aesthetic objects, as a cultural phenomenon, as a social phenomenon, as an institutional product. So that's how *Speaking of Soap Operas* came about. I thought it particularly important to include a consideration of the ways that soap operas had been taken up in a variety of different discourses: that of mass media and sociological research, critical commentary, and industrial discourse. I also thought it important to foreground the historical connection between these discourses and those around the gendered audience for which soaps have been intended since the late 1920s.

I gestured in the direction of the need for a consideration of the experience of soap operas, but I stopped short of engaging in this kind of work myself. I didn't feel very well equipped theoretically or methodologically to do so. Fortunately a number of other scholars—Jan Radway (1984), Ien Ang (1985), Tania Modleski (1979; 1982), and Dorothy Hobson (1982) among them—produced some groundbreaking ethnographically informed studies of viewer experience, all of which were very much relevant to understanding the experience of soap operas and other popular forms.

My current scholarly work also involves creating accounts of the experience of popular culture. I am particularly interested in how we can better understand the historical experience of cinema and in reorienting cinema history in relation to experience. This presents daunting historiographic challenges—few soap opera viewers or moviegoers documented their own experiences as viewers—and so we are left with trying to study the billions of occasions over the twentieth century on which people experienced something they understood to be "cinema" (or soaps) on the basis of very indirect and fragmentary evidence. Despite the theoretical and methodological difficulties, however, I think that understanding culture as experience is extremely important.

When I first began thinking and writing about soap operas in the 1970s, I was eager to highlight what we might call the formal complexity of the U.S. daytime soap opera as an instance of serial narrative and narration, as well as the complexities involved in "reading" them. Both had been flattened in cultural commentary about soap operas and in attempts to study soap operas and their audiences according to the procedures of mass communication research. My account of these complexities was anchored in what seemed to be the stable and enduring logics of commercial broadcast television: domination by a few advertiser-based, national, commercial broadcasting companies and their local affiliates, resulting in a limited range of programming choice.

Speaking of Soap Operas is set in a national and international media landscape that was—unacknowledged by me at the time—being transformed by technological and institutional forces right under my nose: the remarkably swift adoption of the video cassette recorder ("video" doesn't even appear in the index of my book!) in the mid-1980s; the rise of intertwined cable and satellite delivery systems; the explosion in new commercial terrestrial channels across Europe and Asia;, and the multi-directional international circulation of programs and program formats. The essays collected in *To Be Continued: Soap Operas Around the World* reposition the television serial in relation to this global technological and institutional horizon. In doing so, they call attention to other forms of "soap operas" in addition to the open-ended, daily broadcast variety familiar to U.S. viewers. In particular, the "closed" serial form of the *telenovela* had proven to be especially well-suited to regional and global circulation.

As I was writing *Speaking of Soap Operas*, soap opera writers, broadcasters, and advertisers were trying to adjust to social changes as well—changes that would undermine assumptions about the "audience" for soaps that had been in place since the advent of commercial television. Baby Boom women were by the mid-1980s aging out of the prime demographic market for soap operas' traditional advertisers. More women were working outside the home than ever before. Cheaper programming forms—game shows and talk shows—ate into soap operas' share of the dwindling daytime network audience, as did new cable channels. Soaps attempted to broaden their audience appeal, but the number of new viewers was never enough to offset the declining audience base. Cable television was not a hospitable environment for the development of new programs based upon the daytime soap model. It took too long to build an audience for a new soap. The cost per episode was too great in relation to likely audience size. The investment was too open-ended.

Although millions of people will continue to watch daytime soaps for the foreseeable future, in a sense, soap operas are historical victims of their phenomenal success in the deployment of seriality as a means of storytelling and maintaining a viewing audience. Especially from the 1980s forward, serial narration became a feature of many forms of comic and dramatic programming in the U.S., and, in the process, transformed the production of what had previously been thought of as "episodic" programming. The reasons for studying soap operas today are necessarily different from those that applied to my own efforts thirty years ago. There is much more comparative work to be done—examining the role of soap operas and media seriality more generally within national media cultures and across them. We should also ask how the experience of daytime soaps has changed in relation to new technologies of access, distribution, and engagement. But soap operas are still worth studying for their persistence alone—as a form of storytelling, as an advertising vehicle, and as the basis for uncountable hours of experience for uncountable billions of people in the U.S. since 1929.

GROWING OLD TOGETHER
FOLLOWING *AS THE WORLD TURNS'* TOM HUGHES THROUGH THE YEARS

—SAM FORD

> Sam Ford is a research affiliate with MIT's Convergence Culture Consortium and Director of Digital Strategy for Peppercom Strategic Communications. He has written extensively about soap operas in the blogosphere, in online academic outlets such as *Transformative Works and Cultures*, and in his 2007 Master's thesis, "*As the World Turns* in a Convergence Culture." He began watching *As the World Turns* with his mother and grandmother and watched through its 2010 cancellation with his wife and daughter.

Television is an actor's medium. While budgets and schedules have often given movies a greater mastery of grand visual spectacle than television, the on-screen performer has always remained the currency of television fiction. Even today, with television series consistently raising the bar for production values, the actor still holds the most power in connecting with the audience. The value placed on the exploration of character and on-screen performance is more suited to the seriality of television as well. While films visit a character's life for a short time, a television series depicts characters on a regular basis, over a number of seasons, giving viewers a deep history of performances with which to compare the current actor—and her or his character.

In the case of the U.S. soap opera, the exploration of characters may not last for years but decades. Put another way, the soap opera highlights the power of television. With low production budgets compared to primetime fare, soap operas focus on character instead of production values, actor instead of set. Hour-long soaps feature a cast of up to forty, including both full-time stars and frequently recurring actors. The alternation of zoom-in close-up reaction shots from one character to the other through the course of a conversation is the defining series of shots for the genre.[1] Viewers come to know the facial movements and the voices of actors so well that the most minute change in expression or inflection can be meaningful.

Because of the collective memory of the viewing audience, soaps often rely on a historical understanding of a character. And, especially when that character is played by the same actor for a number of years, audiences continually learn more about that character's traits, predicting her or his actions based on past decisions and then revising their understanding of a character based on new actions (Giles 2003). It comes as no surprise, then, that soaps—when they are at their most powerful—value character over plot, reaction over action, and relationships amongst the characters over more episodic "situation" stories. Fans and industry members alike agree that the genre's quality has declined as these differentiating factors are downplayed or even circumvented.[2] This essay demonstrates the power of the history of soap opera texts, characters, and actors through a summary of one character's fifty-year history.

AS THE WORLD TURNS

TeleNext Media, which is funded by Procter & Gamble, produced one of the longest-running soap operas on broadcast television, CBS's *As the World Turns* (*ATWT*), which ran on the network from 1956 until 2010. The soap opera genre got its name on radio because of the shows' sponsorship by soap companies, but P&G remained the only soap company directly involved in funding U.S. broadcast soap operas in the past few decades. While TeleNext and P&G discussed alternate forms of distribution for new original episodes for *ATWT*, no new deal materialized before the show's cancellation (Kiesewetter 2009). From its inception under the supervision of Irna Phillips, largely considered the "mother" of the soap opera genre, *ATWT* popularized many of what are now considered defining elements of the genre. The program initially aired live daily for thirty minutes (it was one of the first 30-minute soaps, along with CBS' *The Edge of Night*, which debuted later the same day on April 2, 1956), breaking away from the shorter fifteen-minute increments of previous soaps. Slow pacing, an emphasis on dialogue, and the now-stereotyped camera angles were all in part pioneered by *ATWT*.

From 1958 until 1978, *ATWT* was unchallenged as the top-rated soap opera, until growing competition in the 1970s ended the show's dominance. Throughout its run of more than half a century on CBS, *ATWT* survived important changes: the switch to color; the conversion from live to taped television in the 1970s;, the shift from thirty minutes to an hour in the late 1970s; and fluctuating ideas about what topics the genre should cover, oscillating amongst family drama, romantic escapist fare, and tackling controversial social issues—or more often some combination of the three.

While its P&G/TeleNext sister show *Guiding Light* (*GL*) phased out many of its longtime characters in the two decades prior to its 2009 cancellation by CBS, *ATWT* retained not only the greatest number of long-term characters but also many of the actors who defined those characters, as compared with other contemporary soaps. For instance, Helen Wagner's performance as Nancy Hughes is listed by *Guinness World Records* (2006) for the longest portrayal of a character by a single actor in history. Wagner spoke the first words on *ATWT*'s debut episode, and her character continued to appear at least a few times a year until her 2010 death, only months before the show's cancellation.

Wagner was joined by several cast members who also remained part of *ATWT* for decades. Don Hastings was the third person to play the role of Bob Hughes, taking the part in 1960. However, he has portrayed Oakdale Memorial's most famous doctor for fifty years. The same year, Eileen Fulton originated the role of Lisa Miller (who went on to become Lisa Miller Hughes Eldridge Shea Colman McColl Mitchell Grimaldi) and who likewise remained on *ATWT* for five decades. These accomplishments were supported by a cast featuring actors who debuted in almost every subsequent era of the show. Marie Masters's portrayal of Susan Stewart began in 1968. Kathryn Hays debuted as Kim Hughes in 1972. Colleen Zenk-Pinter began as Barbara Ryan in 1978. Elizabeth Hubbard started portraying Lucinda Walsh in 1984. Kathleen Widdoes and Jon Hensley joined the cast as Emma Snyder and Holden Snyder in 1985. Scott Holmes took over the role of Tom Hughes in 1987. Ellen Dolan was cast as Margo Hughes in 1989. In addition, Kelly Menighan Hensley began as Emily Stewart in 1992, Maura West as Carly Tenney in 1995, Michael Park as Jack Snyder in 1997, Terri Colombino as Katie Peretti in 1998, and Trent Dawson as Henry Coleman in 1999, and a variety of other actors started on the show during the years 2000–2005. As new actors join and leave a show, the consistent portrayal of these longtime characters listed above act as an anchor for longtime viewers and provide consistency for the show through creative shifts and changes in the featured characters and stories. Further, characters/actors like Anthony Herrera's James Stenbeck (originated in 1980), Ed Fry's Larry McDermott (originated in 1990), and Paolo Seganti's Damian Grimaldi (originated in 1993) were often referenced and brought back to the show for stints, always played by the same actors. Because the show featured actors who started in their roles during every period of the program's history and periodically brought back these other faces, the cast directly indicated the multigenerational storytelling and viewing practices that set soap operas apart from any other fictional television genre.

The most central character in the history of *ATWT* was arguably Tom Hughes. The son of Bob Hughes and Lisa Miller, Tom was born on *ATWT* in May 1961. Miller was, as portrayer Fulton writes in her memoir, daytime television's original "bitch," and her marriage to Dr. Bob was one of the biggest stories of *ATWT* in the 1960s (Fulton 1995). Audiences reportedly cringed at the thought of their Bob marrying a conniver like Miller (Fulton 2006). For almost fifty years, viewers watched Tom mature into a prominent Oakdale attorney. Hughes is the only character in television history to be born on a show and then to be consistently featured for five decades in a daily television text, with viewers able to watch each step of the character's development.[3] Tracing the maturation of the Tom Hughes character can thus provide a lens through which to view both the trajectory of the soap opera genre and the changes in the audience and the culture that surrounds and supports these shows.

SHIFTING PORTRAYALS: THE MANY MEN WHO ARE TOM HUGHES

While daytime television provides viewers with the unique ability to see characters played by the same actor for decades, recasts are much more common on soap operas than on other television series, as a character's pivotal place on a show's canvas is commonly considered more important than the following for a particular actor playing the role. Thus, the duration of so many actors on *ATWT* is all the more impressive because such long-term performances are relatively rare, even in U.S. soap operas. Starting in 1963, when the character was old enough to be consistently portrayed by one child actor at a time,[4] the role of Tom Hughes has been played by thirteen people. In his first few years, Tom was aged more rapidly than real time would allow, and his birth date was revised significantly as the show progressed so that the character would be old enough to allow for certain stories. This aging process was largely accomplished by seven actor changes from 1963 until 1969. After that point, Peter Galman took over the role and played Hughes until 1973. Galman was replaced by David Colson, who portrayed Hughes from 1973 until the end of 1978. After a short-term performance by Tom Tammi, Hughes was played by Justin Deas from 1980 until 1984. Deas was critically acclaimed in his role as Tom, winning a Daytime Emmy Award for his performance. After brief stints by Jason Kincaid and Gregg Marx, Scott Holmes took over on July 3, 1987, and played the character of Hughes until the show's September 2010 cancellation.

As with any attempt to trace the textual or acting history of a particular soap character, a description of this nature may risk either becoming trivial or confusing, but I have included this information to show how complicated discussions of even a single character's history can be on a show as multifaceted as *ATWT*. For some, Justin Deas or Peter Galman may be the "real" Tom Hughes, and another portrayer not true to the "real" character, despite Holmes, for example, having appeared in the role four times longer than any other actor who played Hughes.

The transition from one actor to another often creates a corresponding change in the character's personality, although both writers and actors try to make the shift as smooth as possible. For instance, as I've previously mentioned, the early actor changes were used to age Tom quickly so that he could be used in more complicated stories. Each new actor had to both be true to Tom's history and shift the Hughes character to make him his own. Thus, Scott Holmes's Tom Hughes is different from Justin Deas's portrayal, although Tom reflects past performances as well for any viewers who have Deas's stint as Tom to draw upon from their memory or from viewing scenes on video sharing sites like YouTube.[5]

THE STUDY OF TOM HUGHES

With a character that was on the air consistently for almost fifty years, five days a week with no off-season, tracing the full details of a narrative like Tom Hughes's is impossible. With the ensemble canvas of any daytime drama, soap plots involve so many interactions among a community of intersecting people that a comprehensive study of even one character would be difficult enough to fit into a book, much less one essay. Because soaps are not commonly replayed or re-aired, much of the archive either no longer exists or is incredibly difficult to access.[6] Even if all those episodes were available, watching almost fifty years of *ATWT* featuring the Tom Hughes character would literally take years, and the minute character details revealed over that fifty-year span could never be mastered by one person, even if he or she had the access and the time.

Because of this massive volume of content, much soap opera scholarship has shied away from close analysis of texts or characters, especially when following a character that has developed over decades. For purposes of this essay, I have personally watched most of the years of Scott Holmes's portrayal of Tom Hughes and have also drawn on accounts from online discussion groups, online character guides, and friends and family who have watched the soap opera many more years than I. A collection of *ATWT* history published for the show's fortieth anniversary (Poll 1996) also proved helpful. Soaps themselves have a complicated relationship with their own characters' pasts, as continuity for longtime viewers dictates that the long history of their characters be referenced and names from the past be brought up fairly often.

Writers are often afraid such references will confuse newer viewers, or else the writers are so new to the show that they don't have the personal knowledge to draw on characters or relationships from prior decades.[7] This essay will look briefly at the various stages of Tom Hughes' on-screen life to demonstrate soaps' unique ability to tell a complex story over decades.

CHILDHOOD AND ADOLESCENCE—THE SORASING OF TOM HUGHES

Tom Hughes immediately became a central focus on *ATWT* because he was born to a central couple on the show at the time, Bob and Lisa. The show's writers recognized that only a minimal amount of storytelling could be accomplished with Tommy as a young child. Therefore, he became one of the first victims of SORAS, a disease that now regularly strikes children in soap opera towns. SORAS, which stands for Soap Opera Rapid Aging Syndrome, is a term popularized in the soap opera press and in online fan communities in response to the trend of aging soap opera character (almost always children) much more rapidly than real time would allow.

The early development of Tommy Hughes is one of the most blatant examples of SORASing, as the character was born in 1961 and, by the end of the decade, was serving in Vietnam. The character's birth and early existence was largely a plot device in the dissolution of Bob and Lisa's marriage. Bob, workaholic doctor and son of the featured family of *ATWT*, and Lisa, the ambitious social climber, separated, with young Tommy left stuck in between. He was quickly aged so that he could become an active part of the divorce storyline, acting out because he was resentful of his father's devotion to work. The story was one of daytime's earliest nuanced looks at the social effects of divorce at a time when such issues were becoming prevalent in the social consciousness. Tom grew up in the midst of this struggle between Bob and Lisa, spending periods of time with both parents and also away at school.

Tom was SORASed through constant changes of actors, allowing the baby in 1961 to be in college and then to return from Vietnam by the decade's end. If one were to try to hold fictional Oakdale, Illinois, to a realistic standard of time and aging, Tom's SORASing actually moved his birth date from 1961 back to the late 1940s. SORASing has become an accepted part of soap opera storytelling, and recent *ATWT* head writer Jean Passanante said that all soap writers realize that, while fans may sometimes complain about this aging phenomenon, they almost always accept and even desire it, as aging characters helps create more compelling stories (Passanante 2006). Fans complain about SORASing most when the aging process is either too drastic—as in this case with Tom—or when one younger character is aged while others are not, especially when younger characters get aged so that, as adults, they become older than characters born earlier than they were. With Tom, the SORASing was later reversed slightly through recasting choices, so that the actor playing Tom for the past two decades, Scott Holmes, was born in 1952.

VIETNAM AND DRUG DEPENDENCE

Tom continued to be aged rapidly throughout the 1960s so that writers could use his relationship with Bob and Lisa to examine the generational divide that defined the decade. Tom was frustrated both at his father's place in "the establishment," emphasizing career over family, and at his mother's obsession with maintaining and elevating her class status. In his somewhat justified frustration, Tom befriended his college roommate, who soon facilitated Tom's addiction to speed and involvement in several illegal activities. With his grades failing, Tom revealed to his family that he was thinking of joining the Army and going for a tour in Vietnam. He eventually did so, returning from the war with self-inflicted injuries and an even worse drug dependency. That drug dependency led to Tom being wrongly convicted for the murder of an ex-stepfather, although he was later exonerated.

This period saw the solidification of the Tom Hughes character, growing from being a plot device in Bob and Lisa's story to having a story of his own. By the end of the 1960s, Tom was established as a permanent part of the show's canvas. With *ATWT*'s focus on current social issues, Tom became an outlet through which the writers could examine aspects of the current political climate: the Vietnam War, the motives driving young adults to sign up for combat, the social consequences for soldiers returning from a tour of Vietnam, generational conflicts, and a growing visible drug culture in American society. These issues were addressed through a primary character linked to the central family of Oakdale, the son of two of the show's most heavily featured characters.

CAREER AND MARRIAGE

Tom began the 1970s as a mature young adult, portrayed by Peter Galman from 1969 until 1973. While earlier quick shifts in actors facilitated both rapid aging and the sense of a fractured or shifting character identification for Hughes, the relatively longer period Galman portrayed Tom demonstrated the character's newfound consistency. Viewers may have been reassured to see that the effects of the Vietnam War, the generational divide, and the drug culture that Tom represented in the late 1960s still resulted in a responsible and productive young adult, as Tom moved into a different phase in the 1970s: the search for love and the development of a career.

Several romances led up to Tom's eventual marriage to the demure Carol Deming. Meanwhile, Tom had gained control of his life and decided to use his knowledge of the court system, gained through being wrongfully accused of murder and through his parents' divorce in childhood, to become a lawyer. Tom's career choice was also meaningful to his family because his grandfather and the show's patriarch, Chris Hughes, was a lawyer in Oakdale. However, career and marriage came into conflict, as Tom's focus on law school caused—in an echo of his father's experience—a rift with Carol. This common genre theme of work/love conflict came to a head for Tom when he fell for a client and divorced Carol, becoming one of the divorce statistics he had idealistically hoped to combat when he began law school. In fact, one of the recurring ironies in Tom's story during the 1970s was the many ways in which he became the very person he had rebelled against the decade before, a social consequence that became common in American society as hippies became yuppies.

With the shift to David Colson playing Tom, the character's relationship with his new love, Natalie Bannon, led to a second marriage, which also ended in divorce owing to Natalie's infidelity. Colson's portrayal of Hughes played out these personal conflicts juxtaposed with Hughes's role as young lawyer, an important fixture in a town with as many controversial characters as Oakdale. During Tom Tammi's brief stint as Tom and the transition to Justin Deas, Tom almost married a third time—but this bride-to-be, Barbara Ryan, dumped him at the altar in favor of old flame James Stenbeck.

THE DEVELOPMENT OF A SUPERCOUPLE

Viewers followed Tom through the 1970s as he built a law practice for himself and endured several failed relationships and two failed marriages. But, with Justin Deas portraying Tom, the creation of a soap supercouple was underway. Another central aspect of soap storytelling, the supercouple is what every soap opera producer dreams of: the partnership that viewers can't get enough of, a love story and ongoing relationship that drives ratings and fan reaction. In the late 1950s, *ATWT* created perhaps the first soap supercouple with Tom Hughes's aunt Penny and her boyfriend and husband Jeff Baker.[8] For Tom, however, it was Margo Montgomery (played by Margaret Colin at the time) who would become the love of his life and his partner in *ATWT*'s longest lasting marriage.

As soaps entered the height of their fantasy phase in the early 1980s, Tom's and Margo's love story became one of the greatest examples of the action-adventure and fantasy romance of soaps during this period.[9] Tom's relationship with Margo developed around a story arc that spread across the show, in which a drug kingpin named Mr. Big became involved in the lives of several Oakdale residents. Tom, through his law firm, was investigating Mr. Big, and Margo worked as an assistant for him. However, she was personally involved in the controversy with Mr. Big as well. The adventure with Big led Tom and Margo across the world in a series of action stories, and their romance developed in the course of their travails. Tom and Margo's experiences became the major story of *ATWT* in 1982. At the height of this escapist tale, the supercouple ended up in Europe, at Mr. Big's mercy, in a death trap with the only way out being provided by clues from classic literature. Tom and Margo survived because of their knowledge of a Robert Browning sonnet and ability to reenact a scene from *Alice in Wonderland*.

The couple's escape to France and their fanciful adventures there led to an engagement. Their relationship faced a variety of challenges from social forces: Margo had been involved in recent controversy in Oakdale, and Tom had been dating Margo's aunt. Nevertheless, the couple was married in 1983, in a come-as-you-are spontaneous wedding, with Tom and Margo arriving on a motorcycle. The story also included Margo's career as a police detective, an important facet of the couple's relationships as they played out in the decades since. The popularity of this supercouple was driven by the soap press's revelation that the actors playing Tom and Margo had fallen in love.[10]

Deas and Colin soon left the roles of Tom and Margo. Hillary Bailey Smith took over the role of Margo. Meanwhile, Jason Kincaid's short run as Tom in 1984 was followed by Gregg Marx taking on the role for almost three years. During much of this time, Tom and Margo were on the backburner. However, 1986 introduced a complication in Tom's and Margo's marriage, as a shift in creative focus at *ATWT* epitomized by the arrival of revered soap opera head writer Doug Marland led to a strong focus on family and workplace drama once again.[11] Tom and former fiancée Barbara Ryan began working together, with Tom acting as Barbara's business manager. After a work night when both were drinking, Barbara convinced Tom that they had slept together. This led to Tom and Margo becoming separated and Margo's eventual affair with fellow detective Hal Munson.

TOM'S MATURITY—SCOTT HOLMES TAKES THE ROLE

At this point, Scott Holmes took over the role of Tom Hughes. Tom was out of Oakdale for some time in Washington, D.C., where he was heavily involved in a massive FBI case that involved the Oakdale Police Force. Tom returned to Oakdale to put his marriage back together and began working with Margo on the case. The couple was eventually reunited. Holmes's Tom once again became part of several storylines that sought to renew focus on social issues through personal drama, similar to the stories Tom participated in during the late 1960s. This mid-1980s to early-1990s time period is often celebrated by *ATWT* fans as a glory period for the show, with Marland blending social relevance into a strong writing emphasis on workspace tension and family drama.[12]

For Tom, this new emphasis included a surprise visit from Lien, a daughter born from an affair he had with a Vietnamese nurse during his time in the military decades before. The ensuing drama showed both the personal effects of Tom's discovering a grown daughter and this relationship's impact on his recently reconciled marriage—along with the lasting social effects of the Vietnam War on American society. In particular, the story highlighted the racism the Hughes family had to deal with because of Lien, including a storyline where Lien's high school teacher—another Vietnam veteran—openly showed aggression toward her because of her race.

The other major story for Tom was Margo's discovery that she was pregnant and that the baby was not Tom's but Hal's, the result of the short affair she had during her and Tom's separation. Again, questions of custody and the need for a nuclear family versus the messiness of real human relationships became the focus of baby Adam's birth. Hal eventually decided to be Adam's godfather and to let Tom raise Adam as his son.

At the beginning of the 1990s, Margo's current portrayer, Ellen Dolan, took over the part, and Tom's and Margo's status as a couple cemented with Holmes and Dolan in the roles. Tom and Margo's relationship involved both police drama (because of Tom's role as district attorney and Margo's as a detective), and family drama through which the writers continued to examine major issues Americans were facing. Euthansia was one of these issues. Margo was obliged to unplug her stepfather's life support after he confided in her that he wanted to die naturally, even though he had not made a living will. After Tom and Margo weathered the family controversy surrounding Margo's role in her stepfather's death, the couple had their first child together and named him Casey, after the deceased man.

Later, Margo was attacked and raped while on duty by a man later found to be HIV-positive. Through the next couple of years, the show explored the aftermath of her rape. While she did not contract HIV, Margo became friends with another victim of the same rapist who had contracted the virus and eventually died, with Margo and Tom mentoring the woman's son. When the rapist later broke back into the Hughes home, Tom killed him, and the show examined Tom's ambivalence over whether the murder was self-defense or the unnecessary killing of another human being. Through the end of the

Doug Marland Era, the couple remained an important focus for such debates. These social issues were made even more powerful because of Tom Hughes's positioning at the heart of the show, and by Tom's and Margo's status as the show's supercouple.

In the mid-1990s, Tom and Margo continued to work together, solving local crimes and raising Adam and Casey after Lien left town. Eventually, though, Tom hit a mid-life crisis, returning to some of his early questions about his place in "the establishment" as district attorney. Disillusioned by not being able to make a difference, Tom decided to leave the law and begin a new job as a journalist for *The Argus*, his mother's newspaper. However, his editor at the paper, Emily Stewart, became infatuated with him and took advantage of Margo's growing obsession at work with a troubled youth named Eddie Silva to the point that she convinced Tom that Margo was having an affair. Tom ended up having a one-night stand with Emily, shocking viewers by acting out of character. Although he and Margo reconciled, Tom and Emily's affair led to a child, Daniel, born in the late 1990s.

During this time, Adam discovered that Tom was not his biological father. Angry at Tom because of his infidelity, a SORASed Adam—now a teenager in high school—brought all the issues of Margo's and Tom's separation from the mid-1980s back to the forefront, where they were juxtaposed with Tom's current affair with Emily. Adam left Tom and Margo to live with his biological father Hal and Hal's wife, Barbara. Longtime viewers were reminded of the complicated past of Hal, Barbara, Margo, and Tom, while new viewers were given enough background information to understand the current family drama. In short, the dynamic between these characters worked so well precisely because the story of Adam's discovery that Hal was his father had been woven into the plot—and "resolution" delayed—for years.

Tom and Margo weathered that storm and a later flirtation Margo had with a local sportscaster that led to another separation for Tom and Margo in 2004. The couple—and the viewers—had by that point realized that the two had so much history between them that they could not stay apart. And Holmes, despite being the thirteenth actor to take the role of Tom Hughes, had become the defining actor for the role.

THE VOICE OF REASON

Tom Hughes eventually took over the "voice of reason" role from his father. While Dr. Bob remained part of the *ATWT* canvas, Tom appeared often as a key figure when Oakdale citizens found themselves in trouble. While daughter Lien was no longer in Oakdale (although the character returned for stints),

Tom was actively involved in the lives of his three sons—dealing with Casey's fathering a baby that died at birth while he was still in high school, Casey's later prison stint after stealing to pay off an online gambling debt, and then his decision to follow in Tom's footsteps and become a lawyer; dealing with the death of Adam's biological father, Hal, Adam's subsequent mysterious departure from Oakdale after committing some underhanded deeds, and then his return with a new face and new identity; and taking primary responsibility for son Daniel while his mother battled various legal issues. During this time, Tom shifted back into his role as district attorney to spend more time working with his wife, only to leave the job to go back to private practice when he realized how much his work as a D.A. took away from his personal time. In addition, Tom's character was at the center of family drama when he suffered a heart attack due to the stress of his son Daniel being kidnapped, and Tom and Margo were later stalked by Margo's rapist's brother (whom Margo later killed in the Hughes den, in a scene reminiscent of Tom's murder of the rapist more than a decade earlier). Later, *ATWT* viewers watched the drama of son Adam's return in the guise of Afghanistan war veteran Riley Morgan (complete with a new face and voice, the result of Adam's injuries from an explosion in Afghanistan). Margo first decided to hide Adam's true identity from Tom, to prevent him from having to make moral and legal decisions about whether to turn his son in for various crimes, including the impersonation of a deceased soldier. When Tom eventually discovered the truth, he struggled with the sense that his family didn't trust him but decided to defend his son in court when Adam turned himself in. Alongside the drama of the show's final few years on CBS, viewers watched Tom deal with his parents' health concerns, including multiple strokes for his father and heart problems for stepmother Kim.[13]

Tom and portrayer Scott Holmes, who played the role for more than two decades, remained an essential part of *ATWT*'s fabric even when he was largely playing a supporting role. In soap operas, the interaction among characters takes precedence over the plot-driven day-to-day activities.[14] Tom and Margo were the longest-running couple on the show. For longtime viewers, their relationship with each other and with most of the cast was an important tie that bound together *ATWT* and the show's past, present, and future in a way that no other television couple could achieve.

CONCLUSION: WHAT THIS MEANS FOR SOAPS

The only way to understand the power of soaps and to be able to acknowledge artistry in the soap opera genre is to understand and analyze the texts themselves. If nothing else, this essay proves how difficult a task true textual analysis of a soap opera over time can be and why the nuances of a soap are hard to explain. As with real life, soap opera narratives evolve over long periods of time, resist neat categories, and often involve more characters than can be included in a short plot summary. In some ways, the artistry of soaps remains hidden from anyone outside the fan community.

To trace the character of Tom Hughes is to trace the trajectory of the U.S. soap opera and, to a degree, U.S. television. Tom's career illustrates the soap opera genre's use of SORASing and the supercouple, as well as the constant tug-of-war between the three major strands of soap opera plots: family and workplace drama, social issue storylines, and escapist romance fare. A part of the soap canvas for almost fifty years, Tom Hughes is—in a sense—the history of *ATWT*, and the trajectory of his character marks changes in performers, changes in writing staffs, and changes in audience reception and U.S. society. From tackling divorce to drug culture and Vietnam to living wills and AIDS, Tom's character was involved with many of the controversies that have defined U.S. public discourse over the past few decades. And, for fan communities with lasting memories, his character served as a monument to those social changes and plot turns.

Soaps are always at their best when they blend social awareness with character development and drama, and this brief sketch of Tom Hughes's history demonstrates the power of soaps to create coherent narratives of characters' lives. While many more recent viewers may not have known Tom's rich past (and there is much more omitted than included here), longtime viewers or viewers interested in understanding a show's history were at least broadly aware of the way this character developed over time. Tom's comments and actions in later episodes were often examined and weighed against his past in fan discussions, and his scenes with fellow performers were mined and supplemented with any history Tom might share with others. Dedicated fans internalized the basic narrative outlined here. The fan community's consumption of the text of Tom Hughes's life led to a nuanced understanding of the contemporary character and his complicated relationships both with the variety of characters on the later *ATWT* canvas and myriad characters from the show's history who may have been referenced from time-to-time or who fans believed might return in the future. For fans, mastering this overall narrative is crucial, and several sites and online discussions were dedicated to filling in the blanks in Tom's story as well as those of other longtime Oakdale residents.[15]

Most importantly for the soap opera genre, this essay highlights what U.S. soap operas can do that no other genre can: tell a story over decades and create a rich multigenerational narrative that draws not only on characters throughout the history of the show, but also on the interests of viewers of all ages. Soap opera writers may see this unique attribute of soaps as not only a genesis for creativity but also as an albatross, with a history not welcoming to new viewers and a longtime audience that demands continuity with a show's history. However, this long-term character development and storytelling is the differentiating element of soap opera storytelling and must be fully exploited to draw an audience. As the introduction to this collection emphasizes, the only way for soap operas to gain and maintain viewers is to draw people not just into the narratives, but into the history of the show and the fan community as well. Focusing on stories that do not capitalize on the history of a show and its characters may lead to some short-term viewer gains, but this strategy creates no compellingly different product in which new viewers might invest. In short, soaps' greatest chance for gaining an audience that will stick around is to draw in viewers who are interested in the detective work of piecing together character relationships and complicated pasts, the sort of engagement that soaps downplay when they oversimplify or ignore the history of their shows.

No other television genre could offer a long-term viewing experience like watching the life of *ATWT*'s Tom Hughes. Finding ways to highlight and draw from that history and to connect viewers with not only historical footage, but also longtime viewers who can help act as "tour guides" and resources for the history of these characters, is the best way to draw new viewers who can become actively engaged with a show and its characters, thus developing a long-term commitment to the show. On *ATWT*, Tom Hughes rarely had a significant front-burner storyline in the show's final decade, a fact that longtime viewers often highlighted in fan discussions. While the show remained committed to including the character, his marginalization to supporting status (and sometimes functionally—as a lawyer—rather than in a role drawing on his rich history on the show) muted the power of *ATWT* and sometimes wasted the potential of having such deep character history.

The show treated many of the other longtime characters listed earlier in the essay similarly, despite the fact that many of the actors who portrayed them are recognized as veterans and masters of their craft because of their ability to draw on decades of history. Waste of soaps' differentiating factors—longtime characters and multigenerational casts—is not specific to *ATWT*, even though this particular show was blessed with the deepest contemporary ties to its longstanding history and a dedication among the producers to hang on to key actors and characters. For soaps to survive in the modern media landscape, it is crucial that the genre fully embrace what makes it unique—this history—in ways that are accessible to new viewers, connecting them to the history of the show and to other viewers who have watched for years.

NOTES

1. For more on the camera work in soaps, see Timberg and Alba's essay in this collection.
2. See pieces by Scardaville, Scodari, Liccardo, and Metzler in this collection for more on this perceived decline in quality.
3. The only characters on television longer, such as Nancy and Bob Hughes and Lisa Miller, were already adults at their debut.
4. When soap opera children are babies, they are rarely shown on screen and were even more rarely shown during the early days of soaps, when the programs aired live. Because soaps tape so frequently, babies are switched every few days, so that viewers have come to accept that the baby's looks will change constantly.
5. See Webb's essay in this collection for more on YouTube's impact on soap opera viewing.
6. See Wilson's essay in this collection for more on institutional soap opera archives.
7. For instance, characters will refer to illnesses or phone conversations with former faces from the show because certain off-screen characters remain an integral family link to multiple characters currently on-screen. These references are a reward for long-term viewers designed to be small enough not to confuse newer viewers.
8. When Jeff's portrayer left the show, his character was killed abruptly, causing a nation of television viewers to mourn. This story is often cited as one of the most powerful examples of viewer identification with television characters.
9. See more on this shift in Liccardo's essay in this collection.
10. Justin Deas and Margaret Colin were eventually married.
11. See further comments on Marland's *ATWT* work in the pieces by Casiello, Harrington and Brothers, and Scardaville.
12. *ATWT* fans refer to Marland even more often than to creator Irna Phillips as an exemplar for current writers.
13. This account does not reflect stories from the final few months of *ATWT*'s CBS run leading up to its 2010 cancellation, as this book was in production during this period.
14. For more on the relationship between the aging of characters and their portrayers, see Harrington's and Brothers's essay elsewhere in this book.
15. See, for instance, the "Who's Who" sections and the family tree sections on each soap at the Soap Central Web site, located at http://www.soapcentral.com/.

PERSPECTIVE
WRITER KAY ALDEN ON WHAT MAKES SOAPS UNIQUE

BASED ON AN INTERVIEW BY SAM FORD

> Kay Alden is currently co-head writer of *The Bold and the Beautiful* and former head writer for *The Young and the Restless*, a show for which she wrote from 1974 until 2006 and won four Daytime Emmys and two Writers Guild of America awards. She has also worked as a consultant to ABC Daytime. Alden has been a lifelong fan of daytime soaps and completed academic work on soaps while at the University of Wisconsin-Madison.

The soap opera is historic in terms of the quantity of material that has been produced within the genre. For instance, *CSI: Crime Scene Investigation* (CBS) bemoaned the fact that William Peterson was leaving the show after some one hundred-plus episodes. I'm writing the newest daytime drama currently on the air, CBS's *The Bold and the Beautiful* (*B&B*), which celebrated its twenty-third anniversary in March of 2010. We are currently writing episode 5,830 at the time this is going to press. CBS's *The Young and the Restless*, which has been airing for thirty-seven years (since 1973), is currently writing episode 9,410. *Guiding Light* (CBS), which broadcast its final episode in September 2009 after more than seventy years on the air, closed out its historic network run with episode 15,762. CBS aired its final episode of *As the World Turns* in September 2010, after more than fifty-four years on the network, capping off another historic broadcast run. In all, this venerable daytime giant aired more than 13,800 episodes on the network. To create and produce that amount of material is nothing less than remarkable. No other genre can compare in terms of the sheer number of hours of entertainment programming. The daytime serial is a phenomenon—a phenomenon that's no longer as bright a shining star as it once was, but it's still in a category of its own.

This is a daily medium that relies on viewers tuning in tomorrow. These are shows that people invite into their homes, ideally five times a week. Why do they do that? What is it about? Why do people keep tuning in, day after day, year after year? On *B&B*, four core characters have been portrayed by the same actors for twenty-three years, since episode one. Stephanie, Eric, Ridge, and Brooke (portrayed, respectively, by Susan Flannery, John McCook, Ronn Moss, and Katherine Kelly Lang) have been available to *B&B* viewers every day for nearly a quarter of a century. That is astonishing. Think of Susan Flannery, whom viewers have watched daily for the past twenty-three years. This beloved character has literally aged from gorgeous leading lady to gray-haired matriarch on our television screens, and we have been with her every step of the way. Who in Hollywood endures like that? It is a tribute to the actress that she has not only endured; she has embraced her character's maturation. In so doing, she has earned the undying love and loyalty of the audience who "tuned her in" each day.

There is something beautifully reminiscent of life and the way we grow and the way we age in daytime television. There is no vehicle other than the American soap opera that focuses on life in such an experiential way, literally making viewers a part of the day-to-day growth process. Daytime drama respects and elevates the process of living. The basis of the drama evolves from a focus on relationships and everyday life.

The Internet provides a new platform for the daytime audience to be *heard*. I find what's being said by soap viewers online quite fascinating. I think these forums are helpful; I don't think we've really had the opportunity to figure out just what role those responses should play. I do think there is a tendency to overreact, to say, "Oh, my gosh, they're not liking this! I've got to jettison this." You have to weigh your commitment to a story against what kind of reaction you're getting. I absolutely believe in looking at audience response; this is valuable information to have. But we need to be cautious how strongly we react to it.

On *B&B* Web sites, comments are sometimes so harsh that I can hardly bear to look at them. Some viewers are frankly abusive to us as writers, to our characters, and to our performers. Sure, they're words, and you can say they're not going to hurt anybody, but you can't deny that it gets into your head. That kind of input, I tend to feel, is an argument for not reading online message boards every day. It's good to let a few days go by and then see how fans are reacting to what they're seeing. Keep in mind, we've already written several weeks or sometimes months past what fans are watching. So, even if the fans don't like it, we still have a long time to go from what viewers are seeing and reacting to on air. If we aren't careful as writers, intensely negative viewer response can create a crisis of confidence, shaking our faith in our storylines.

I believe daytime soaps still do the best job of all television genres of cutting into the characters and the relationships. We have no time limit; no one tells us how many days we will tell a story or how long until we get to a certain plot point. We are not limited in that sense. If stories are interesting, we can take our time letting them unfold. When the viewers care about the characters, which is our task as writers, the audience will stay involved for the journey. On the other hand, attention spans aren't what they used to be. It's a delicate balance. If we spend too much time exploring motivation, the audience may lose interest. The art of daytime television lies in finding that delicate balance.

Some have argued that primetime serialized dramas may be doing a better job of attracting and holding an audience than we are in daytime. However, a primetime show has to find a "tune in again" moment once a week; in daytime, we need to find that compelling moment each and every day. That is our constant goal: to make the viewer feel compelled to watch and see what happens tomorrow. Rather than being intimidated by the growing popularity of primetime serialized shows, I am encouraged by it. As viewers tune in to *Gossip Girl* (CW) or even *Survivor* (CBS) and *Big Brother* (CBS), I see hope for the survival of the daytime drama. Primetime dramas and many reality shows reaffirm the power of the serial form, and daytime drama remains the master of this format. People inherently want to know what happens next. It's part of human nature. Understanding this desire, as well as how the daytime audience relates to the characters they know and love, is key to a genre that asks viewers daily, more than 250 times a year, to "tune in tomorrow."

PERSPECTIVE
SCHOLAR NANCY BAYM ON SOAPS AFTER THE O. J. SIMPSON TRIAL

BASED ON AN INTERVIEW BY ABIGAIL DE KOSNIK

> Nancy Baym is associate professor of communication studies at the University of Kansas. She is the author of *Tune In, Log On: Soaps, Fandom, and Online Community* and writes regularly about online fan communities at her blog, *Online Fandom* (http://www.onlinefandom.com/). She started watching soaps as a teenager when she had a summer job cleaning hotel rooms; the woman who trained her also taught her all of the characters and storylines on *General Hospital,* which she watched regularly (in addition to the rest of the ABC lineup) for the following decade.

I guess I'm a very distant observer of soaps—I haven't watched a whole episode of a soap opera in a long, long time. It was hard to get back into soaps after the 1995 O. J. Simpson trial, which came at a time when I had been so intensively writing about soaps and reading about soaps, and I was overexposed to the genre. It was good to get a break from soaps at that time. The storylines that I was a total sucker for were these long-arc romances where the couple can't be together and the parties don't realize they love one other, even though each knows they love the other person. Getting together takes forever, and both characters think the love is unrequited.

After the O. J. trial, those plots were gone. They were too long. Everyone on soaps went from being strangers to being married in six weeks and then got divorced. Before the trial, everyone on soaps used to eye each other and *not* kiss. There was a plot on NBC's *Days of Our Lives* (*DOOL*) where Shane had amnesia and didn't remember his wife Kimberly, but finally it all came rushing back to him, after months and months of waiting for that moment. Jack and Jennifer on *DOOL* had to play out their romance over years. On *One Life to Live* (ABC), the show managed to get everybody into the storyline of Billy being gay. Everybody was implicated, and such intricacy takes a long time and careful plotting. I thought that storyline was brilliant. Stories like that took advantage of the wide, long canvas to do things that other genres couldn't.

A lot of the drop in the audience that occurred around the time I quit watching also had to do with people realizing that they could in fact get through your week just fine without their soap. Picking up a five-hour-a-week habit again just didn't seem like a good idea after the O. J. trial. It's like smoking: a compulsion. Once you wean yourself from it, you want to stay off. Regular soap viewing requires so much investment, so much more than other kinds of media. So people thought, "Maybe I'll watch some primetime." There were always soapy elements on primetime. The relationships were not left hanging in the same way, but primetime shows like NBC's *Emergency!* dealt with family trauma in the 1970s in a manner that mimicked soaps.

However, soaps never get their due. People say of *Lost* (ABC)—"It's ambiguous; fans have to talk about it to make sense of it! Television is revolutionized!" But I say, "Really?" None of that is new. However, viewers now have less time for and less commitment to soap operas. There's so much more media now available in shorter doses. Whether you watch soaps is all a question of lifestyle. My kids are at school in the day, and, when they're at school, I need to be working. At nighttime, I have to make dinner. However, if something should change in my life, if I should spend a lot of time in bed sick, with an illness, I'd be back watching soaps in a heartbeat. I know I'd be watching soaps if I were homebound. I can also see myself watching soaps later in life.

When I do turn on the soaps now, as I do from time to time, I don't really watch, but when I see someone I recognize from when I used to watch, I find that experience very powerful. It's an incredibly powerful kind of continuity. The endurance of these characters, their continued existence—I feel admiration for these actors and actresses who live this double life, who spend a whole lifetime playing one other person. That's really a weird career.

These people have been there my whole life. It's a parasocial relationship—a kind of family—and I do feel emotionally attached to those people, even if I never liked their characters that much. Monica Quartermaine on ABC's *General Hospital* (*GH*) was always stiff and uninteresting to me, but she's been on my radar almost thirty years now, longer than anyone I know. And then there are those young people—the ones who were children on these shows and are now women, like *GH*'s Robin Scorpio.

It's like comfort food. It's nice to have it, but it's not like anything goes wrong if you don't.

OF SOAP OPERAS, SPACE OPERAS, AND TELEVISION'S ROCKY ROMANCE WITH THE FEMININE FORM

—CHRISTINE SCODARI

> Christine Scodari is a professor of communication and multimedia studies and a women's studies associate at Florida Atlantic University. She has written numerous pieces of scholarship on issues of gender and age in soap operas, including her book *Serial Monogamy*. She has been a cafeteria viewer of numerous daytime soap operas since her teen years, including *All My Children, General Hospital, One Life to Live, Days of Our Lives, Another World,* and *Guiding Light,* and began engaging with them more intently because of her academic work.

INTRODUCTION

In 1987, Fiske catalogued television's gendered attributes, observing that the masculine brand of TV programming marginalizes women, the private sphere, and committed relationships, instead focusing on "lone wolf heroes" or "hero teams" that bond within a context of hierarchy and goal orientation. In terms of narrative structure, masculine TV relies upon action to advance plots that reach episodic closure and punctuates them with celebratory triumph. Feminine television—epitomized by the soap opera genre—features personal settings, interactions, and relationships, ensemble casts with a solid female presence, and a preference for character-driven storytelling. Structurally, the feminine narrative favors process and the deferred resolution of multiple staggered and serialized storylines.

Scholars such as Brown (1994) and Nochimson (1992) have theorized that serialization opens the soap opera text to resistive interpretation, serving to rescue this feminine genre and its fans from unwarranted critical denigration. However, as Mumford (1995) and I (Scodari 2004) have demonstrated through textual and/or ethnographic audience analysis, such a claim is qualified, as individual soap opera storylines do eventually achieve closure, and the story content of such closures can retroactively undermine emancipatory readings.

Serialized structure had long been suspect for its association with soaps and their fans. It was not until the 1980s that American primetime drama series began dabbling substantially in the feminine form. (Noted exceptions include the 1960s self-proclaimed soap opera *Peyton Place* [ABC] and CBS' 1965 *As the World Turns* [*ATWT*] spinoff, *Our Private World*.) Programs that dared to resurrect supposedly one-off storylines and/or characters or blatantly attenuate some of their narratives were lauded by critics such as Newcomb (1985) as the "champagne of TV" and were, in due time, showered with Emmys. The unbridled romance with serial storytelling in such primetime dramas would, however, be short-lived. What factors have emerged and/or coalesced to constrain and/or interrupt the use of such storytelling in both nighttime and daytime dramas?

This essay adopts a macroscopic perspective in pondering the ebb and flow of serialization and other feminine narrative qualities as theorized in terms of American soap opera and impacted by audience segmentation and other economic contingencies. It considers how the gendered narrative elements of soaps have been absorbed into, handled cautiously in, or cast out of primetime genres such as science fiction "space operas" and reworked in light of the challenges daytime dramas currently face. It explores how soaps have adapted by peppering their narratives with primetime conventions. Ultimately, it argues that the commercial imperatives and gender biases of television conspire to circumscribe the feminine form and, along with it, the genre that started it all.

PRIMETIME: LATHER AND RINSE

Wittebols (2004) contends that, in the fifteen years or so prior to the publication of his book, the "soap opera paradigm" governed multiple genres of television, such as news, sports, and reality. The "market critical" aspects of this paradigm include serialization (or "seriality"), real-time and audience-omniscient storytelling, "seeming" intimacy and semiotic play, and chaos and conflict as identifying themes (2004, 34–9). His point is well taken, particularly in linking strategies tied to the consumerist *raison d'être* of American soap opera to trends in other genres. However, many, if not most, of the series and genres he cites are, at best, scattershot in their incorporation of Fiske's (1987) elements of feminine narrative, such as complete serialization and emphasis on private sphere issues. Moreover, despite Wittebols's thesis, the political economy of American commercial television does not necessarily lead to the presence of such traits but sometimes leads, instead, to their containment or outright absence.

In Wittebols's (2004) rendering, the predominant gender of the target audience and the fact that young males are a more highly prized audience demographic[1] are not key factors in the industry's profit-seeking equation. In fact, going too far in the direction of feminine narrative is unlikely to attract this elusive audience segment. News may over-dramatize and serialize political storylines, as Wittebols (2004) suggests. However, these stories still center on the public sphere, thereby retaining at least some masculine narrative traits and, theoretically, male viewers. Similarly, professional wrestling may wallow in "backstage intrigue" and other soapy flourishes, but the narrative is still driven by physical action, and the endgame is ultimately one of masculine triumph and resolution.

In terms of primetime drama, Wittebols and others tend to conflate soap operas with serial episodic series.[2] While programs such as *Melrose Place* (Fox) and *Dawson's Creek* (WB) certainly meet the definitions of soap opera and feminine narrative *à la* Fiske, cop dramas such as *Cagney and Lacey* (CBS) or *NYPD Blue* (ABC) and medical dramas such as *ER* (NBC) or *Grey's Anatomy* (ABC) are actually semi-serialized gender hybrids rather than purebred soaps, in that some of their storylines continue and others conclude within a given episode. In the case of these particular programs, the main action occurs in the public sphere and involves both public and private sphere issues. Personal storylines are almost always serialized, while many of the professional plots—such as the medical cases on *ER* or *Grey's Anatomy*—include one-episode characters and wrap up by the end of each installment.[3] Contrast this with ABC Daytime soap opera *General Hospital* (*GH*), in which the medical cases generally involve regular or recurring characters entangled in ongoing personal crises.

Indeed, in recent times, primetime television's embrace of feminine narrative and, particularly, serialized structure, has been tentative. As viewing options mushroomed and program ratings continued to decline during the 1990s, narrowcasting by gender and other demographics became commonplace, and networks looked to vehicles that would be cheap to produce and/or profitable in syndication, and neither proposition is true of dramas with more than a modicum of serialization (Thompson 1996, 180; Turow 1997). "Unlike soapier shows," Havrilesky (2003, 1) explains, those that allow a viewer to "pick up in the middle" and understand what is going on—whether an episode is first run or rerun—are much more lucrative in the syndication market. Contemporary lifestyles hinder "appointment" viewing, and depending on serial narrative content to enhance a program's enjoyment consequently diminishes audience retention.

When considered in light of the huge investment associated with primetime drama, serialization has been increasingly limited to premium cable or to top-rated or smaller network shows meant for women—or else gingerly employed and/or dispensed with entirely. Yes, the success of semi-serialized programs aimed at female viewers, such as *Desperate Housewives* (ABC), has allowed similar "dramedies" and dramas to be greenlighted on major commercial networks, but only a few runaway hits have stuck around for any length of time.[4] And, yes, *24* (Fox) features an over-arching, serialized narrative, but among other masculine traits, the show inserts "mini-closures" in virtually every episode, allowing viewers to miss a week or two and still understand the narrative logic of a given episode upon their return. As with comparable shows touting a healthy male viewership, such as *Lost* (ABC) and *Heroes* (NBC), *24* is vigorously promoted and aired in uninterrupted blocks—in this case starting in winter rather than autumn.[5] According to Jenkins, female viewers of these traditionally male-oriented sub-genres are regarded by producers and creators as icing on the cake. He points to *Lost*'s ubiquitous character flashbacks as elements of serialization appealing to female fans intrigued by ongoing relationships, while "fanboys"—in masculine, puzzle-solving fashion—utilize the flashbacks' character clues to unravel the conundrums present in the show's "real-time" narrative (Jenkins 2006a). But, as Kustritz (2007, 2) argues, such gender hybridization is fraught with risk and requires "delicate negotiations between producers, writers, and various groups of fans who often find fulfillment in a program for vastly different reasons."

It is not surprising, then, that many science fiction and other series explicitly privileging a young male audience have been hamstrung in any proclivity to hybridize masculine and feminine narrative. *The X-Files* (Fox), *Farscape* (Sci Fi), and *Firefly* (Fox) serve as three cases in point. As my previous research (Scodari and Felder 2000) chronicles, *X-Files*'s innovative yet schizophrenic structure interspersed standalone episodes in which Agents Mulder and Scully investigated and resolved a paranormal mystery in each installment with "mytharc" (mythology arc) episodes that advanced the series' continuing narratives and highlighted personal relationships. The latter type of episode was used sparingly, however, at just three junctures during a season (sweeps months) to ensure enhanced ratings and promotion.

This narrative structure produced somewhat disharmonious gender hybridization and contributed to the emergence of two warring camps among the show's fans: one that read romantic development into the agents' relationship, accentuating events in the mytharc, and another that hoped the pair would remain strictly platonic. This second, mostly male group argued that any hint of romance would render the show a soap opera, pointing to creator Chris Carter's promise that love would never bloom. In fact, according to actor David Duchovny (Mulder), at one point Carter fretted that a slight touch of hands between Mulder and Scully could come off as "too soap opera-y" (Schilling 1999, 5). Nonetheless, a proponent of a romance—echoing its mostly female advocates' reasoning—claimed on the alt.tv.x-files Usenet newsgroup that if Carter is "trying to write the anti-soap, he's failing miserably as far as the mythology is concerned" (Scodari and Felder 2000, 246). The mytharc's feminine dollop continued to be policed throughout the show's nine seasons, as securing young male viewers was Fox's key objective.[6] While there were nebulous hints of a love story as the series wound down, one was never overtly depicted until the second X-Files feature film, *I Want to Believe* (2008), was released in an altered political and economic context well after the series ended.

Created for cable's Sci Fi Channel, *Farscape* (1999–2003) became the first critically acclaimed program in the niche network's stable. Sci Fi executive Bonnie Hammer touted it as a model for attracting fans of both genders (Lucas 1999) and claimed a mandate to broaden her network's repertoire beyond a "very boy representation of science fiction" (Beale 2001). This "space opera," as it is often dubbed (Lavigne 1998), follows the escapades of a wise-cracking American astronaut, John Crichton, who is stranded on a massive, living ship in a remote sector of the universe with a motley crew of alien refugees after being sucked through a wormhole. This would seem to be a classic, sci-fi scenario appealing to the genre's characteristically young male enthusiasts. However, the series cultivated an audience of both sexes and a semi-serialized pattern that fueled affinities among regular and recurring characters that progressed, regressed, and were interwoven with and impacted by other episodic and/or extended plotlines. As in real life (and soap opera), the untangling of one predicament inevitably led to another, and little rest was afforded the heroes. This approach prompted the following commentary from a wavering male fan on the alt.tv.farscape Usenet newsgroup: "*Farscape* has been one of my favorite shows since I first saw it, but I can understand someone not liking it. They never win! Crichton especially, they are constantly traumatised [. . .] and rarely get that relax-at-the-end-of-an-episode moment [.]"[7] The remark clearly reflects expectations associated with Fiske's (1987) masculine television conventions. Creators did toy with Captain Kirk's "girl on every planet" formula, placing Crichton in tantalizing poses with females of myriad exotic species, but he seldom followed through. Instead, he undertook a rocky but resolute romance with a gun-toting shipmate, the alien soldier Aeryn Sun.

In light of the perils of gender blending discussed above, it is fair to say that *Farscape*'s soap operatic elements made its situation precarious from the outset. Then, as the third season wrapped, the series was offered a two-year renewal on the condition that it improve its ratings and demographics (Owen 2003, 30). A year later, the show was axed, despite glowing notices from professional reviewers such as *TV Guide*'s Matt Roush (2004), whose timely online editorial praising the series and admonishing the network for canceling it was read to the cast and crew by executive producer David Kemper after he informed them of the bad news (Kemper 2006). Adrienne Crew (2003, 2) of *Salon* notes attributes such as "strong female characters, feminist storylines, and the sexual tension between human John Crichton and his alien flame" in her complimentary post mortem (2003, 2). But it was just such elements that spelled the show's doom, making it too serialized and complex for those viewers who, lacking a penchant for "appointment television," engaged in sporadic viewing and then, frustrated with missing key plot developments, ceased watching entirely (Owen 2003, 30).

Overall ratings, cost, and expectations that it would perform better with young males (males in general constituted only half of the audience) were also factors in the show's demise, with the Sci Fi network giving little thought to approaching advertisers eager to reach the more gender- and age-diverse audience already in place (Owen 2003, 30). Claudia Black, who played Aeryn, observed that the network "didn't understand the power of the romance between Crichton and Aeryn" (Crew 2003, 2). Kemper disavowed network promos depicting Crichton as an omnipotent space playboy, but creators did tweak the show's formula, trying to boost viewership as season four commenced. Core fans were put off by the alterations, while new audiences never materialized (*SFX Magazine* 2003). By the time *Farscape* left the air, the network had begun to stake its future on *Stargate SG-1*, an established sci-fi show with a masculine narrative, and to repudiate its niche by jumping on to the reality TV bandwagon with affordable series such as *Scare Tactics*. In 2004, Sci Fi launched another semi-serialized space opera with its critically acclaimed, "re-imagined" version of the late 1970s program *Battlestar Galactica*. This show, too, was destined to fade out at the end of its fourth season after similar audience makeup and retention issues emerged, and its creators made the decision to preemptively conclude the program upon completion of a major story arc (Martin 2007; Miller 2007, 2).

This conventional deference to short-term ratings success is a hindrance to creativity and character-driven storytelling. *Firefly*, a space opera similar to *Farscape*, was picked up by Fox in 2002 on the reputation of its creator, Joss Whedon of *Buffy the Vampire Slayer* (WB, UPN) and *Angel* (UPN) fame (Gilbert 2002, N1). Whedon was ordered by Fox to shelve a planned premiere recounting the characters' intricate back stories and replace it with a plot-driven, action-packed installment more easily digested by the desired audience: young males (Freeman 2002, 3). The shelved opener capped off the series three months later, when fans protested that, had it aired first, it might have lured enough additional viewers to avert cancellation (Gilbert 2002, N2).

If *Farscape* and *Firefly* were considered the epitome of fiscally unviable soapiness at the dawn of the twenty-first century, reality series such as *Survivor* (CBS), *American Idol* (Fox), and *The Bachelor* (ABC) were less problematic. Such shows are semi-serialized and melodramatic, as Wittebols attests, but they also include weekly mini-closures within a finite number of episodes, making it easier for viewers to either watch religiously or drop in and out at will. Moreover, these programs are as potentially effective in attracting young female viewers as the more costly primetime soaps and serial episodic dramas they began to outnumber on the major networks over a period of several years in the early 2000s.

Thompson's (1996) *From* Hill Street Blues *to* ER: *Television's Second Golden Age* chronicles the period of the 1980s through the mid-1990s in which "quality" dramas, most of which were serial episodic shows, reigned in primetime. Audience fragmentation and other contingencies began to intervene, leading to narrowcasting and other strategies designed to circumvent the loss of advertising revenue, a phenomenon dissected in Turow's (1997) *Breaking Up America*. This environment exacerbated already existing concerns about serialization, leading to steadily increasing curbs on feminine narrative conventions in drama during the mid- to late- 1990s and early 2000s. By the end of the 1990s, one consequence of this change was the emergence of a plethora of temperate, cost-effective procedural programs with masculine, single-episode narratives, such as those populating the Law & Order (NBC) and CSI (CBS) franchises. After lamenting the demise of her serialized favorites, Schlossberg (2002, 2) observes of these non-serialized alternatives: "Watching these shows is like having a series of one-night stands with the same fun guy. You have a good time but wouldn't be devastated if it all ended." Women characters do partake in the episodic triumph over crisis and conundrum dominating procedurals—but usually under the patronage of seasoned male bosses. On *Law & Order*, the female principals (assistant district attorneys played by Jill Hennessy, Carey Lowell, Angie Harmon, and three others in the course of fourteen of the show's seasons) come and go with more frequency than male leads. It is as if any female actor will do as long as she is attractive and a generation younger than her male co-star. A dearth of feminine story tidbits and the near absence of character arcs—especially ongoing romances between core characters—sharpen a division between the mostly static regular characters and the shorter-term suspects and victims with whom they deal.

So, in the 1990s and early 2000s, as primetime lathered up and then rinsed off its soapy excesses, what *was* the daytime drama to do?

DAYTIME: MIX AND MATCH

In the 1990s and beyond, ratings for afternoon soaps tumbled as their producers and creators battled audience segmentation and other threats to the genre's business model (Kennedy 1993). Inexpensive tabloid talk and news shows proliferated in daytime, the latter stimulated by coverage of the O. J. Simpson case, which erupted in 1994 and, for months, preempted the daytime lineup while garnering high ratings. These insurgent genres simultaneously heightened and satisfied appetites for sensational storytelling. At the same time, more and more women entered the workforce and thereby ceased to be part of a fungible daytime audience (*Soap Opera Digest* 1995).

Rather than primetime taking its lead from soaps, soaps stepped up their sampling from primetime. One such case was the Matt and Donna romance from the now defunct NBC soap opera *Another World* (*AW*), which I analyzed extensively in previous work (Scodari 2004). This older woman/younger man coupling initially took the form of 1940s screwball comedy film romances and their primetime progeny in hit sitcoms and "dramedies" of the 1980s and early 1990s such as *Cheers* (NBC), *Moonlighting* (ABC), and *Northern Exposure* (CBS). Transpiring primarily between 1992 and 1996, the story was not intended to be long term. Its genre mixing was likely meant to forestall any permanent investment in the couple since, on soaps, serious romances are, well, *serious*. Many fans familiar with primetime incarnations of romantic comedy love stories had developed a more complex set of expectations. But, alas, the commercial imperative to free up a young "hunk" for romance with a nubile heroine, and thereby attract soap opera's most desired audience— young women—meant that the storyline had to end sooner or later. When it did, the story's tone shifted from comic to tragic, and its empowering message was undercut by eventual reproduction of the crude stereotype of the insecure, desperate older woman. Moreover, this was a hard closure, much like those punctuating masculine narratives. In 1999, when the show left the air, the story had been swept so far under the rug that everyone was paired off in the final episode except for Donna and Matt, who shared nary a word. This case demonstrates that, as Mumford (1995) has also maintained, soap opera serialization doesn't always produce resistive readings; rather, the resolutions of particular plotlines can set hegemonic meanings in stone, regardless of how a story might once have been imagined. Johnson concurs, observing that, "{W]hether through interpretive, legal, or narrative measures, fan activity is discursively dominated, disciplined, and defined to preserve hegemonies of cultural power" (2007, 299).

Beginning in the new millennium, the popularity of competitive reality series such as *American Idol* (Fox) and *The Bachelor* (ABC) prompted integration of musical and other contests into daytime. In the autumn of 2003, *One Life to Live* (*OLTL*; ABC) featured its fictional "Search and Destroy" vocal competition, while *All My Children* (*AMC*; ABC) conducted a "Sexiest Man in America" contest, with hopefuls profiled during commercial breaks. Viewers' votes granted a cameo appearance on the show, among other things, to the winner. Such stories bring with them a smattering of the masculine by inserting relatively quick finishes to break up the extended soap opera narrative, a pattern similar to the periodic or episodic mini-closures of many primetime reality programs and male-skewing dramas.

The paranormal plots popularized in cult series such as *X-Files* and *Buffy* have also been influential. Building on the camp sensibilities and gothic storylines that came to define *Days of Our Lives* (*DOOL*; NBC) and *Passions* (NBC), and following their 1960s antecedent, *Dark Shadows* (ABC), other traditional soaps took a cue. In 2000, *Port Charles* (ABC), then a recent spin-off from *General Hospital* (*GH*), temporarily fended off cancellation with vampires, vampire slayers, and angels. It also adopted the structure of Latin American *telenovelas* by carving its storylines into finite, thirteen-week arcs or "books," thereby inching away from the feminine extreme on the continuum of gendered narrative. In the same vein, the CBS soap *Guiding Light*'s (*GL*) resident diva, Reva Shayne, traveled backward in time from 2001 in a quest to recapture her once and future true love, only a few years after having been cloned.

The ascendance of reality crime shows, police procedurals, and similar primetime series prompted soaps to tap into the masculine and foreground crime and law enforcement stories, many of which supplied masculine plot elements and mini-closures. Beginning around 2000, following the success of HBO's *The Sopranos, GH* shifted its attention to characters and storylines based on organized crime. In 2007, this soap depicted a hostage crisis at a hotel by aping the "real-time," hour-by-hour storytelling format popularized by *24*. On *GL*, elements of the CBS hit procedural *CSI* were expressly mimicked during the week of March 31, 2003, in the depiction of a hostage situation, police stakeout, and subsequent emergency and operating room activities. Shortly thereafter, Dr. Rick Bauer, a longtime fixture in Springfield's Cedars Hospital, suddenly became the medical examiner and began showing up at crime scenes wearing a Crime Scene Investigator (CSI) jacket reminiscent of the primetime series. Storylines also focused on the minutiae of police work, as when Detective Harley Cooper talked her troubled former stepdaughter down from a suicide attempt. In early 2007, an episode of *GL* literally transformed into reality television when the actors stepped out of their fictional roles and, in news documentary fashion, worked to help rebuild Gulf Coast dwellings in the aftermath of Hurricane Katrina as the news cameras followed.

In the summer of 2007, *OLTL* capitalized on the popularity of the Disney Channel's hit movie *High School Musical* with a storyline revolving around casting and rehearsals for a high school play that was actually written for and ultimately performed on the show. This was not only a shorter-term arc, but a blatant effort to ape primetime and corral the teen and preteen audience at home for the summer recess.

The soaps borrowed from these primetime genres in other ways, ushering *cinéma vérité* style and other such departures into the daytime panoply. In 2006, *AMC* began digitally manipulating its product using FilmLook software, which mimics the higher resolution, less mechanical visuality of film and many primetime dramas. After both *AMC* and *ATWT* experimented with the use of handheld cameras, *GL* completely switched to this technique in 2008 and escalated its exterior location shooting to resemble candid reality programs such as MTV's *The Hills*.[8] Such newfangled tactics are designed to appeal to a younger, savvier audience, but they have been met with mixed reviews from both critics and fans, along with even lower ratings (Cortez 2008).

Such incorporation of traditionally masculine components—as well as superfluities like abrupt inter-cutting of scenes (*GL*), entire episodes dedicated to the perspective of a single character (*GL*), and musical installments (*OLTL*)—sprinkle daytime soap operas with gender-diverse traits. Indeed, there is little evidence that the industry is as apprehensive about integrating masculine narrative elements into traditionally feminine genres as it is about the reverse. However, this does not alter the fact that these programs are still largely serialized and air five times a week. Today's American woman-on-the-go, whether working outside the home or not, engages in such an array of activities during the day that the commitment these soaps demand becomes onerous. This is also true of teen girls, who engage in online social networking and other activities that have supplanted afternoon soap watching. These prospective viewers might rely on the soaps' notorious redundancy or use cable's SOAPnet, DVR, or online viewing to time-shift their "stories" to the evenings. But, in the current system of exchange, with advertisers skeptical of DVR users who wait to play back and/or skip through commercials (Levin 2009), and the sheer quantity of primetime fare and other diversions with which DVR or SOAPnet viewing must compete, such alternatives are unlikely to sufficiently boost the soaps' profitability quotient. As long as their producers and advertisers eschew older viewers—those more often at home in the daytime hours—their attempts to rejuvenate soaps could prove fruitless.[9]

In the last decade or so, six network daytime dramas—including the long-running *AW* and its replacement, *Passions*, as well as the seventy-two-year-old *GL* and the fifty-four-year-old *ATWT*—have ceased airing on major networks. Rumors of cancellation continue to plague a multitude of other shows. If American public television (PBS) opened its embrace to popular cultural forms—as Ouellette (2002) recommends in her book, *Viewers Like You?*—it might portray American soap opera in transformative ways, abandoning glitz and glamour for gritty realism in the manner of the BBC's long-running serial, *EastEnders*. But, for the reasons Ouellette reports—such as the conviction that PBS should program "highbrow" alternatives to commercial television genres, rather than alternative *versions* of such genres—such a shift in mission is unlikely. Perhaps one solution is fewer and less costly soaps dividing the available audience, a change which is already occurring, plus enticements for advertisers who covet such viewers. Web streaming and other alternate distribution mechanisms might also provide respite. Whatever their fate, the continuing trials of afternoon "sudsers" are symptomatic of the industry's skittishness when it comes to feminine narrative.

DISCUSSION AND CONCLUSION

Wittebols (2004) is correct that the "soapification" of other television genres can distort, cheapen, or—as in the case of news—divert them from their critical role in a democratic society. However, the features of feminine narrative Fiske (1987) associates with soap opera are not welcomed with open arms by the industry; neither are they the most automatic or egregious offenders in terms of the effects Wittebols (2004) recounts. In fact, serial narrative, despite the fact that it can serve as a vehicle for hegemonic meanings, has the capacity to enrich television programming, especially scripted sitcoms and dramas. The old school, non-serialized drama series—populated by male heroes whose serial romances with disposable female characters are obliged to begin and end within an episode, and whose nemeses are just as swiftly dispatched—whose formula was established before 1980, are trumped by those programs with at least some measure of serialization and diverse protagonists who encounter complex, long-term challenges and liaisons and evolve with their experiences. Despite its excesses, the afternoon soap opera is the progenitor of the latter, more sophisticated, more feminine formula.

However, the current political economy of television confounds this formula in telling ways. Narrowcasting by gender, in which feminine narrative elements can be more liberally employed in shows geared to women, but are constrained in series targeting males, is a key factor. Turow's (1997) concern with the impacts of narrowcasting and audience fragmentation is borne out by this impasse, in which gender blending is limited or enjoined by institutional apparatuses that cultivate consumer tastes. Anything targeted to young males that inserts an iota of romance or sentiment, or denies celebratory triumph to its heroes on an episodic basis, can be—as this essay demonstrates with regard to series such as *X-Files, Farscape,* and *Firefly*—maligned as "soap opera" by networks, creators, and fans.

The plot/character dichotomy inherent in such gender distinctions also tends to limit what counts as action in a text. Character details can be dismissed as "outside the plot," yet character arcs are plots that incorporate a range of events, not the least of which are interior. Mental activity is consequently devalued. Hence, genders become culturally linked with particular narrative textures, thereby segregating the sexes according to dichotomous roles that can limit experiential resonances and, consequently, hinder relationships. Even as they respond to economic exigencies by adding a double dash of the masculine to their recipes through the insertion of shorter term storylines and other masculine elements based in primetime reality, crime, and occult series, among others, the future of afternoon soap operas remains uncertain. But it is still true that the artistic vision of television at large continues to be encumbered by a love/hate relationship with the feminine form traditionally embodied in this trendsetting, yet still belittled genre.

NOTES

1. For more on these trends, see Hauck (2006).

2. Others, such as Sconce (2004, 96–7), seem to blur the distinction between non-serialized, episodic dramas and serial episodic or "cumulative narrative" dramas.

3. Thompson (1996, 70) notes that creators of *Hill Street Blues*, which in the early 1980s heralded the heyday of the serial episodic drama, were ordered by NBC to insert at least one story each episode that would begin and end within the episode. This suggests that, from the start, networks were skeptical about all-encompassing serialization in primetime drama.

4. As of March 2008, *Grey's Anatomy*, having been frequently ranked in the top five of the Nielsen ratings in terms of both total viewers and those eighteen to forty-nine, and attracting more than twenty million in overall audience, is the most visible example of what is required for a female-skewing program with a serialized narrative to remain in a major network's good graces beyond a couple of years.

5. *Heroes*, a TV comic book, began dividing its autumn and winter/spring runs into distinct chapters that could easily be viewed as separate seasons. *Lost* began airing its abbreviated fourth season in the winter of 2008 to avoid interruptions in its weekly serialized storytelling on account of holiday reruns and preemptions.

6. The show's eventual placement as part of a Fox Sunday night lineup that included NFL broadcasts and *The Simpsons* was due to the network's desire to optimize its appeal to that coveted, young male demo (*USA Today* 1996).

7. SwitchDawg, comment on "Something against *Farscape*," alt.tv.farscape, March 28, 2003, http://groups.google.com/g/81b7fefd/t/7bf01c5547df3273/d/ec70a0db28c14ac7.

8. For more on *GL*'s transformation, see Erwin's essay in this collection.

9. For more on older viewers, see the essay by Harrington and Brothers in this volume.

THE IRONIC AND CONVOLUTED RELATIONSHIP BETWEEN DAYTIME AND PRIMETIME SOAP OPERAS

—LYNN LICCARDO

> Lynn Liccardo is a longtime soap opera critic who currently writes about the genre on her Red Room member blog (http://www.redroom.com/blog/lynnliccardo). Her articles on soaps have appeared in *Soap Opera Weekly,* including a demographic analysis of soap opera viewing entitled "Who Really Watches Soap Operas." She watched *As the World Turns* on CBS from its 1956 premiere to its 2010 cancellation and was a longtime viewer of *Guiding Light*. She has watched numerous other soaps and is a participant in online soap opera fan communities.

I've always believed that, the higher one's tolerance for ambiguity, the more fully one is able to appreciate soap opera. Perhaps, then, it's oddly fitting that the run-up to what many consider daytime soaps' greatest triumph—the iconic wedding of Luke Spencer and Laura Webber on ABC's *General Hospital* (*GH*)—points to many of the reasons why the genre finds itself in its current sorry state. The 1981 event transcended anything daytime had ever seen: *Newsweek* featured the couple on its cover, legendary movie star Elizabeth Taylor made an appearance, and more than fourteen million viewers tuned in for the two-day event. To the delight of ABC, six million of *GH*'s newly gained viewers were teenagers, whom soap execs logically saw as the future of daytime (LaGuardia 1983, 178–86).[1] The future didn't unfold at all as they had envisioned, though.

The road to the wedding actually began in 1978, when Gloria Monty was named *GH*'s executive producer in a last-ditch effort to save the show from cancellation. When Monty asked Anthony Geary to join the cast as Luke Spencer in 1978, he famously said, "I hate soap opera." Monty reportedly replied, "Honey, so do I. I want you to help me change all that" (LaGuardia 1983, 181). The exchange raises two obvious questions: What did they both hate about soap opera, and exactly what did Monty want to change? Since neither Geary nor Monty ever elaborated on their famous exchange,[2] what they really meant when they said they hated and wanted to change soaps is open to speculation. Were they being ironic, or maybe a little hyperbolic? Were Geary and Monty expressing the unspoken feelings of others in daytime when they articulated their "hatred" of soaps? Whatever the meaning, this idea that the soap opera is something that needs to change in order to not be hated has become deeply embedded in daytime producers' collective psyche during the past thirty years.

I'm not the only observer to believe this hatred of the soap genre has been internalized among many of those responsible for making soap opera, and that this self-loathing lies at the heart of soaps' decline. For critic Patrick Erwin, "It all boils down to this: There came a time when the industry decided it was going to be ashamed of what it did."[3] Erwin's sentiment is echoed by a former writer for CBS' *The Young and the Restless* (*Y&R*), Sara Bibel (2008).[4] In a 2008 blog post, Bibel comments:

> [A] lot of what's wrong with soaps today stems from the inferiority complexes of many of TPTB [or "The Powers That Be," which is online shorthand for the network executives, producers, writers, directors and others who oversee daytime drama]. So many people behind the scenes seem to be ashamed of the fact that they're working on soap operas instead of primetime/cable/movies/theater/etc. Why else would they be constantly trying to make soaps more like other mediums? During my career as a soap writer, I encountered people in positions of authority (no I won't name names) who used the word "soapy" as a pejorative. In interviews, producers and headwriters often tout changes they are making in the show as being "more like primetime," as if that's automatically superior.

It's not hard to see the Geary-Monty exchange of three decades ago resonating in today's "inferiority complex of many of TPTB," and contemporary daytime producers and writers "constantly trying to make soaps more like other mediums." Monty wasn't kidding when she said she wanted to change soap opera. For instance, in 1981, she oversaw an action-mystery-fantasy for *GH* that centered on a global weather machine powered by an enormous diamond called the Ice Princess, which kept Luke and Laura on the road and away from *GH*'s home base of Port Charles for months. Not only was the fantasy subject matter far from the usual soap fare, the focus on the Ice Princess plot had a considerable impact on the rest of the cast. As LaGuardia explains, "Actors who were once the core of [*GH*] were now coming to the Gower Studio a few times a week to say four or five lines about Luke or some other young character yet were obviously not needed for any of the stories" (1983, 185).

Imitation being the sincerest form of flattery—not to mention increased ratings—it was inevitable that other shows would follow. It's worth considering here that the enthusiasm with which other shows began imitating *GH* may indeed reflect that "hating soap opera" was not limited to Anthony Geary and Gloria Monty, but shared by others in the industry. *GH* was on the verge on cancellation in 1978 because the show was losing its time slot to NBC's *Another World*. While the ratings in 1978 were nowhere near what they were when soap viewing peaked in the mid-1960s, overall, most soaps were relatively healthy at the time of *GH*'s transformation.[5]

Nevertheless, as one fan said, soaps "started trying to outdo each other with attractive couples in impossible situations and that eventually led to the corruption of the genre."[6] For instance, an online poster noted, "[W]e had Tom and Margo trapped in a booby-trapped castle by Mr. Big on ATWT [CBS' *As the World Turns*] (Loved that story, BTW - but it was anything but typical soap),"[7] While that fan may have loved the Mr. Big story on *ATWT*, my mother—a longtime viewer of that soap—most emphatically did not; in fact, she hated it. As with *GH,* long-term *ATWT* characters such as Chris and Nancy Hughes (played by original cast members Don McLaughlin and Helen Wagner, the latter of whom continued in the role until her 2010 death at age ninety-one), their son Bob Hughes, and his ex-wife Lisa Miller were pushed to the side while their son/grandson Tom Hughes and his girlfriend went off on adventures far from Oakdale.[8] Sometime during the Mr. Big saga, my mother had enough and abandoned the show she had watched and loved for more than twenty-five years. Even when Douglas Marland became head writer of *ATWT* a couple of years after the Mr. Big story and brought neglected longtime characters back to the forefront of the show through multi-generational stories, I couldn't coax or nag my mother into tuning back in. Patrick Erwin spoke for many disenchanted soap fans, including my late mother, when he said in 2007: "Gloria Monty did wonderful work when she changed *GH* and updated the look and feel of the show, but that change irrevocably broke the genre away from what I think made it special: everyday people leading extraordinary lives."[9]

Prior to the Luke and Laura phenomenon of the 1980s, soap operas were primarily about domestic and interpersonal relationships. Conflicts originated within families, romantic entanglements, or close friendships, and plots were driven by the emotional complexity within each character, not the other way around. For Luke and Laura, daytime soaps adopted a variation of the soap narrative in which the conflicts that affected main characters were not domestic, but rather originated outside the couple and the community. Instead of telling stories about a woman falling in love with the wrong man, or a child acting out after a divorce, soaps turned to tales of omnipotent villains who threatened to wreck havoc on a global scale. That Luke and Laura prevented the Ice Princess from freezing the whole world was indeed extraordinary, but everyday? Not really.

While daytime's super couples were enjoying their excellent adventures in the early 1980s, primetime underwent some changes of its own. The one that would have the most impact on daytime was the 1978 premiere of CBS' *Dallas*.[10] Described on the Internet Movie Database (IMDb) as "the soapy, backstabbing machinations of Dallas oil magnate J. R. Ewing and his family,"[11] *Dallas* captured the public's fancy. The *Dallas* phenomenon culminated in the classic November 21, 1980, "Who shot J. R.?" episode, which garnered stratospheric ratings—eighty-three million, according to one source (Cagle 1990).

Dallas was often called a "primetime soap opera," but was it really a soap? Not according to Larry Hagman, who played the charmingly villainous J. R. Ewing and who formerly starred on CBS' *The Edge of Night*. According to Hagman, "Dallas was a cartoon" (Hagman 2008). Hagman's description was apt: On *Dallas*, there was never any character development, just neverending scheming and betrayals amongst characters who were perennially either good or bad. This formulation was a far cry from daytime soaps at the time. I don't recall the soaps I grew up watching ever exploring the "backstabbing machinations" of the filthy rich. My memories of daytime soaps echo those of former Amherst College professor William H. Pritchard, who described soap operas as "a seemingly endless process by which people talk themselves into and out of happiness and misery" (1986, 31), a "process" that resembled most viewers' everyday lives far more than the Ewing family's wheelings and dealings.

In a description of *Dallas* on IMDb, one poster, Tad Dibbern, was careful to distinguish between *soap opera* as a noun and *soap opera* as an adjective. By calling *Dallas* a show that followed "a soap-opera style,"[12] Dibbern acknowledged the serial format of the show but stopped short of actually calling it a soap opera. This distinction was largely lost in the aftermath of *Dallas* and the success of shows such as ABC's *Dynasty* and CBS' *Falcon Crest,* which premiered after *Dallas* and embraced its format. In the mind of much of the U.S. public, the adjective "soapy," while accurately describing primetime serials like *Dallas* and *Dynasty*, became synonymous with the noun *soap opera*. *Soapy* has become shorthand for everything from "a cheaply melodramatic plot twist" (Ford 2008b), to "surreal, campy, and wayyyy out to the left of reality,"[13] and is often linked to the descriptor "guilty pleasure," an oxymoron whose formulation recalls the nation's Puritan roots. Further, "soapy" in the pejorative sense describes how the public, and many of daytime's decision makers (as the Bibel excerpt earlier in this essay emphasizes), have come to perceive daytime soap opera.

The success of *Dallas* and *Dynasty* also led daytime soaps to shift the sorts of characters they focused on. Soap operas were traditionally made up of two families: one middle-class, one well-to-do. The men were professionals, physicians and attorneys. With the exception of the occasional secretary or nurse, women were stay-at-home wives and mothers. And "well-to-do" probably meant that the father of the family was a judge, a hospital chief-of-staff, or a newspaper publisher. Some soaps expanded this narrative emphasis over time to focus more fully on the efforts of working-class—and Irish, Italian, Jewish, Polish, and African-American—families to improve their lives. For instance, when Agnes Nixon created ABC's *One Life to Live* (*OLTL*), one story focused on how Larry Wolek's siblings helped him achieve his dream of becoming a physician—of course, not without some resentment on the part of his brother, Vinnie. Meanwhile, Claire Labine and Paul Alva Mayer continued the trend of featuring working-class/white-collar conflicts within families in 1975 with ABC's *Ryan's Hope*, where the children of bar owners Maeve and Johnny all became professionals. For viewers living in the small-town suburbia of the 1970s, soap operas were populated with characters whose aspirations to upward mobility and domestic struggles seemed familiar.

But, beginning in 1979, denizens of small Eastern and Midwestern soap cities like Pine Valley, Llanview, Oakdale, Springfield, and Genoa City were suddenly joined by oil tycoons (the Lewises on *Guiding Light* [*GL*], the Buchanans on *OLTL*), high-powered CEOs (Lucinda Walsh on *ATWT*, Victor Newman and John Abbott on *Y&R*, and Palmer Courtland and Adam Chandler on *All My Children* [*AMC*]), and European royal heirs (James Stenbeck on *ATWT*) who inexplicably set down roots in the small, provincial towns in which soaps have always been set. So, instead of seeing soap characters who looked like their neighbors, or who at least resembled the physicians and attorneys who lived on the "richer" side of town, middle-class suburban soap opera viewers were now watching the lives of the incredibly rich and famous, characters whose lives would have been available to viewers only in gossip columns or the society pages. While these larger-than-life characters have never completely replaced more traditional characters, the soap landscape had clearly changed. This shift in soaps' focus in the 1980s, along with the concurrent phenomenon of adventure-mystery narratives begun by Luke and Laura, further distanced daytime soap viewers from the characters they were watching on the screen. Daytime soaps were beginning the process of becoming the same kind of "guilty pleasures" as their primetime counterparts.

The early 1980s also ushered in the era of what has come to be called "quality television" or "television's second golden age," the heart of which was a new kind of serialized show (Thompson 1996, 19–35). That daytime soap opera began abandoning its traditional storytelling conventions at the same time as the next generation of primetime serials adopted some of the very qualities that daytime soaps were discarding is beyond ironic. It would be difficult to overstate the importance of *Hill Street Blues* (*HSB*)'s NBC premiere in January of 1981. When it left the air in 1987, *Los Angles Times* critic Howard Rosenberg was calling it "the TV show of the 1980s" (Rosenberg 1987). While *HSB* was not the first primetime serial, it was the primetime serial that set off what Robert J. Thompson called "the quality revolution." Thompson lists twelve characteristics of "quality television," among which are several descriptors of what I am calling traditional (i.e., pre-1980s) soap (1996, 13–6):

- Quality TV tends to have a large ensemble cast
- Quality TV has a memory
- All quality shows integrate comedy and tragedy in ways Aristotle would never have approved
- Quality TV is self-conscious
- The subject matter of quality TV tends toward the controversial [...] (quality series) frequently included some of television's earliest treatments of abortion, homosexuals, racism and religion
- Quality TV aspires toward "realism"

In the wake of *HSB*'s success, a string of critically acclaimed primetime serials followed, including subsequent NBC series *St. Elsewhere*, *L.A. Law*, and *ER*, and ABC series *NYPD Blue*. However, despite their serial format and the debt Robert Thompson argued these shows owed to the soap opera genre (1996, 35), they were rarely, if ever, referred to as "primetime soaps." In the minds of viewers and critics alike, the title "soap opera" was reserved for the over-wrought melodrama of *Dallas* and *Dynasty*.

More recently, quality television has evolved yet again. For instance, debuting in the fall of 2002, NBC's *American Dreams* centered on the Pryors, an Irish-American Catholic family in early-1960s Philadelphia. Through a diverse set of characters, the show deftly integrated the divisive issues of the 1960s—the changing roles of women, liberalized sexual mores, race relations, etc.—into a series of subtle and intimate family portraits. The show's storytelling was the antithesis of the melodramatic machinations of *Dallas* and its ilk. *American Dreams* was the first primetime serial to completely capture the ethos of what daytime soaps consistently were pre-Luke-and-Laura: ensembles of fully developed, multi-generational, lower-middle-class or middle-class characters shown in open-ended, inter-connected, intimate stories, where the actions of any one character reverberated for all. While *American Dreams* was cancelled in 2005 (along with two other shows featuring the same kind of storytelling, CBS' *Joan of Arcadia* and the WB's *Jack and Bobby),* these shows perhaps paved the way for the fullest embodiment of this trend.

In 2006, NBC's *Friday Night Lights* (*FNL*) premiered to the kind of critical acclaim reserved for shows such as *HSB*. *TV Guide's* Matt Roush spoke for many critics when he called this show about a high school football team and the community surrounding it, "a side of life you don't see anywhere else on primetime TV" (Roush 2007). The show's focus on high school football might give the impression that it was designed solely for male viewers, but *FNL* isn't really about football; football provides the vehicle through which emotional stories are told. As Connie Britton, who plays *FNL* female lead Tami Taylor, put it, "What people respond to about our show is that it's really about human relationships" (Britton 2008). As I've said, this is what soap opera, both daytime and primetime, is all about.

For daytime viewers fed up with the fact that their soaps have both dropped their focus on history and multigenerational families and become overrun with over-the-top melodrama, primetime shows such as *FNL* have provided refuge. Many longtime fans complain that today's soaps are "hokey and over the top,"[14] "stripped of the structure, dialogue and look that made the genre popular" (Bibel 2008) and deride the overall "corruption of the genre."[15] In other words, soaps today often seem even more cartoon-like than *Dallas* ever did. According to *Snark Weighs In* (2008), "[M]any primetime fans are former daytime fans who now stick exclusively with primetime, because it's the only place they can get anything resembling the socially aware, character-driven, serialized storytelling they used to get from soaps."

TPTB of daytime soaps have always been quick to list the challenges daytime faces: the increase in competition for daytime audiences with the introduction of cable and satellite programming; more women working outside the home than in soaps' heyday; residual effects of the O. J. Simpson trial, which interrupted soap broadcasts for months in 1995 and contributed to viewers permanently tuning out; and too few younger viewers. However, by placing responsibility for soaps' declining viewership on these factors—true though they may be—TPTB have avoided taking a cold, hard look at the damage their misguided decisions have wrought upon the quality of daytime soaps. The sad reality is the damage has been so profound that, even if every daytime soap could revert to its previous glory in terms of narrative structure and intergenerational casts, this might well be too little, too late to attract either former or new soap viewers. Put another way, the future of soaps on the broadcast networks may already be sealed, owing to the decline in quality of the daytime soap genre over the last three decades.[16]

As was beginning to be clear by 2008, many of those same factors were threatening the future of primetime serial drama on broadcast as well, driven in part by the continuing shift of viewers to cable. This decline has led the networks to create less expensive programming that can draw equal if not larger audiences; hence, the proliferation of reality shows. NBC upped the ante in 2009 when they turned over the last hour of primetime Mondays-Fridays to Jay Leno for a program similar to his longtime late night staple, *The Tonight Show*. (However, the experiment ultimately failed, and Leno returned to his traditional late night slot in 2010.) It is also discouraging to note that CBS, "the most watched network" on broadcast television, has not even a single "soap opera" style serialized drama in its primetime lineup (de Moraes 2008). Meanwhile, the fate of *FNL* has been a constant source of discussion for the television industry, as the show was saved from cancellation after its second season only by a deal struck between DirecTV and NBC, whereby the satellite provider underwrote a portion of the production costs in return for the right to air the show first exclusively to DirecTV subscribers. *FNL* is now guaranteed to last through at least a fifth season.

Perhaps most disheartening was the 2009 cancellation of ABC's *Dirty Sexy Money* (*DSM*). Premiering in the fall of 2007, just before the Writers Guild went on strike, *DSM* was a hybrid, marrying the wealth, glamour and family dysfunction of *Dallas* and *Dynasty* with the fully developed, multi-generational storytelling of *FNL* and the soaps of old. Because the strike hit so soon after the 2007 fall season began, *DSM* was not able to solidify a fan base, and ABC's decision to wait until fall 2008 to bring the show back certainly didn't help matters.[17] But, when it returned for its second season, *DSM*'s unique storytelling had been eviscerated by the network and replaced with simplistic cartoon characters skipping from plot point to plot point—not unlike the changes *ATWT* endured. Fans lamented the loss of the emotionally complex characters they had come to love; indeed, many denounced the changes as "soapy."

Meanwhile, the three decades that have passed since Gloria Monty asked Tony Geary to help her change soap opera have taken their toll on the genre. Not every show has abandoned the qualities that previously defined soap opera, but holding on to them consistently is a struggle. In a 2008 interview, *ATWT* head writer Jean Passanante described the show's latest attempt to reinvent itself: "We're playing more of a story in a day, and then we're not playing it the next day. We're creating more of a unit within a single episode, so that it feels like it has a beginning, a middle, and an end (*CBS Soaps In Depth* 2008)."

What Passanante is describing is storytelling that's essentially episodic, rather than serialized, even though serialization is commonly cited as the one element all soap operas have in common. Additionally, this model not only stripped the storytelling of the emotional depth and complexity viewers craved, but allowed *ATWT* fans to watch only the stories that interest them without risking missing anything, a fact driven home by this fan comment on *ATWT*'s Media Domain message board: "The self-contained episodes on ATWT & viewing spoilers for the week makes it so much easier for me to know when I will watch ATWT again. Not sure how or why TPTB have come up with this new concept but it sure is working well. I think I've watched a total of one or two episodes in the last two weeks. The story lines are not capturing any interest out of me even from the characters & actors that I like[.]"[18]

Despite the declines of the past three decades, common sense has prevailed every few years when a soap opera improves by returning to the style of storytelling that made soap operas successful in the first place. In 1985, as mentioned earlier, Douglas Marland restored *ATWT* to its previous glory by drawing upon the show's long history to create richly textured, multi-generational stories involving the show's core families. However, not all alienated viewers return when a show corrects course. As I mentioned earlier, my mother was one disgruntled fan who did not return to *ATWT* to witness Marland's transformation of the show. Therein lies the paradox underlying the soaps' current sorry state of affairs and why the current strategies continue to put the future of daytime at risk. And, as the numbers sadly illustrate,[19] there are never enough new viewers to make up for those who've been lost forever The latest self-destructive manifestation of soap operas' thirty-year inferiority complex—the quasi-episodic storytelling adopted by *ATWT*—is therefore all the more disturbing. The show went down this road in the past, but TPTB seemed not to pay attention to the lessons of history: *ATWT* has been cancelled.

As a genre, however, daytime soap opera is nothing if not resilient, and common sense prevailed on another soap when longtime *OLTL* writer Ron Carlivati, who took *OLTL* back to the Basics of Soap Opera 101, was named the show's head writer in 2007. And what exactly are the Basics of Soap Opera 101? In 2008, a Media Domain poster summed them up perfectly:

> *OLTL* is a perfect example of how good writing and an understanding of the soap technique (and it is a separate technique from most TV writing) can lift a genre out of the doldrums and make it soar. All ages of characters, all good actors with story, story building out of story, interwoven and yet with new twists, characters staying in character and pulling story thru that, beautiful sets, appropriate costuming, the now-and-then outdoor shot ... well, it is all coming together to create new excitement and interest in soap fans.[20]

While some storylines generated under Carlivati's reign have created controversy, and while not every story or character has resonated with viewers, the renewed *OLTL* has given rise to much passionate, thoughtful, and contentious discussion among fans—a sure sign of successful soap storytelling, day or night.

When I first began thinking about the relationship between daytime and primetime soaps, I argued, "[D]aytime would do well to understand what is working on primetime soaps, because it's what used to be working on daytime" (Liccardo 2008). At the time, I believed the future of daytime soaps would lie in the shows' willingness to reclaim their past. But that was then. While I still believe that daytime soaps would be much improved by reclaiming what they've discarded—*OLTL* certainly has been— I'm just not sure this change will be enough to save the genre. In the wake of *ATWT's* cancellation by CBS in September 2010, rumors persist that *OLTL's* run could soon be drawing to a close as well.

It's the rare mainstream media article on any aspect of daytime soap opera that doesn't mention the genre's declining ratings and imminent demise.[21] And there are days when I'm among those who believe that the damage inflicted on these shows over the past thirty years has been so profound that, even if every daytime soap could revert to the genre's previous glory with respect to its style of storytelling and its breadth and depth of character interaction, the daytime soap opera as a whole may well still fail. However, as this volume points out, soaps are just now beginning to take advantage of alternate viewing platforms.[22] Many shows are now available online. Some can be seen on ABC's cable network, SOAPnet (until its 2012 cancellation). On-demand cable is a possibility yet to be explored. If these alternate platforms can be fully monetized, daytime soaps may well stick around. But all the fully monetized alternate viewing platforms in the world cannot help the soaps survive if they don't first help themselves. And that means dealing with the storytelling. Since good storytelling begins and ends with creators who love, not hate, what they're doing, the future of daytime soaps hinges on the genre's ability to stop hating itself. Will that happen? Tune in tomorrow.

NOTES

1. Regarding the numbers for the wedding, Wikipedia puts the number at thirty million but cites no source. See http://en.wikipedia.org/wiki/General_Hospital#Broadcast_history.

2. Gloria Monty died in 2006. If she or Anthony Geary ever elaborated on the exchange, I've not been able to find it.

3. This quote is from Erwin's comment on Marlena De Lacroix's September 21, 2007, piece "OJ Didn't Do It, Kill Soaps That Is," on *Savoring Soaps*, which is no longer available online. Note that Erwin has an essay elsewhere in this book.

4. Bibel has an essay elsewhere in this collection as well.

5. For an overview of the ratings, see http://en.wikipedia.org/wiki/List_of_US_daytime_soap_opera_ratings#1980s_Ratings.

6. BeetFarmGirl, comment on "The Moment All of Daytime Jumped the Shark," Television Without Pity (TWoP), October 29, 2007, http://forums.televisionwithoutpity.com/index.php?showtopic=3161153&st=0.

7. TWoP Snarkian, comment on "The Moment All of Daytime Jumped the Shark," Television Without Pity (TWoP), November 2, 2007, http://forums.televisionwithoutpity.com/index.php?showtopic=3161153&st=45&p=9156876&#entry9156876.

8. See Ford's essay in this collection on the history of *ATWT*'s Tom Hughes character.

9. See end note #3.

10. *Dallas* was not the first serialized primetime drama by any means. ABC's *Peyton Place*, which debuted in 1964, was most clearly related to soaps, followed by an *ATWT* spin-off, CBS' *Our Private World*, in 1965. However, *Dallas* was the first show of its kind to capture the public's fancy in a big way.

11. See Tad Dibbern's description of *Dallas* on Internet Movie Database (IMDb) at http://www.imdb.com/title/tt0077000/plotsummary.

12. See end note #11.

13. Greg Jenkins comment on *Dynasty*, IMDb, undated, no longer available.

14. See end note #6.

15. See end note #6.

16. Soap blogger Mark Harding, who in his professional life is a behavioral scientist and statistician, http://markhsoap.blogspot.com/, feels that the decline in soaps' quality "is likely one element in a multi-causal chain that brought us to this point. Even there, because 'quality' is so subjective, I am not sure that is the right word. What we're talking about is the whole host of stylistic and substantive changes that Tom [Casiello, current *Y&R* writer and contributor to this book] and others have been cataloging (disregard of older characters, quick cut scene structure, short arc stories, emphasis on silly plots that twist characters to fit the beat-of-the-day, etc.). I'm not sure how much 'variance' these kinds of things explain, but they are surely part of the equation" (Harding 2008b).

17. For more on the strike's impact on soaps, see essays from Bibel and Metzler in this collection.

18. Unfortunately, Media Domain, http://www.mediadomain.com/, like many soap opera boards, does not archive posts, and this one is no longer available. This lack is a problem for scholars that will only grow worse with time.

19. See an overview of soap opera ratings at http://en.wikipedia.org/wiki/List_of_US_daytime_soap_opera_ratings#1980s. The annual *ATWT* ratings in 1977–1978, before Luke and Laura, was 8.6; in 1982–1983, post Luke and Laura, it was 7.6. By 1986–1987, with Doug Marland writing the show, the rating average was 7.0. In 1993–1994, after Marland's death and before the O. J. Simpson trial, the rating was 5.8. In 1995–1996, after the trial, the rating averaged at 4.4. By 2007–2008, *ATWT*'s average rating was a 2.4. For more on soap opera ratings, see Harding (2008a) and Harding (2009).

20. See end note #18.

21. See, for instance, Lidz (2009).

22. See Levine and Gonzales' contributions, for example.

PERSPECTIVE
SCHOLAR LOUISE SPENCE ON COMPARING THE SOAP OPERA TO OTHER FORMS

BASED ON AN INTERVIEW BY ABIGAIL DE KOSNIK

> Louise Spence is professor of radio, television, and cinema at Kadir Has University in Istanbul, Turkey. She is author of *Watching Daytime Soap Operas: The Power of Pleasure*. Spence began watching soaps on a snowy, black-and-white television set as a child at her mother's feet. The two watched *Search for Tomorrow* and *Love of Life* as the mother fed her infant daughter.

I need to confess that I haven't been watching soaps since my book, *Watching Daytime Soap Operas*, went to press nearly five years ago. I know I could probably get right back into them if I wanted to start again. I did take an almost ten-year hiatus once before and had no problem starting up again. But my concentration in the book was as much on methodology as on soaps themselves. And following this methodological approach, when contemplating the continuing appeal of soaps, or the lessening of that appeal, one might want to start by looking not only at daytime television viewing habits and the number of women who are at home during the day, but also at the popularity of other texts that represent suffering—melodrama, opera, sentimental song, addiction memoirs, abuse memoirs—and at the popularity of other ongoing texts.

Certainly, the interest in television serials has not waned. They seem to be more popular than ever. When I was still living in the States, *The Sopranos* (HBO), *Desperate Housewives* (ABC), and *24* (Fox) were the rage. There were serial aspects to other esteemed primetime shows: for example, *The Shield* (FX), *The Wire* (HBO), and *Lost* (ABC). And, of course, there is another form of popular TV, the sprawling genre of actuality-based television entertainment programming (what the British industry calls "popular factual entertainment" and what is generally called "reality TV" in the U.S.), and the more recent genre of "docusoaps." One might consider these serials as well. And, with the ability to time-shift and purchase season-long DVDs, much "accumulating" television could be enjoyed in the quiet of the afternoon. Yet we know little of the appeal of the serial form. What are the pleasures of accumulation? Of regularity and permutations? Of waiting? Of anticipation?

I am not ready to imagine soaps dead. Sure, the troubled economy in the U.S. will probably affect advertising revenues. But if the current global financial crisis is more than simply an economic crisis—that is, if it is a systemic crisis—then we ought to look at more than simply TV economics. For a start, we should look at rates of broken family relationships. During the Great Depression, divorce went up. Have family relationships been breaking down now, and how does that affect living arrangements? Do women live alone? Are they performing paid work in the home? (These days, women are losing their jobs and having their hours cut back faster than men.) And how does that impact their viewing habits?

In *Watching Daytime Soaps*, I tried to argue that you cannot look at viewership in isolation from other relationships. My own mother said that she lost interest in soaps after her mother died. I certainly watched soaps much more when I was talking to people about them. Does following soaps on the Internet count? Maybe not to the TV industry. Women who follow their stories on the Internet or via print sources are "viewers" whose attention time cannot be sold to Proctor & Gamble. But how do scholars account for those people? Is the number of people participating in chat rooms and discussion boards declining? Are soap performers still profiled in the Sunday supplements? Do they appear on talk shows?

Can the notion of "soap opera" still have cultural currency, as there are fewer U.S. soaps on air? Tim Cole, a British historian of the Holocaust, referred to the 1978 U.S. TV miniseries *Holocaust* as a "soap-opera style" program, "portraying the 'Holocaust' through the experiences of one imaginary family" (1999, 13). When is the last time you saw a soap, even a British soap, with only one family? Cole wasn't talking about the seriality of the series. Nor is there any reason to believe that he was speaking of hyperbole, excess, or melodrama when he used the term. He was talking about the fact that it is a domestic drama. The term "soap opera" seems to still have cultural meaning, even if that meaning is not the same to everyone.

As the Cole case demonstrates, whether we are referring to the essential features of the genre or not, we are still able to communicate when using the term. And, in most cases, what we are communicating is that the film, TV show, story, or telling of an event that is referred to as "a soap opera" is in unsuccessful competition with some other narrative form that is privileged and valued by the comparison. This tendency to contrast other narrative forms with soap operas draws on deeply entrenched dissonances in our discourses on popular culture and asks us to consider the worth we place on originality and realism in telling stories. Are they saying that the ways these soap-like stories are told take us away from the real? No, they are they saying that they take us away from realism "understood as an attempt to duplicate a world we think we know" (Wood 2004). This is what is being revered. Sometimes, it seems like soaps are the premiere "other" against which all alternative forms must compete. And, sometimes, this competition takes a truly perverse turn, that other narrative style is appreciated not because of its own craft, but simply because it is not soap-like—or because it is not popular!

PERSPECTIVE
SCHOLAR JASON MITTELL ON THE TIES BETWEEN DAYTIME AND PRIMETIME SERIALS

BASED ON AN INTERVIEW BY SAM FORD

> Jason Mittell is associate professor of American studies and film and media culture at Middlebury College. He has written on issues of television genres, including the soap opera, in *Genre and Television* and the textbook *Television and American Culture* and writes regularly about television narratives on his blog, Just TV (http://justtv.wordpress.com). He became a regular viewer of *General Hospital* with his mother at the height of the Luke and Laura phenomenon of the early 1980s.

I think that, just like a lot of soap researchers, my research interest in narrative complexity on primetime television shows stemmed first from my personal tastes and fannish investments. I remember a number of programs that I was watching around 1999–2001 that seemed distinctive in their narrative strategies, offering new possibilities for primetime programming. Some specific examples include *The West Wing* (NBC), *Buffy the Vampire Slayer* (WB/UPN), *Angel* (WB), *Six Feet Under* (HBO), *Alias* (ABC), and *24* (FOX). All of these shows seemed invested in expanding the vocabulary of primetime television, both by incorporating serial form (which had been a growing trend for two decades) and by embracing more overt narrative experimentation: temporal manipulation, moments of "retelling" the same scene from a different perspective, "reboot" scenarios that change the course of the series quite significantly, and an overt acknowledgment of narrative mechanics (like *24*'s "real-time" structure and title or *Six Feet Under*'s "death-of-the-week" norm). And I'm gratified that, since I published my first essay on the topic (Mittell 2006), many more shows that fit the mode have aired—although per the logic of television, most have failed to last beyond a single season.

The question of how soap operas have influenced primetime narrative complexity is a big question for my work, and I think my answer won't be popular. But it's not arrived at casually or dismissively.

Certainly, the soap opera plays an important role in the history of serialized TV, as it has been the centerpiece of the serial form for decades. If we look at the history of primetime serialized programming, soap operas are a common reference point. In the 1960s, *Peyton Place* (ABC) had much in common with soaps, with a focus on a web of relationships within a community, an assumed female audience, multiple airings (two or three episodes) each week running year-round, and even some production norms common to soaps (although it was shot on film, not live). The producers of the show refused the soap label, highlighting its novelistic source material over the lowbrow assumptions tied to the daytime genre. (This link between the televised novel and soaps became more prevalent later through the Latin American telenovela form). Despite the show's success, other primetime serials failed throughout the 1960s.

The next wave of primetime serials that debuted in the 1970s were explicitly tied to soaps, but through genre parody: *Mary Hartman, Mary Hartman* (a syndicated series) and *Soap* (ABC) both were successful serials that acknowledged their soap opera roots but explicitly mocked the genre. *Soap*'s success helped change a lot of network assumptions: that audiences won't be able to follow a weekly serial; that viewers are too inconsistent for serials in primetime; that viewers will drop out over the summer; that serials are only for the traditional daytime audience; that men won't watch.

These shows set the stage for the 1980s serial boom of the primetime melodramas (*Dallas* [CBS], *Dynasty* [ABC], *Falcon Crest* [CBS], *Knots Landing* [CBS]) and the "quality" serial hybrids (NBC's *Hill Street Blues*, *St. Elsewhere*, *Cheers*, and *L.A. Law*; and ABC's *thirtysomething*). This also coincides, of course, with the boom in both viewership and legitimacy of daytime soaps in the 1980s, so serialization starts to lose many of its lowbrow assumptions. By the 1990s, I'd argue that serial form existed as an independent concept from the soap opera genre. Thus, even a show that I believe explicitly references and comments on soap norms—*Twin Peaks* (ABC)—was not framed within the context of soaps by most critics; if *Twin Peaks* had aired in the 1970s, I think it would have been viewed primarily through the lens of soap operas, much like *Mary Hartman* was.

All of this history simply presages a point that I imagine some readers of this book will find controversial: I don't think the contemporary primetime narrative complexity that I write about has much in common with or influence from soap operas, except through their common connections to 1970s and 1980s primetime serials. They are distinctly different in production method, scheduling, acting style, pacing, and formal structure. In reading interviews with, and talking to, primetime creators, I've never seen any reference to soap operas as a point of inspiration or influence. Likewise, there is almost no crossover of creative personnel between daytime and primetime drama.

So what are the shared features? Seriality, of course, and often an investment in melodrama. Otherwise, the way daytime and primetime handle serial form and melodramatic writing and performance are so different that I don't see this as a particularly compelling link.

This is not to say that there aren't primetime shows that are soapy. I think *Friday Night Lights* (NBC), *The O.C.* (Fox), and *Dirty Sexy Money* (ABC) are all examples of recent shows with a tie to soap operas' mode of melodrama (and all of which I find quite enjoyable). But these shows aren't particularly invested in the form of narrative complexity that I'm studying. They lack the self-aware storytelling mechanics, the play with temporality and subjectivity, and the commitment to a long-term accrual of plot clues and mechanistic interconnectivity that I believe is central to the mode of "narrative complexity."

One thing we need to remember is that an average primetime serial produces around twenty-two hours each year, or even less for cable series (without subtracting out commercial time). A soap opera produces more than ten times that amount each year, a stark difference in the sheer amount of storytelling volume. How is that volume filled? Here are my working hypotheses:

- Soaps spend much more time talking about events that have happened rather than showing them, while primetime serials show events more frequently than talking about them.
- Soap dialogue includes the names and relationships of characters more frequently than on primetime.
- The amount of narrative change that happens over one week of a soap opera is less than one episode of a primetime serial.
- The amount of narrative change that happens over one year of a soap opera is less than a season of a primetime serial.
- Soaps involve more interwoven characters than primetime, where separate storylines have fewer interactions.
- Individual episodes of primetime have much more defined boundaries and distinctive features than on daytime.
- Individual storylines on primetime serials are introduced and concluded far more quickly than on daytime, with the exception of major plot arcs and mythologies, as on ABC's *Lost*.
- Narrative events have far more emotional and character repercussions, both for an individual character and the community at large, on daytime versus primetime.
- Missing a week of a soap opera would cause less confusion than missing a week of a primetime serial (assuming the viewer does not watch the "previously on" recaps

on primetime) because daytime incorporates far more recapping into the dialogue than on primetime.
- A published "recap" of an episode on a fan Web site is far more likely to focus on character reactions to information and events on daytime versus the actual events themselves on primetime recaps.

To be clear, I'm not trying to suggest that these differences should be measured in terms of quality—there are different appeals, pleasures, and aesthetics at work here. We also have to account for the differences in viewing strategies. I believe that soap viewers are less likely to watch an episode straight through with their full attention aimed at the screen than primetime viewers are. Soap viewers use recording devices to time-shift and fast-forward through plotlines or characters they don't care about, frequently multitask while watching, and rely on recaps from their community of viewers or paratexts like *Soap Opera Digest* or fan Web sites. Certainly, primetime viewers do all of these things, but I would guess with less frequency. I think these viewing tendencies reinforce textual norms that encourage daytime repetition and slow pacing, versus primetime speed and moving forward.

The primetime shows I'm most interested in explicitly discourage such viewing strategies and textual redundancies. Their plots and enigmas are constructed to reward viewers who watch every episode carefully, and they refuse redundancy and repetition for dramatic effect. For instance, *The Wire* (HBO) plants some seeds of narrative backstory in the first season— such as Clay Davis's bribery or Cedric Daniels' history of corruption—that do not blossom until later seasons, even five years later in some instances. The moment of recognition that accompanies the connection to this deep backstory depends on that lack of redundancy, rewarding your careful attention over years rather than having foreshadowed and gestured toward a turn of events explicitly. I would guess that such moments in soaps are more about backstory relationships, rather than dangling loose ends of plot. Further, since (for many soaps) the fans know more about the backstory than the producers, such moments might be more frustration than revelation!

I think that serial storytelling taps into a widely shared and even primal set of narrative pleasures: for instance, investment in an ongoing storyworld, relationships with characters, and long-term interest in what will happen next. I think back to my own brief soap opera viewership: I got hooked in because of the "what next?" question. But the obstacles for regular viewership were too high, so I abandoned it for the more convenient (although costly!) serial form of comic books. I gave up comics around the same time that I got invested in primetime serials (*Dynasty* and *St. Elsewhere* were two of my favorite 1980s shows), and I've continued to be an ardent primetime serial viewer for twenty years.

Many changes in television production and consumption privilege this mode of seriality: more flexible technologies of viewing; a lower commitment threshold; and more choice allowing for the flourishing of serials across genres and channels. So, when my kids get interested in seriality, why would they turn to soap operas? I don't think that there's enough of an appeal there without the intergenerational community to nurture soap fandom. So, as viewership dwindles, so does the next generation's exposure and investment—a recipe for making soap fans endangered, if not extinct.

What remains unique about soaps are the daily ritual and amount of textual material. For many, these are core pleasures that a primetime serial cannot replace, even with paratextual exploration online via fanfic, fan Wikis, blogs, online vidding, etc. Is it enough to sustain the genre? Hard to say—I could imagine a mode of online storytelling that could effectively match that ritual, but I haven't seen an example that's worked yet. But we're still in the primitive era of online video, so I wouldn't be surprised if someone comes up with a smash hit online serial in the next few years.

I definitely see soap opera as a genre, while "serial drama" is a format. My concept of genre is as a cultural category bearing assumptions and associations—and probably no television genre is as laden with assumptions as the soap opera! Some of those associations are tied to textual form, such as production style, performance, and narrative mode. Others are more operative within the industry and audiences, such as the assumed viewer base, low cultural value, and norms of consumption. Thus, I don't think that primetime serials are still considered soap operas, or—at most—they bear the adjective "soapy" while acknowledging that they're not the same as daytime.[1] This is a change since the 1970s, when the producers of *Soap* couldn't imagine serial television apart from the soap opera.

I think it's important to detach seriality and melodrama from the specific genre of daytime soaps. Obviously, soap operas have been prime sites for both serial form and melodramatic television for decades, but, today, there's a wider range of these formal and tonal elements across genres—including nearly every primetime fictional genre, reality TV, sports, and even news, where both melodrama and serial form seemed to bubble up quite a bit in the 2008 election. But this doesn't make these texts "soap operas," any more than when a soap opera incorporates a crime plot does it become a cop show.

I know other scholars who disagree with me on this point concerning the influence of soaps on primetime. They argue that the presence of serial melodrama in primetime suggests the expansion of soap opera outside of daytime, and assert that this influence suggests that soaps should be seen as more legitimate and central to the cultural values of television. But I just don't see that argument applying today, as the majority of both producers and viewers of primetime serials have never watched soaps and probably wouldn't particularly like them if they did. Soap operas did not invent serial form or melodrama—they just happened to be the dominant locale for both throughout the bulk of television history. I see today's primetime serials as much more influenced by other serial formats, like comic books and nineteenth-century novels, than by soap operas. Thus, today's primetime serial is cousins with the soap opera, sharing common ancestry from the nineteenth-century novel, but very few primetime shows seem to be directly influenced by daytime traditions.

There is a strategic cultural politics to asserting the importance of soap opera to the more legitimate cultural realm of primetime serials, but I see a dangerous side-effect to this claim. If you argue that shows like *Lost*, *The Sopranos* (HBO), and *The Wire* are indebted to the soap opera genre, it's pretty easy to rebut that these programs are "more evolved" than daytime. The aesthetic criteria that celebrate primetime serials will judge soap operas as cro magnon relatives at best and thus worthy of evolutionary extinction! If you're a soap fan, it's a dangerous game to suggest that the primetime serial can be regarded as a mutated soap opera, because soaps are going to lose this aesthetic evaluative comparison for the vast majority of critics and viewers. I see it as both more accurate and politically useful to maintain that soaps possess their own unique values, aesthetics, and rewards distinct from primetime serials, forcing us to evaluate daytime on its own terms. And, based on the soap opera's own assumptions and aesthetic terms, most primetime serials fail to measure up.

As for what would be lost if soaps died out—there's certainly a sense of sadness that would accompany the extinction of storyworlds that have persisted for so long. But I'd be curious to know whether soap fans today are truly invested in the shows as they are now, or more nostalgically holding onto a series as it once was. How many viewers are watching out of habit and hope for a potential return to a golden age, versus how many still find the shows rewarding? American television is so invested in the notion that cancellation equals failure that people mourn the loss of long-running primetime series even after they've lost their luster. I think the cultural place of shows like *ER* (NBC) and *The Simpsons* (Fox) might be even higher if they'd ended in their prime, rather than going on for years after their core fans had moved on. This "infinity model" of TV is certainly most prevalent for soaps, where cancellation is quite rare and lamented. But the idea that a story can go on forever just doesn't make much sense if you think about it and really has few precedents in other narrative traditions.

If you're looking for advice on how to reinvigorate soap operas, I'd say producers should focus on what the genre does best: immersive, slow-paced, dialogue-driven melodramatic storytelling that rewards long-term accrual of character knowledge. If there's no longer a market for this, then the genre will disappear—and perhaps there does need to be some contraction to concentrate the audience who still want these pleasures to choose from only a few shows. But I'd argue that the problems daytime is facing can be solved best by trying to be more like soap operas, not more like primetime.

NOTE

1. For more on the contested meanings of "soapy," see Liccardo's essay elsewhere in this collection.

PRESERVING SOAP HISTORY
WHAT WILL IT MEAN FOR THE FUTURE OF SOAPS?

—MARY JEANNE WILSON

> Mary Jeanne Wilson is a doctoral candidate in critical studies at the University of Southern California's School of Cinematic Arts. She is also currently an employee of ABC Network's research department, working with daytime programming. She is co-author of the essay "Soap Opera Survival Tactics" with Ellen Seiter, and her dissertation work explores issues surrounding archiving soap operas and fan pleasure in archived episodes. She started watching soaps as a kid with her mom and has watched at least twelve different soaps since, with *General Hospital* being her longest watched.

INTRODUCTION

Television is, by its very nature, an ephemeral medium. Without home recording or playback technologies, a television broadcast disappears from a viewer's reach as soon as it airs. This lack of permanence creates a constant struggle to capture and preserve historical television texts, and viewers and researchers alike can be regularly frustrated by a lack of access to programs not being currently broadcast or rerun in syndication. The daytime soap opera is a particularly "unruly" genre when it comes to accessing past programming.[1] Unlike other narrative television, a soap opera episode is seldom broadcast more than once.[2] In addition, while the commercial sale of television on DVD is becoming a more readily available option, the sheer volume of episodes that constitutes a soap opera has thus far rendered these shows unsuitable for DVD release thus far, a reminder that the content available for sale in this format is only a narrow slice of the scope of television history.

With few commercial options available for accessing historical soap episodes or storylines, major U.S. television archives might seem a viable alternative; however, the soap opera texts available in these institutions are surprisingly fragmented. Without access to the soap opera in reruns, in the commercial marketplace, or even in the collections of prestigious televisions archives, how can a fan, scholar, or producer access the rich history of television's longest running genre? By exploring these questions surrounding the accessibility of soap opera history, I hope to expose how important issues of preservation, or the lack thereof, may be to the future of the soap opera genre.

The soap opera holds a prominent place in television history, with a presence on the air from the earliest days of broadcast television in the U.S. ABC's *General Hospital* (*GH*) is the longest-running scripted program on television today, and CBS' *Guiding Light* (*GL*) was the longest-running drama in broadcast history, lasting from its radio debut in 1937 until its final episode in 2009 (Hyatt 1997, 490). Not only are soaps still a permanent fixture of the broadcast day, they have also played a prominent role in television scholarship. In the 1980s, the soap opera was one of the first genres feminist scholars examined in their attempt to reevaluate popular media labeled "women's genres." The works of Dorothy Hobson (1982), Christine Geraghty (1991), Ellen Seiter (1982; Seiter, Borchers, Kreutzner, and Warth 1989), Robert C. Allen (1983, 1985), Tania Modleski (1979, 1982), and others established the soap opera as a key genre within television studies, in part because of its culturally devalued status due to its association with the feminine. With the intense academic focus on the genre during the rise of television studies, one might assume that soap operas would have become a priority for acquisitions during the formation of television archives. Unfortunately, neither their popularity and continued broadcast presence nor their prominence in academic circles has garnered soap operas the space or attention they deserve in television archives.

Part of the motivation behind the development of television archives in the 1960s and 1970s was to raise the still-derided medium's cultural status. In trying to elevate television's standing, archives often sought out what would be recognized as "quality" television thought to have lasting cultural relevance. Soap operas fell far outside cultural notions of what would be considered quality TV. The soap opera's association with a female target audience, as well as female creators and writers, marked soaps with low cultural and artistic status and directly affected their place in official television archives.[3] This development is particularly disconcerting considering that archives are often key institutions in determining a canon for any art form. Brunsdon explains that, despite resistance to the development of a sanctified canon in television studies, judgments are still implicitly made about a television canon by sources of institutional power even while they are theoretically rejected by the academic field (1997, 129).

To be fair, the magnitude of U.S. television production leaves domestic television archives with the unenviable task of selecting texts to be saved for future research while having to reject other texts, which will then most likely be lost as part of television history. This means that much local, off-network, and non-"quality" programming is already lost or will be lost if not actively preserved. Unfortunately, many soaps have already suffered this kind of erasure from television history because of the lack of attention they have received from archives. In this study, I examine the state of soap opera collections in three major U.S. television archives: the University of California Los Angeles (UCLA) Film & Television Archive; the Library of Congress Motion Picture, Broadcasting and Recorded Sound (MBRS) division; and the Paley Center for Media. By exploring the relationship of each archive with the soap opera genre, we can see not only how these institutions have and will continue to shape soap history but also what these issues of preservation and the loss of soap history might tell us about the future of soaps.

THE RISE OF THE TELEVISION ARCHIVE

Kompare's *Rerun Nation* (2005) traces the emergence of U.S. television archives in the 1960s and 1970s, elevating the television text to that of a valid object of scholarly research worthy of preservation. In 1965, the Academy of Television Arts and Sciences Foundation sought to establish a national television library to be maintained by Columbia University, American University, and UCLA (Library of Congress 1997, 156).[4] The Library of Congress established a separate division for American television and radio with the 1976 Copyright Revision Act, which also prompted a huge increase in television registrations to the U.S. Copyright Office. The Museum of Broadcasting (later renamed the Museum of Television and Radio and again renamed The Paley Center for Media in 2007) also opened its doors in New York in 1976 (Kompare 2005, 102–14).[5] While the establishment of these museums and archives allowed greater access to television's past, their founders, their institutional structures, and their selection policies would greatly influence and shape the prevailing television canon and narratives of television history.

The particularly low cultural status and unusually large broadcast volume of the soap opera makes it perhaps an extreme example of the unwieldy and contentious practice of television archiving. In addition, the extremely serialized nature of the soap opera compounds difficulties in archiving the genre, with financial and spatial limits on acquisitions in direct conflict with the long-form narrative requiring that viewers watch numerous sequential episodes to follow even one of many simultaneous plot lines. For instance, while there is extensive academic work on soap operas as a television genre, little attention is given to individual or particular sets of episodes, as opposed to many primetime series.

This lack of scholarly emphasis on individual productions coincides with the lack of commercial and archival availability of these shows and the subsequent inability for researchers to watch historical runs or view individual episodes within the context of the episodes that came before and after. What is saved today will determine what future academics, fans, and media industries professionals will know of television's history; therefore, we must examine the reasoning behind the selection of particular soap opera titles and episodes for preservation. Television archivist Terry Cook (2006, 169) warns fellow archivists that "they are determining what the future will know about its past, which is often our present.... We are deciding what is remembered and what is forgotten, who in society is visible and who remains invisible, who has a voice and who does not." When looking for soaps in major U.S. television archives, we discover just how much of their history has sadly met this fate of invisibility.

UCLA

The UCLA television archive contains more than 100,000 titles, favoring American commercial broadcasting and cable. The strongest holdings are in fiction programming of the 1950s through the 1970s, including a variety of soap opera episodes. In order to draw conclusions about the archive's soap holdings, the holdings must be contextualized within the collection and the archive's larger mission. UCLA generates a formal television collection policy articulating its goals and acquisition procedures, which I used—along with staff interviews—to explore the makeup of the collection.[6] The policy breaks down their acquisitions into three tiers:

- *Television Collected Extensively*: entertainment programs judged as milestones of popular culture or of "significant" historical or cultural interest, as well as news or public affairs programs dealing with topics of significant historical or cultural interest.
- *Television Collected Broadly:* programs that have historical significance, cultural impact, or artistry. These programs are judged on several factors: popularity; quality as judged by critics, historians, and scholars; influential subject matter; or their contribution to popular culture, both positive and negative. The collection proposes saving around 30 percent of the episodes from series in this category, such as each season opener and finale, episodes of special note, and a sampling of more typical episodes from each season.
- *Television Represented by Sample Shows:* programs for which the archive acquires a limited number of sample shows to represent the whole. This is the only

description that actually lists specific genre headings such as game shows, soap operas, talk shows, reality shows, cartoons, commercials, infomercials, and music videos. The probable formula for this category is no more than 10 percent of a show, with two or three episodes from each season.

While archives are often forced by limited resources to restrict their holdings, these small "representative" samples in the third category will greatly influence future characterizations of entire genres. The soap opera is the only ongoing fully scripted genre that falls in this "sample shows" category, and, while the daily style of these shows may often be similar, the content of each episode for each individual soap is always different.

A representative sample of soap operas doesn't permit full exploration of the cultural impact of particular shows or of the genre itself. Particularly troubling is the likelihood of losing those soaps that may not have reached critical or financial success, but nonetheless articulated alternative expressions of this so-called "formulaic" genre. Many terminated soaps never achieved the long-running status of those still on the air but are still significant because of their innovation or experimentation within the genre. For example, UCLA holds only ten of the 470 episodes of NBC's *Generations*, the first soap featuring an African-American core family, which ran from early 1989 to early 1991.[7] The archive holds only four of the 352 episodes of ABC's *The City*, which ran from late 1995 to early 1997. The show was a spin-off of ABC soap opera *Loving*, which tried to incorporate heavy location shooting and the latest production styles of primetime television in hopes of giving the failing soap new life.[8] Shows such as *Generations* and *The City* may not be representative of the soap genre as a whole, but their considerable contributions to the genre run the risk of slipping through the cracks of soap history if they are not more strongly marked for preservation (McNeil 1996, 320).

While UCLA may have somewhat sparse numbers of certain soaps, they also hold one of the largest blocks of consecutive episodes of any one soap opera in a television archive. The archive received a large gift from ABC in the early 1990s of approximately 24,000 shows, including large runs of both daytime and primetime programming (UCLA Film & Television Archive Undated A). A portion of this gift contained the complete run of *GH* from its premiere in 1963 until 1970, consisting of nearly 2,000 individual episodes. Typically, the archive would not accept such a large number of episodes of a program that is part of the "represented by sample shows" category; however, due to the nature of the gift, the archive was required to accept the donation in total. While 2,000 episodes is an impressive number, *GH* entered its forty-fifth year of production as of April 1, 2008. With approximately 250 episodes each year, the total run will be more than 12,000 episodes by the end of

2010. Still, 2,000 consecutive episodes of a program is a rare grouping when compared to the rest of the archives' soap holdings. With a large run of consecutive episodes, fans and scholars will actually be able to access the unique narrative qualities of the soap opera genre, which depends on long narrative builds, as well as the specific characteristics of *GH* during that particular time period. This large chunk of consecutive episodes is the type of archival holding that will allow both television scholars and fans to properly revisit a soap's history in great detail.

After 1970, however, the cataloged holdings for *GH* total approximately sixty episodes dating from 1971 to 2000, and most of these episodes are Emmy nomination screeners.[9] The archive's second largest soap holding is *The Bold and the Beautiful* (*B&B*), with 261 episodes of a show that has been on-air since 1987 and which will have produced almost 6,000 episodes by the end of 2010. Compare this holding to only ninety-one episodes of *GL*, the longest running of all U.S. soaps. These disparate numbers persist throughout UCLA's soap opera collection and demonstrate that the block of *GH* episodes is an anomaly in a collection that otherwise pays little attention to the strong importance of continuity across episodes in soap opera storytelling. These numbers also reflect the cultural hierarchies in television preservation, whereby the least collected "sample show" tier includes a genre that has both sustained popularity and broadcast presence and one in which the idea of sporadic episodes ignores the essential nature of the long-form continuing narrative to the genre.

LIBRARY OF CONGRESS

The Library of Congress MBRS division operates within a very different context than a university-affiliated archive such as UCLA. The bulk of the MBRS television collection was built by retaining deposit copies registered for copyright protection. All "original works of authorship" qualify for U.S. copyright protection once they are "fixed in a tangible form of expression," and registration with the U.S. Copyright office is not a requirement for copyright protection (United States Copyright Office Undated, 2). Registration of television material for copyright is really up to individual production companies. Some companies want to register every episode, and some only send in a few. All episodes are protected by copyright regardless of whether they are registered with the copyright office. While most MBRS television holdings come from the U.S. Copyright Office, only relatively small selections of the thousands of programs registered in the Office are chosen to remain part of the MBRS collection (Spehr 1992, 50).[10]

At the 1990 symposium, "Documents that Move and Speak," former MBRS chief Paul Spehr (1992, 52) discussed the overwhelming number of considerations that his archivists faced when selecting television programs for their collection:

> Over the years many different standards have been proposed by the staff of the Library [...] quality, popularity, program content, innovation, artistic achievement, technical achievement, documentation of the times, social and or political impact, tastes and/or trends, quality productions by or about important persons and programming that has caused controversy [...]. Each one of these has merit and we have used them all at various times over the years. The problem is that if one uses many standards, it opens the flood doors and we are once again overwhelmed with material.

MBRS acquisition policies are thus constantly shifting, along with the unpredictable nature of copyright deposits. Until the mid-1960s, the Library acquired very few television programs, emphasizing quality programming and collecting extremely small samples of entertainment programs: "The Library simply underestimated the social and historical significance of the full range of television programming. There was no appreciation of television's future research value [...] Of those registered for copyright during these years, the Library chose only an occasional sample of entertainment series [...] and the so-called "quality programs" (Library of Congress Undated)." Spehr (1992, 52) also notes that, at different points in the collection's history, the Library used very tight limitations, eliminating entire genres, including the soap opera.

Librarian Rosemary Hanes (2006) told me the MBRS often adopts a similar system to UCLA's "television represented by sample shows," taking a sampling of shows they feel are formulaic and of which they already have a reasonable number of episodes. She cited Fox's *Cops* and *America's Most Wanted* as examples of such programming. Similar to the soap opera, these reality shows have a low cultural status and are seen as being interchangeable and thus easily represented by any selection of episodes. Soaps, with their prolific production and their reputation for being indistinguishable from one another, again find themselves at the bottom of the priority list when it comes to preservation.

MBRS's soap opera holdings reflect the uneven and changing nature of their copyright policy and collection emphasis. Compared with the number of episodes of a long-running soap like *GH* with UCLA's collected episodes, the MBRS holdings are smaller and more sporadic. The latter collection contains approximately 1,200 episodes of *GH*. The MBRS has a complete run of the program's premiere year in 1963, and then not a single episode for the period 1963–1980 was archived. The collection does include an almost complete run of episodes from 1980 through 1984. This set of episodes is followed by only five shows from 1985 and nothing again until 1988.

The irregular nature of MBRS television episode holdings has as much to do with the emphasis production companies put on copyright registration as they do MBRS selection policies. ABC Daytime, which produces *GH*, was often diligent about registering their soaps during certain time periods (as indicated above with *GH*), when CBS and NBC, by comparison, registered very little (Hanes 2006). These uneven registration practices suggest that large chunks of significant shows, or even whole programs, may be forgotten, a lapse that stymies any attempt to characterize the specific nature of soaps aired on different networks or created by one production company versus another.

Take for example the recently cancelled NBC/DirecTV soap opera *Passions*. Known for its use of supernatural plotlines and camp sensibilities, *Passions* aired from 1999 until 2008. Less than ten years is a relatively short run compared to those enjoyed by many U.S. soaps; however, because of regular copyright deposits by NBC Studios, the MBRS holds copies of more than 500 individual episodes of *Passions*.[11] In comparison, the MBRS holds only five individual episodes of the only current NBC broadcast soap, *Days of Our Lives*, produced by Corday Productions, Inc., a show that premiered in 1965 and has currently been on the air five times longer than *Passions*. While having such a high number of episodes for an experimental show like *Passions* is fortunate, the disparity between the holdings for this show versus another will make characterizing the success or failure of one soap compared with another on NBC during the 1990s and 2000s very difficult.

THE PALEY CENTER FOR MEDIA

The Paley Center for Media, formerly the Museum of Television and Radio, has a different mission than the other television archives examined here, in that it operates as a public museum rather than as a traditional archive.[12] Archives at publicly funded institutions like UCLA and the Library of Congress are both accessible to the public, but The Paley Center specifically tailors its holdings to reach general and industry audiences. Founded by CBS chairman William S. Paley, the museum's main mission has been the cultural legitimization of television and shaping the public image of the medium. The Center cites public access and exhibition as the main impetuses for founding the museum: "[T]he history of—and the history captured by—television and radio was not being exhibited and interpreted for the general public. Outside of university level media/communications courses, the significance of television and radio was not being analyzed and interpreted in public settings." (Gibbons 1997, 227) Paley's vision was to elevate the medium by presenting and preserving what he and his industry colleagues considered the very best of television's offerings. While the Center's selection policies have expanded, much like those of other television archives, its close relationship to the industry is reflected in the collection. For instance, rather than drawing on librarians or academics, Paley insisted that the museum's board consist of network executives and industry colleagues (Kompare 2005, 113). Additionally, the organization has long-standing contractual agreements with studios, networks, and production companies to donate, routinely, programming selected by the Center (Gibbons 1997, 227). Perhaps these industry ties account for the Center's emphasis on popular as well as the critically acclaimed programming and may help to explain why the soap opera genre is so well represented in this collection, in comparison with other archives.

While the Center does not have complete runs of any U.S. daytime soap operas, their collection of soap operas and material about daytime serials is comparatively large. Unlike the other archives, the Center has a more wide-ranging soap opera collection, including recordings related to daytime soap operas in addition to regular episodes. The Center's selection process is shaped substantially by its relationships with various soap opera production companies (such as TeleNext Media/Procter & Gamble Productions and Bell-Phillip Television Productions), as evidenced by the volume of some programs in the Center's archive.[13]

The Center's daytime drama holdings can be roughly divided into three major categories. The first group consists of a selection of regular daily episodes for a wide variety of soaps, similar to those collected by UCLA and MBRS. Rather than random episodes from donations or copyright registration, these episodes were chosen to highlight major events or characters within the narrative (Simon 2002). These episodes contain special events like births, weddings, and deaths or other major plot points in soap opera narratives, including infidelities, kidnappings, and murders. This group also includes many series premieres and finales. While the Center holds a fair number of these single episodes, each must still be viewed out of their narrative context. These climactic events result from months—even years—of complicated storylines and character development. While these episodes may provide a useful glimpse into the narratives and characteristic styles of particular soaps, they represent only a small portion of the complicated histories of the programs as a whole.

The second group of holdings might be classified as anniversary episodes. This group consists of both primetime specials and regular daytime episodes that celebrate an anniversary or the history of the program by revisiting favorite storylines and rewarding longtime viewers with series of clips from the past. Primetime specials are more often reflections on a specific soap opera as a whole, and include interviews with cast members and producers alongside longer montage sequence. The daytime anniversary episodes, on the other hand, are often contained within the current narrative of the program. The daytime episodes in this group also include many holiday episodes, wherein core families of the show gather and reminisce about family history and where viewers are treated to "memories" of the characters in the form of historical clips. This group of the Center's holdings suggests the importance of narrative history to viewers. These episodes allow for an abbreviated look at the history of the program while skipping the complex storytelling surrounding these revisited moments.

The third group of programs could be classified as the Center's original programs or special events. The organization has produced several public programs and seminars about the soap opera genre, recorded exhibitions and seminars held at the Center and featuring soap operas and their production. For instance, in 1998, the organization held a special series entitled "Worlds Without End: The Art and History of the Soap Opera," which featured events focused around the current daytime soaps and particular tropes or attributes of the genre, such as "The Supercouples," "The Divas," and "The Rituals of the Soap Opera." Most of these seminars included discussions with cast and crew, and almost all featured elaborate clip reels—like those seen in the primetime anniversary shows—to introduce the show or topic of the night.

These programs are particularly notable for their emphasis on an individual as the "author" of a soap opera, with attributes of a particular series associated with that show's creator or producer at the time, such as in "The Agnes Nixon Seminars" in 1988.[14] This emphasis on authorship is unusual for a genre such as soaps, as authorship is usually only associated with "quality" television.[15] Soap opera production actually depends on the lack of visible authorship, so that one writer can be replaced by another over the years without the change being too jarring for the audienc. But the Center recognized that certain writers and producers stand out, despite the industry's need to suppress an authorial signature for the sake of continuity. In addition to their original programs, the Center's "She Made It: Women Creating Television and Radio" initiative honored the important achievements of 150 women in television and radio, including legendary soap opera producers and writers Nixon, Irna Phillips, Gloria Monty, and Claire Labine. This recognition of individual women and their contributions to the soap opera genre is indicative of the particular attention the Center has put on soaps, especially their connection to women, which is not seen in other television archives.

Recognizing authorship in individual soap operas opens up the possibilities for the examination of differences between specific soap operas and the contributions of specific soap writers, producers, and other creative talents. However, any kind of in-depth scholarship on a individual creator's or producer's influence would be impossible with just a few episodes of a particular show, and this difficulty suggests the need for more complete archival collections of all soaps.

Special events celebrating the soap genre highlight the differences between the Center and archives like the Library of Congress or UCLA. The Center has a more significant investment in soaps than the other archives, paralleling its attention to public taste, rather than to the critical opinions of scholars and archivists. Fortunately, the elevation of television's public image has not caused the Center to eschew the popular and discard a genre with such low cultural status. In recognizing soaps' longevity and popularity, the Center also acknowledges the importance of women in soap opera production and the power of the female audience that sustains the genre, a focus that dovetails with soaps' prominent place in the rise of television studies.

THE FUTURE OF SOAP OPERA HISTORY

What does the preservation of soap operas in archives have to do with the future of the soap opera form? What does it mean for soaps' future if we cannot reassemble its past? For scholars, the loss of these early episodes makes it extremely difficult to discuss soap opera content and performance in the early days of television. Television archivist Samuel Suratt (1995) laments the loss of many early kinescopes due to cost cutting management, pointing out that "virtually none of the early soap operas have survived, the soap manufacturers, which owned them, being even more parsimonious that the networks which broadcast them." Archival research reveals that, despite all that is written about early soaps on television, only a handful of these early recordings actually remain. While watching the content may not be essential to all scholarly work about soaps, one certainly must make numerous assumptions about the genre with so little content left as supporting evidence—especially considering how inadequately "representative samples" provide context for the complex narratives of this genre. The lack of complete runs of particular soap operas may also prevent a scholar from accurately characterizing the unique qualities of their object of study. Preservationist Robert Rosen (1997, 452) discusses the implications of archival acquisition policies that have neglected to see the value in archiving the entirety of a show's episodes:

> The soap opera, of no particular interest to the student of television as an art form, is of enormous import to the social historian. One may be tempted to say we only need a sample of a particular series. But very often the heart and soul of that series lies in the formula that evolved over time [...] The single most important criterion for the archivist in making selections is humility, not precluding the possibilities for future generations to discover for themselves the value of these material.

The "heart and soul" of a soap opera is very much the everydayness that becomes part of its viewers' lives. The ritualistic nature of soap viewing is again a favorite topic of television studies, which pays scant attention to the specific content that enters the viewer's home on a daily basis. With no way to view the evolution of a soap over time, much of the discussion about soap opera conventions and the so-called formulaic nature of the genre lack historical evidence to properly support these claims. Furthermore, this lack of individualization among soaps also risks limiting the ways we can analyze the genre.

Examining the archival holdings of daytime soap operas can do nothing to recapture lost episodes, but it can inform how we study what remains. When conducting research on any television text, attention must be paid to not only its production and viewing context, but also to the context in which the text is accessed by the researcher. If it can only be seen in an archive, how did it end up there? How did the character of the institution or their collection policy influence what programs, and even which episodes of a program, they chose to preserve? How did business relationships or copyright regulations help determine what we will be able to study as part of television history? Answers to all of these questions will have serious effects on the ways scholars study the soap opera genre in the future. While the physical and financial limitations of television archives will obviously continue, scholars need to be aware of how these restrictions affect the field. The "representative samples" that have been preserved of an individual soap opera must be ruthlessly interrogated as to how "representative" they actually are.

As I've emphasized throughout this essay, if soaps are simply archived as a "genre" rather than as distinct programs, this arrangement precludes scholars from comparing particular series, creators, networks or episode runs within one series to the same elements of another. Dedicated soap fans often follow the particular styles and careers of soap opera producers or writers, but the material has to be archived in order to trace these characterizations and make comparisons in the future. Without sufficient content, it may be impossible to discuss the different styles of particular soap opera production companies or different networks' soap opera line-ups. The ability to review historical content and analyze how storylines coincided with previous ratings trends might be of particular importance to contemporary soap opera production. While the vast decline in soap opera ratings is commonly attributed to increases in competing programming and home technologies, content-based analysis between historically successful storylines and contemporary ones may prove to be a valuable research tool for both television producers and scholars.

Industry-focused research on soap operas would also benefit from analysis of the coinciding content and how it reflects industrial or technological shifts. Having sufficient content from a specific soap available in public television archives would open up tremendous research possibilities. Building stronger relationships between soap opera producers and institutional archives would also insure that soap operas achieve the recognition they deserve as part of television history. While public archives may not be actively seeking out soaps for preservation, soap opera producers are currently grappling with ways to capitalize on their own archives in the digital era. Platforms for soap opera reruns such as SOAPnet (which will go off air in 2012), the PGP short-lived online Classic Soaps channel, and the availability of episodes of cancelled soap *Passions* on NBC.com, all indicate that producers are eager to find ways to maximize the profitability of their own archives, as well as fans' desire to rewatch older soaps. While these offerings may not be motivated by ideas of historic preservation, they still hint at the idea that daily soaps may have an afterlife after all.

Soap opera fans have already responded to the lack of accessibility of soap history in the commercial marketplace by creating their own personal taped off-air archives or posting their collections on public video sharing sites like YouTube, but these collections don't have the security of preservation and public access that they would in an official archive. Fan clips are always in danger of being taken down because of copyright infringement, and personal fan archives rarely meet the physical requirements to properly store a video collection— not to mention the fact that fans have no legal purview to do anything with their collections other than view them for personal use.[16]

Fan archiving may help to save some of soap history for the future, but without the institutional weight of a public archive, will soap history be given its proper place within television history?[17] All these complications highlight the need to increase archival emphasis on the soap opera genre, with its unique long-format narrative structure recognized and accounted for, rather than used as an excuse to treat a singular episode as a generic example of the whole. While the ability to examine the daily texture of many soaps is already lost, careful attention to the archiving of current soaps will lead to greater resources for studying and revisiting the pleasures of the daytime soap opera in the future.

NOTES

1. I use the term "unruly" with reference to Rowe's (1995) work on women and excess. As soaps are gendered as feminine texts, their ever-expanding size seems to fit Rowe's theories.

2. Soaps will occasionally rerun episodes on holidays or to celebrate a show's anniversary or memorialize an actor. This does not include the recent phenomenon of daily rebroadcasts of soap operas on cable channel SOAPnet (which will go off air in January 2012).

3. For more on the low cultural standing of soaps, see Nasser's essay in this collection.

4. All items were later consolidated at UCLA.

5. See Kompare (2005, 111–4) for a more detailed history of these institutions' origins.

6. These categories come from the UCLA Film & Television Archive (2003) collection policy.

7. BET Cable Network purchased all 470 episodes of *Generations* after its cancellation.

8. Profiles of *Generations* and *The City* appear in Museum of Television and Radio's *Worlds Without End.* (1997).

9. UCLA has a donation relationship with the Academy of Television Arts and Sciences (UCLA Film & Television Archive Undated B).

10. The Library can also demand programs that have not been registered if they are commercially traded, and the Library has the right to record programs off-air if necessary.

11. A production company may register an episode for copyright at any time within the life of the copyright, so there is a possibility that many more episodes will be added to the collection in the future.

12. I am still examining the Paley Center as a museum, even though they have taken the word out of their official title.

13. The Paley Center's Los Angeles Board of Governors currently includes William J. Bell Jr., President of Bell-Phillip Television Productions, Inc. Bell-Phillip Television Productions solely produces *B&B* and produces CBS' *The Young and the Restless* in conjunction with Sony Pictures Television. TeleNext Media/Procter & Gamble Productions were the producers of *GL* and *As the World Turns* for CBS, among various other soap operas.

14. Agnes Nixon created ABC's *One Life to Live* in 1968 and ABC's *All My Children* in 1970. See Williams's piece featuring an interview with Nixon elsewhere in this collection.

15. See Feuer (1984).

16. The role of fan as soap opera archivist and fan archives falls outside the purview of this article, but is a topic I intend to cover in future projects.

17. See Webb's essay in this collection for more on soap video archiving practices on YouTube.

DID THE 2007 WRITERS STRIKE SAVE DAYTIME'S HIGHEST-RATED DRAMA?

—J. A. METZLER

> J. A. Metzler runs a boutique marketing and communications consultancy. He took part in a daytime writer development program in 2003 and has been a lifelong soap opera viewer, particularly of CBS's lineup of daytime serial dramas.

I was raised from a young age on a diet of CBS network daytime dramas. In fact, one of my earliest television memories, dating from when I was nine years old, was of a drunken Katherine Chancellor—a character on *The Young and the Restless* (*Y&R*)—driving her cheating husband Phillip off a cliff, simultaneously depriving Phillip's pregnant, working-class mistress, Jill Foster, of any type of happy future, and sealing my fate as a long-term closet fan of the serialized genre. My parents were understandably worried by their child's interest in adulterous, alcoholic, and sociopathic characters. However, I was fascinated by worlds where fifteen or more characters, flaws notwithstanding, could be entwined and interconnected in a slow, intimate, continuous, and evolving storyline with no set beginning, middle, or end. Today, there are six network soap operas left on the air, down from thirteen twenty years ago; daytime has lost more than half its available viewing audience in the same number of years. CBS cancelled the longest-running soap on the air, *Guiding Light* (*GL*), in 2009, and the subsequent longest running show, *As the World Turns* (*ATWT*), in 2010. Owing to declining ratings, many of the soaps that remain are on what the industry calls the "bubble" of cancellation.[1]

Fans and critics noted that *Y&R*, the flagship of the CBS daytime lineup, struggled creatively after the death of its creator and longtime show runner, William J. Bell, in April 2005. In an attempt to contemporize and update the classically slow-moving yet most popular soap in order to appeal to new viewers, Sony Pictures Television (the production company that co-owns *Y&R* with Bell-Phillip Television Productions, William J. Bell's family corporate entity) hired a well known veteran primetime scribe and producer to join the show. Ex-*Knots Landing* (CBS) and *Homefront* (ABC) writer Lynn Marie Latham came aboard as a creative consultant in late 2005 and was quickly promoted to co-head writer alongside William J. Bell's long-term collaborative writing team, led by Kay Alden and Jack Smith. Soon after, Latham took over as both sole head writer and executive producer of the show, and she was essentially the only creative executive guiding *Y&R* for the next year. Latham soon started replacing key members of Bell's experienced writing and production teams, many of whom had worked together at *Y&R* for the better part of three decades.

Since childhood, I have always regarded *Y&R* as a kind of televised comfort food—familiar and always satisfying. I relied upon *Y&R* to consistently feature the same strong characters that had been on the show's canvas for twenty years or more. After Bell's demise and Latham's takeover, the show's canvas became littered with new, unfocused characters that had no history or family ties. Under Latham's pen, the slow buildup to stories—a William J. Bell trademark since the show's premiere in the early 1970s, and a primary reason why I had been a longtime viewer of the show—was replaced by too-quickly developed plots that concluded in anti-climactic and unsatisfying payoffs. Some stories began and then were abruptly terminated, leaving viewers hanging. The pace of the show quickened; there were more scenes per act, but they were shorter and frenetic, bearing little resemblance to the slow-cooked pacing of traditional daytime, which more closely mimics the pace of real life. Latham's faster, choppier *Y&R* resembled primetime programs much more than soap opera.

Y&R was not alone in changing its pacing, narrative style, and cast. Every other soap had made similar changes in an attempt to survive amid the thousands of programs playing on more than 120 channels in the post-network era of television. But I was still surprised when Latham's *Y&R* jettisoned the style and structure that had made the show the top-rated soap opera for years. With its new primetime feel, *Y&R* just didn't seem the same to longtime viewers like me. It was almost as if I were watching a completely different show, a show that—within months of Latham assuming production control—I no longer chose to watch on a daily basis.

On November 5, 2007, as Latham was instigating major change at *Y&R*, The Writers Guild of America (WGA) voted to strike against the Association of Motion Picture and Television Producers (AMPTP). The last time the WGA went on protracted strike was nearly twenty years before, in 1988. The 1988 strike lasted a long twenty-two weeks, or close to six months—the longest WGA strike in the guild's history. Because soaps are the only *continuous* dramatic genre on television, with new episodes each weekday year-round, the producers and network executives overseeing the twelve soap operas on the air during the 1988 strike decided to hire scab (non-union) writers to script the twelve soaps. The scab writing teams were most likely comprised of college interns, advertising agency personnel, and production secretaries, and soap fans like myself began to think that almost anyone behind-the-scenes was being enlisted to write scripts. In other words, for longtime viewers of daytime drama, scab-written shows meant trouble. By the end of the 1988 walkout, with most of the scab-written soaps suffering creative challenges, one show—ABC's *Ryan's Hope*—was issued a death notice. A few other serials experienced backstage upheaval as, following the conclusion of the strike, experienced soap staff writers were ousted in favor of the cheaper, less experienced scabs who had written in their stead.

As the most recent WGA strike dawned in November 2007, it did so amid countless media predictions about the death of the daytime soap genre, and close on the heels of an expensive Latham-produced *Y&R* story arc and promotion entitled "Out of the Ashes" that was largely ridiculed by viewers as a ratings stunt. I thought to myself, "Great . . . a writers' strike is just what daytime needs right now to help plan its wake." During the WGA strike, I watched each daytime soap, paying attention to how the remaining scripts penned by shows' staff writers played out. I kept an eye on the opening credits of each show so that I could note the day when the "real" writing teams' names disappeared, indicating that the scabs' work was being featured. At that point, I gloomily predicted, the quality of the scripts would get worse, and soap viewership would decline even more. Rumors pegged *Y&R* as the first soap that would run out of original scripts. In late December, nearly two months after union writing had stopped, *Y&R*'s staff writers' names no longer appeared in the credits. I watched that first scab-penned *Y&R* episode and was tremendously surprised that the show did not get worse, as I had long anticipated, but got, even in that first episode . . . a little better.

In that (and subsequent) episodes written by scabs, aspects of *Y&R* that I had thought "broken" and unfixable under Latham's writing team began to be remedied: long-term characters made sense again; scenes lasted longer than a blink; the history of the show was referenced rather than ignored or contradicted; old friendships among various characters were recognized. It was starting to look, and more importantly, to *feel* like *Y&R* again.

Could the genre of daytime drama be saved by just letting scabs write the shows, I wondered? From what I was witnessing on my television screen, the potentially inexperienced scabs seemed to be able to write better scripts for *Y&R* than a seasoned primetime writer such as Latham. Could this herald a new renaissance for the fading format? My mind could barely wrap itself around this possibility. Then, when the new "written by" credits rolled on that late-December episode of *Y&R*, I saw that the show had not been scripted by scabs. There were actual writers listed, experienced, well known daytime drama writers Under union rules, scab writers cannot be credited, but *Y&R* was being written, not by unknown scabs, but by Josh Griffith, a former head writer for *One Life to Live*, and Maria Arena Bell, the daughter-in-law of creator William J. Bell, who had apprenticed with him on both *Y&R* and *The Bold and The Beautiful*, and wife of Bell-Phillip Television Productions President Bill Bell, Jr., the production company that co-owns *Y&R*.

During the 2007 strike, strikebreaking writers weren't scabs, as in the 1988 strike. They were professional WGA members. They were not primetime writers who had no familiarity or history of professional involvement with the soap genre. They were experienced daytime writers—with a twist! A new term entered the lexicon of soap viewers, one that wasn't in circulation during the 1988 writers' strike: *financial core,* or *fi-core*. Under federal law, members of a union may legally resign from the union during a strike and opt to become "financial core non-members." As a financial core non-member, a writer is bound by union rules and, hence, can work—and get paid—during a union-sanctioned work stoppage or strike. Experienced daytime writers such as Griffith and Bell chose to go fi-core rather than let their viewer-challenged shows be written by non-union scab writers, a controversial decision but one supported by some fans who had seen the strike as the end of daytime soaps.

In the union world, a fi-core is really only one step removed from a true scab, reviled by his or her picket-line-walking peers. But, unlike scabs, credited financial core members are up-front about their strikebreaking. In 1988, the fi-core option existed for WGA members. At that time, however, soap operas were network cash cows, and there was very little chance of the entire genre disappearing during the course of the strike. Soap writers could afford to walk the picket lines with their peers. The state of the industry was much different in 2007. Daytime drama writers likely felt that, if there was a prolonged strike, and, if soaps' quality declined substantially during the strike, even more viewers would lose interest than already had. Post-strike, there might not be any shows left for them to come back to write.

I'm sure a few fi-core writers had legitimate financial hardships and therefore chose to cross the picket line and face the ire of their striking compatriots in order to earn a steady paycheck. However, the romantic in me likes to think most of these fi-core writers were putting themselves on the firing line and risking the condemnation of their peers in the hope of saving daytime drama itself. I think that at least some of the soap writers who chose to become fi-core were motivated by a strong desire to give the soap genre as good a chance as possible of surviving the strike. In the case of at least one fi-core writer, this concern for the future of soaps was probably quite personal: Maria Arena Bell may have been trying to protect her family's and William J. Bell's legacy, which had been threatened by the changes made under Latham's tenure.

For close to two months—the length of time that fi-core writers were scripting the show—Y&R viewers were treated to what had become rarities during Latham's stint as head writer, but which longtime fans regarded as the components most responsible for the show's reputation as the greatest soap on-air: historically rich scenes; character friendships long-since abandoned by the previous writing team; small but poignant dramatic moments; and the show's classic slower pace. From my perspective—at least in the case of Y&R—the 2007 writers' strike was a hit!

The strike officially ended with a new contract ratification between the WGA and the AMPTP on February 12, 2008. Strike-written scripts would continue to be aired for another few months. However, with the end of the strike, the backstage drama eclipsed the fictional onscreen drama. Y&R's production companies immediately opted to continue with their new popular fi-core-led writing and with family member Maria Arena Bell at the head, instead of their pre-strike writing team. Over the next several months, Bell slowly built on the foundation she and Josh Griffith had created during the strike. Several unresolved Latham-era storyline threads were put to rest, and new storylines focused on long-term characters familiar to Y&R viewers. As writer contracts came due, several writers who had been hired by Latham were let go, and new writers—including at least one from William J. Bell's fired team—were brought on, adding to the cadre of scribes who knew the show's history.

As this book goes to press, Y&R is still holding on to its number-one daytime ratings slot. The show's viewers have seen the results of Maria Arena Bell's storylines, with their more traditional pacing, and the Latham era has faded from viewers' memories. GL featured significant experimentation with the show's format and look, and ATWT reinvented its storytelling structure to try to appeal to new viewers prior to CBS' cancellation of both shows, but Y&R proved that going back to the basics of soap opera—character-driven storylines that draw on a show's rich history and unfold slowly over many months—just might be a viable strategy for keeping the genre alive.[2]

In November 2008, I once again found myself watching Katherine Chancellor, played by the same actress I watched as a child (Jeanne Cooper has been in the role since 1973), in a front-burner storyline. In a twist that could only happen in a soap opera, what viewers thought was the "ghost" of Katherine was attending her own funeral, where the ghost reveals that she—and the body in the casket—are actually a lookalike character. "Katherine" is surrounded by grieving family members, many of whom are characters who departed from the Y&R canvas long ago but reappeared for this special occasion, played by the same actors that originated them in decades past. As I watched that show, I found nostalgia creeping in from my childhood—and a feeling of warmth driven by familiarity—and I realized that this feeling was what motivated me to watch continuing daytime dramas.

While no writing team that is producing more than 250 hours of television a year can be 100 percent perfect, *Y&R*'s writing team, led by Bell, managed, in the months during and following the writers strike, to capture the essence of why some of us watch soap operas day after day, year after year. With a creative and legacy stake in the continued quality of the show, and with such an intimate knowledge of their show's themes and characters, Bell took the changes she started to implement with the writers strike and began to truly demonstrate how the intimacy—and passion—of a show's creative team can make a dramatic change that helped bring renewed commitment and investment to even a cynical and disappointed fan base. That cynicism and disappointment is, unfortunately, fueled by frequent lapses in consistency and long-term vision, a characteristic that plagues even the best creative teams and has plagued *Y&R* again during the six to nine months prior to going to press. The additional pressure to deliver a show that will satisfy long-term viewers and bring in a new audience, all while undergoing massive cuts to production budgets, only add to the pressures placed on writers.

At their best, soaps supply us with a sense of continuity. They feature the same communities, all intertwined with one another's lives in various complicated ways over decades. They give us the chance to watch a group of people experience life at roughly the same rate that we experience it, and evolve and age as we evolve and age. Great soaps provide familiarity in a world that for many is filled with the unfamiliar. Thanks to the 2007 WGA strike and fi-core writers, *Y&R* once again became my television version of comfort food.

NOTES

1. For more on the precarious state of soaps, see Bibel's essay in this collection.
2. For more on those two shows' experiments, see Liccardo's essay and Erwin's essay in this collection.

SECTION THREE
EXPERIMENTING WITH PRODUCTION AND DISTRIBUTION

"THE RHETORIC OF THE CAMERA IN TELEVISION SOAP OPERA" REVISITED
THE CASE OF *GENERAL HOSPITAL*

—BERNARD M. TIMBERG AND ERNEST ALBA

> Bernard Timberg is a writer, documentary filmmaker, and university professor, most recently at East Carolina University. His essay "The Rhetoric of the Camera in Television Soap Opera" was among the first scholarly examinations of television soap opera production. His interest in examining soaps came at the suggestion of doctoral advisor Horace Newcomb, and was intended to complement Timberg's work on daytime talk shows, news, and advertisements.
>
> Ernest Alba is a graduate student in anthropology at the University of Texas at Austin. His work in soap opera studies can be found at *MIT CMS: The American Soap Opera* (http://mitsoaps.wordpress.com/). His essay "The Effect of the Youtube Phenomenon on the Soap Opera Text" was published through the MIT OpenCourseWare initiative. He grew up watching a variety of telenovelas.

INTRODUCTION

In Spring 2009, ABC's *General Hospital* (*GH*) underwent a makeover of its seventh floor central nursing station set after an on-screen disaster/fire. The set has been a constant on the show, with modest updates, for nearly half a century. The fire cleared the way for other changes on the show as well: moving into the use of high-definition cameras, for example.[1] *GH*, like other venerable soap operas discussed in this book, was facing the challenges of the twenty-first century.

Levine (2001, 66) points out that cultural studies scholars such as Stuart Hall (1980) and Richard Johnson (1986/7) have called for attention to cultural production as the "integration of production analyses and studies of texts, audiences, and contextual influences." Most work has focused on the decoding side of the encoding/decoding equation: analyses of texts as social documents and sociological or ethnographic studies of the audiences that consume them. Published journalistic or scholarly accounts of camera technology or behind-the-scenes production in television have been much rarer, including Arlen (1981), Cantor (1971), Cantor and Cantor (1992), Gans (1979), Intintoli (1984), Newcomb and Alley (1983), Shanks (1976), Timberg (2002), and Tuchman (1979).

Timberg (1981) does not provide a production study per se, but rather extrapolates production technique from semiotic features of soap opera that emerged from the way these shows were produced: camera angles, lighting, sound, and narrative themes as they were embodied in soap opera camera work and direction. Like the purloined letter of the Edgar Allen Poe story, Timberg suggests, soap opera camera work seemed too obvious, too prosaic, simply too formulaic to merit much discussion at the time. With occasional exceptions in television soap opera critical literature, that omission remains true today. Applying Timberg's work on *GH* and ABC's *All My Children*, Alba (2008) analyzed CBS's *As the World Turns*. Surprisingly, after such a long period of time, many of the same principles of camera rhetoric held true. Alba (2008b) summarized these principles as follows:

1. use of close-ups and extreme close-ups to create an intense, intimate camera style
2. eye-level camera angles used within scenes versus low or high shot angles
3. use of Z-axis alignment with images in foreground and background emphasizing relationships between characters
4. truck-ins or pans to establish or refocus physical relationships of characters in the scene
5. meaningful fades and dissolves at a scene's end
6. camera moves designed to capture stylized expressions of emotion: pity, jealousy, rage, self-doubt
7. transitions indicating connectivity between situations or people
8. a dance-like camera choreography that constantly shifts point of view, following or circling characters and discussions

How did these principles develop? Where did they come from? Will they continue to guide visual styles of story as soap operas move into new forms of production and new technological venues of reception for their audiences? Can new technologies and camera techniques save soap operas in their present state of crisis and make them more relevant to contemporary audiences?

When looking at where the soap opera may be going, it is useful to look at where it has been. With that in mind, this essay examines the camera work of the first episode of *GH* on April 1, 1963, as positioned within the show's thirtieth anniversary in 1993, and as compared to a 2009 episode of *GH*.[2]

THE CAMERA WORK OF *GH* IN 1963

THE JUXTAPOSITION BETWEEN 1963 AND 1993

The thirtieth anniversary show of *GH*, directed by five-time Emmy nominee Peter Brinckerhoff,[3] intercuts scenes from the 1963 show as memory flashbacks.[4] It is a normal day in the life of Port Charles, though the staff is celebrating the hospital's thirtieth anniversary as well. The history of the show itself is recognized in two places. At the beginning, *GH* head doctor Steve Hardy steps out of character to address the viewing audience directly, walking down a long, empty corridor as he recalls the history of the show. At the episode's end, Hardy concludes the show, turning to walk down the same long corridor.

The main focus of the thirtieth anniversary show in 1993, a party being prepared to celebrate *GH*'s thirtieth anniversary as a hospital in Port Charles, is intersected by a second story that harkens back to the show's first episode. This secondary story begins with the surprise visit to the hospital that day of Angie Costello (now Collins), Dr. Hardy's first patient. She had been in the hospital thirty years earlier because of a disfiguring car accident, and returned that day to await the birth of her first grandchild.

The show features a series of flashbacks to these Angie's and Dr. Hardy's interaction in *GH*'s inaugural episode. Each flashback begins and ends with the iconic extreme facial close-up that demonstrates the "inner look" of a reminiscing character, with Dr. Hardy remembering Angie as his first patient and how difficult she had been. These flashback scenes provide viewers with a fascinating intersection of television history and televisual soap opera styles and experimentation. There are obvious contrasts between the 1963 and 1993 shows: black and white vs. color; camera movement; the number of cameras covering a scene; the use of extreme close ups. An analysis of the retrospective clips from 1963 aired on the thirtieth anniversary episode shows us how far camera work has changed in "telling" soap opera stories.[5]

THE 1963 OPENING

The opening sequence of the 1963 show, used again as the "cold opening" of the thirtieth anniversary show in 1993, runs only seventeen seconds, with not a single word spoken. It unfolds in one long take, accompanied by the lush piano score composed for *GH* by Kip Walton. The music, as well as the visual images, "brand" the show. The images are shot in crisp black and white and seem primitive to an eye used to the modulated lighting and muted electronic colors of contemporary soap operas. This copy appears to have been struck from a 16 mm film negative. We see white dust spots and streaks on the film, indications that the negative had been kept in less-than-pristine conditions. The black and white images and the sharp shadows struck by the blunt single-light setups placed around the set bring us intriguingly into the world of *GH* in 1963.

The ending of the opening sequence is particularly striking, as the elaborate choreography of blocking, character motion, and camera movement come to a sudden halt in a freeze frame. As the show's main title dissolves over the freeze, a negative-reversal of blacks and whites occurs, and the bold white capital letters of "General Hospital" appear. The typography will remain essentially the same for the show's next thirty years on the air. A deep-voiced announcer proclaims: "*General Hospital*, brought to you by . . .," and the show begins.

The first sequence starts with the camera pulling back from the traditional serpents-on-a-staff medical symbol on the back wall to reveal the entire seventh floor reception area. (When the new set of *GH* came on screen in 2009, that same symbol appears located in the background, with the visual message clear: some forms of continuity on a soap opera of this kind are sacrosanct.) The establishing shot that follows the camera pullback brings us clearly into an earlier era of television cinematography. We pick up the movement of a nurse in a gleaming white uniform winding in from the far left back of the set. She passes several nurses' desks and comes to the central receiving desk, carrying a tray with a pitcher of water. When Nurse Brewer is almost in the precise center of the scene, just as she is passing the receiving desk, and just before the end of the seventeen-second sequence, the shot freezes and dissolves into a negative image of itself. Blacks turn into whites and whites into blacks.

This opening scene employs an experimental film technique akin to the freeze-frame flashback technique used by director Sidney Lumet and film editor Ralph Rosenbloom in *The Pawnbroker*, released in 1964, a year after *GH* was first broadcast. This particular freeze-frame moment on *GH* seems anything but random. It was, along with the music, an important "branding" moment for the show. What did this unusual negative reversal freeze frame *mean*? It is certainly, first of all, an establishing shot, in the fullest meaning of the term. It gives us a powerful schematic, a floor plan of the set we are about to experience—the same set that, for forty-five years, will represent the heart and core of *GH*. It is "frozen" in our minds with the opening musical chords of the show. However, the freeze frame does more: It provides us with a condensation, a thickening of the action, before it has begun. We are asked to pause, to think.[6]

What we are presented with here is also, in effect, a visualization of the principle of melodrama itself. Melodrama depends on sudden reversals, twists and turns in the lives of its characters. These reversals can develop gradually or break suddenly in the course of a normal soap opera day. Abrupt turns and revelations of plot or emotion are captured by the camera here in one embracing image of change and reversal, turning white to black and black to white in, literally, frozen time.

The freeze frame also casts in relief the gender configuration of *GH*: the personal and professional relationships of doctors, nurses, and patients in the seventh floor reception area. The hot whites of the nurses' uniforms, which have been dominant in the scene up to this time, dissolve to black and now stand out like the black silhouettes of antique family portraits. As the starch white uniforms and caps of the four nurses in the scene turn dark on the screen, the images of the males all dressed in dark clothes turn to stark white after the freeze frame dissolve. One institutional symbol, the switchboard, also appears as a dark object now turned white. The dark/then-white male images frozen into position are: a doctor leaning casually over a nurse's desk in the background; a family member in earnest conversation with the nurse at the receiving desk; and a third figure, the largest and most prominent in the scene, of a man lying on a stretcher, his dark head of hair now a dominant white splotch in the middle of the screen. The patient's gurney is next to Nurse Brewer, frozen in her own tracks, about to pass and move into the first dramatic scene of the show, frozen in her own tracks of motion.

These three men and four women are isolated but also in clear positions of relationship and juxtaposition. Since there were only seven or eight characters on the show at launch, the opening scene features virtually the entire cast. The only characters we have not seen are the ones who are about to appear in the following scene: Dr. Steve Hardy and his patient, Angie Costello, whose room—and emotional turmoil—Nurse Brewer is about to enter.

GH'S FIRST DRAMATIC SCENE

What is remarkable about the hospital room scene from the 1963 show featured in these flashbacks—the first dramatic scene of the series—is how closely it is modeled on "The Eye of the Beholder," a 1960 episode of CBS' *The Twilight Zone* (*TZ*). In this *TZ* episode, a woman, whose face is also seen only in bandages, is cared for by a doctor who reassures her that the plastic surgeon is doing everything he can. The atmosphere is dark and ominous in the *TZ* episode. Though the lighting is still expressionistic on the pilot episode of *GH*, it features a much friendlier, more business-like therapeutic setting. The room is established, without words, in pools of light, with the blinds tightly closed. The first black-and-white images we see are bare legs and a hospital gown, with feet dangling a few inches above the ground. The feet abruptly hit the floor and begin to move through shafts of light and shadow. The camera tilts, and a hand spasmodically grasps the Venetian blinds, pulls them down, and then closes them violently. The camera pulls back and tracks with the bare legs across the room, showing a hand as it grabs a towel and reaches over a dresser to hang it up (over the one mirror in the room, as we later learn) before moving to a shot of a fist smashing in frustration on a table. All of this is covered in a single traveling shot, with eerie piano music playing in the background. So far, we are in a world not far removed from *TZ*—a world in which we perceive objects dimly, quasi-subjectively, in a very spooky atmosphere. The sequence has lasted for forty-five seconds without a single word, but the mood changes abruptly as Nurse Brewer enters the room.

> Nurse Brewer: Oh, Angie. Not again. This is getting positively childish. (She opens the blinds and removes the towel from what is now clearly a mirror.) You know, if you realized it, you're a lucky girl—lucky to be alive. (She pauses to fold the towel). You're so young. (She moves to the patient in the corner, still facing the wall). Honey, why do you do it?
>
> Angie Costello: You know why! (The camera cuts to Dr. Hardy in the corridor, walking toward the room; he meets Nurse Brewer in the hallway as she leaves, and then he enters).
>
> Dr. Hardy: Look, your face will be healed in a few days, and we'll take off the bandages. We'll know how we stand after the plastic surgeon has a look at it.
>
> Angie Costello: I don't want anyone to see my face. I don't even want to see it myself! What's to be alive with . . . with a face like this? (She goes back to bed.) I wish I were dead!

The dialogue quoted above gives us a sense of the dramatic content of this scene. But what about the visuals? How do they "tell" the story? Are any of the rules of the "camera rhetoric" of soap opera, established by Timberg (1981) and distilled by Alba (2008b), in effect in this scene at the very beginning of *GH*'s history?

1 - *Use of close-ups and extreme close-ups* There has clearly been a significant development between 1963 and the 1993 show: *There are no extreme close-ups in the 1963 show*. We know that extreme close-ups were used in other forms of early television drama, such as the 1956 *Playhouse 90* (CBS) episode *Requiem for a Heavyweight*; however, in the 1963 *GH* episode, the camera keeps its distance in establishing shots, two-shots (two characters in medium shots in the frame), and occasional close-ups. The extreme close-up is not yet part of the vocabulary of *GH* in 1963, and yet it becomes *the defining characteristic of soap opera camera work.*

2 - *Eye-level camera angle vs. low or high angles* The angles are somewhat nuanced in this scene, as the two cameras in shot/reverse shot patterns shift position, but direct eye-level angles prevail.

3 - *Use of Z-axis alignment* This is a prominent feature of the scene, with Costello primarily in the foreground and Hardy in the background. However, the patient is constantly (and dramatically) traversing the Z-axis in wild actions and strides as the scene progresses.

4 - *Truck-ins and pans* Both kinds of camera movement occur in both the opening/establishing shot of the show and in camera movements that track Angie Costello around the room.

5 - *Fades and dissolves for ending scenes and transitions* Dissolves are used to illustrate key moments of transition, here from the freeze at the opening of the show to the resumed movement of the nurse in the hallway and from interior scenes in the room to actions occurring in the hallway outside.

6 - *Stylized expressions of pity, jealousy, rage, self-doubt etc.* Here, because Costello's face is bandaged, "less is more" as we intuit all the emotions of a patient whose face we cannot see.

7 - *Transitions indicating connectivity between situations or people* We experience the shifting and circling perspectives of Dr. Hardy and his troublesome patient from the shifting choreography of the two cameras used in the scene to connect them.

CHANGES IN CAMERA/EDITING TECHNIQUES FROM 1963 TO 2009

Television soap opera camera production has changed dramatically since the first soap operas were broadcast on television in the early 1950s. Digital, high-definition color has replaced the black and-white standard format shots of the early shows. Changes in production of *GH* have mirrored these and other technological changes over the years. Contrasted with the first grainy opening shot in 1963 episode, we now see on our televisions—and, increasingly, on the

Internet and in various screen interfaces with the Internet—high-definition, widescreen, color, and digital images. Despite these monumental advances in the production of television soaps, the basic semiotics of camera production on *GH* in 1963 (and in 1993) remain intact in 2009—with one fundamental difference. In 1963, emotion was conveyed primarily through actors. In 2009, emotion is still conveyed of course by the actors—but also, increasingly, through the camera. The chief example of this change is the extreme close-up, which constitutes new vocabulary that allows every emotion and facial quality of the actor to fill the screen. A character's emotion is conveyed precisely through the completeness with which the actor's face and the character's emotions are visually displayed.

Other prominent changes are noticeable. The 1963 scene examined earlier in this essay makes little use of cutting from shot to shot—or, in as-if-live television shoots, camera switching. Instead, the camera is trucked from place to place in elaborate following and tracking camera moves. Furthermore, there are no extreme close-up shots. To contrast, let us compare the scene to a June 17, 2009, *GH* scene in which central characters Nikolas Cassadine, the son of iconic *GH* character Laura Spencer, and Elizabeth Webber, the granddaughter of Dr. Steve Hardy, discuss a dream Nikolas had about his recently deceased ex-wife, Emily Quartermaine, and his current love interest, Rebecca Shaw, Emily's twin sister.[7] We see an establishing shot that includes both Nikolas and Elizabeth in which a camera is placed strategically over Nikolas's shoulder, allowing for a full shot of Elizabeth, who happens to be the focal point of the conversation at first. This shot is followed by medium establishing shots for both characters. As the conversation continues and it becomes clear that Nikolas is bothered by Elizabeth's claim that a relationship with Rebecca is "a terrible idea," the camera shots begin to reflect the shift in focus of the conversation from Elizabeth to Nikolas. As Elizabeth continues to try to convince Nikolas to give up on his pursuit of Rebecca by painting her as a liar who portrays herself as a "sweet, misunderstood victim," she is given a clear close-up shot without Nikolas in the picture at all. Significantly, it is not an extreme close-up, which is being saved for Nikolas, the focal point of this conversation. As he begins to contemplate Elizabeth's characterization of his relationship with Rebecca, however, Nikolas moves from being framed in a medium shot to a close-up shot and, finally, an extreme close-up shot.

The differences in framing between the medium shot, the close-up, and the extreme close-up are subtle but telling. The camera maintains its distance from Nikolas as Elizabeth tells him a relationship with Rebecca is a bad idea. He, in fact, seems partially to agree. When she personally attacks Rebecca, he gets defensive ("How are you so sure about that?"), and the camera moves in for a close-up. When she tells him that he needs to let Emily go, he becomes silent and introspective, and the camera moves in for the extreme close-up. He knows that Elizabeth is right about his still missing Emily, though she is wrong about his feelings for Rebecca, and he needs to let Emily go if he is going to have a relationship with Rebecca. The extreme close-up allows us to see the subtle interplay of those two emotions on Nikolas's face: the sadness he still feels from his loss of Emily, and his love for her sister. The timing of the shots is significant as well. Camera switching is frequent (in an accelerated pacing) toward the end of the scene to create a kind of tension we would not sense from the long-take camera style of 1963.

The relatively fast pacing of camera switching in a modern soap opera is reflected not only by this scene but also by the scenes of intimate dialogue that precede and follow it. In the preceding scene, which lasts one minute and twenty-five seconds of screen time, there are nineteen shots for an average of 4.7 seconds per shot. In the scene that follows, which lasts a minute and six seconds on the screen, there are fifteen shots, for an average of 4.4 seconds per shot. In the Elizabeth/Nikolas scene described above, there are ten shots over a fifty-second sequence, for an average of five seconds per shot. Not only is the narrative effect of the camera-switching pattern similar in all three of these scenes, with emotion constructed in reaction shots and cross-cutting patterns in the dialogue, but the average length of a shot in these sequences is relatively consistent amongst the scenes: about five seconds per shot.[8]

In contrast, the camera-switching patterns of dialogue and intimacy in *GH*'s first episode are quite different. Intimacy scenes are held mostly within one camera shot, and cuts usually signify action (e.g., Dr. Hardy or the nurse coming in and out of the room, or Angie getting up or out of the bed). A brief tantrum sequence in which Angie throws her food to the floor and Dr. Hardy and the nurse come running to the room, includes three shots that are each no longer than two seconds in length, followed by a forty-three-second shot in which Dr. Hardy lectures Angie about her behavior. Even with the tantrum sequence, the total average for the 1963 hospital room scenes, which last seven minutes and thirty-six seconds and are all scenes in which intimate dialogue and confrontation take place, is 15.2 seconds per shot. This is three times the average length of shots in the 2009 episode for similar scenes, a dramatic difference.

The rhetoric of the camera in 2009 on *GH* reflects the fact that electronic editing is now much easier than it was in 1963, and the director's ability to switch between cameras and shots plays a much greater role as a primary vehicle for emotion in nearly every scene. The camera helps tell the story, as it seems to be tunneling into Nikolas' mind. Significantly, this is a scene in which Nikolas is aware of his true feelings—sadness at the loss of Emily, but love for her sister Rebecca—and the internal conflict that his feelings create. In the

1963 episode, camera shots are generally standardized in long-take moves that circle characters' emotions. The actors and their actions are the primary vehicles of emotion, e.g., in the meaningful pauses and movements of Dr. Hardy in and out of Angie's room. The one time an overt camera expression of emotion is used in 1963—a speedy zoom-in on Angie's face—it is an extremely self-conscious shot that calls attention to itself. In 2009, camera shots seem much smoother (here, fluid-head tripods and the hydraulics of modern pedestal cameras certainly play a role) and more adept at conveying emotion. In a typical episode, we see some of the same basic elements used in shots: the close-up, the eye-level camera angle, the use of Z-axis alignment, the capture of stylized expressions at the end, etc. Yet the composition and pacing of each scene have shifted into patterns of camera switching, and scenes rely on extreme close-up shots to go "into" the face and reveal the characters' deepest thoughts and emotions.

This shot-by-shot formal analysis makes a larger point. In order to create drama, the work of conveying emotion in the story is increasingly in the hands of those who tape and edit the scene. While the actors and an elaborate *TZ*-inspired mise-en-scène create a dramatic space where viewers can come to understand the desperate Angie in 1963, camera-switching techniques and an extreme close-up shot on Nikolas are more prominently used to enhance the power of the character's emotions in 2009, and the crisp images of digital cameras emphasize inner rather than atmospheric emotion.[9]

CONCLUSION

It is evident from this brief comparison that the production values and camera technology of soaps have evolved: for example, in the way that *GH* uses high-definition cameras to deliver high-quality visuals in 2009. However, it is no great secret that soap operas are being forced to cut back on costs due to dwindling audiences and profits. Part of the response, at least at *GH*, seems to be greater reliance on intimate sets and camera work to create tension and drama on the screen. This reliance on technology is exhibited as well in the online component of the show. *GH* routinely uploads behind-the-scenes vignettes and hosts question-and-answer sessions with actors on their Web site. Other essays in this volume document this effort to use technology to reengage audiences. It is clear that technology is a tool with which soap opera producers hope to reignite audience participation, make soap operas more relevant to contemporary audiences (especially young audiences), and rescue their productions from cancellation.

Despite these technological changes and dislocations in how emotions, confrontations, and conflicts are portrayed on the screen in soap opera today, it is surprising how little has changed in the semiotics of the camera. The basic soap opera camera conventions described by Timberg (1981) remain largely intact. Though there have been instructive and significant shifts in some patterns of production—in the acceleration of switching shots to create an emotional nexus between characters, or in a new set to expand the arenas of action, for instance—many of the techniques we examined from the 1963 scene still dominate the semiotics of the camera in *GH* in 2009.[10]

There is a sense of intimacy, history, and continuity[11] not just in the narrative outlines and verbal texts of soap opera, which establish certain aural/verbal codes that allow listeners and viewers to follow stories even when they look away from the screen or leave the room. We also find these same principles in soaps' visual codes. When we look closely and follow the texture of a soap opera's camera codes and switching patterns and analyze a soap opera scene-by-scene and shot-by-shot, as we have done with *GH*, we find forms of intimacy, history, and continuity in soap opera production techniques and visual codes that span 45 years of the show's history, from its pilot to its most recent episodes. Yet these central, highly identifiable features of soap opera's visual codes and semiotics have been either taken for granted or neglected by scholars of soap opera. We hope to have shown here how and why these visual forms are as important to the long life and staying power of soap opera storytelling as the verbal, literary, and sociological features that have been more frequently discussed and examined.

NOTES

1. For more, see Green's essay in this collection.
2. A bookend comparison of two episodes of a genre of television has been instructive in other instances. For example, the first newscast of veteran anchorman Walter Cronkite for CBS in 1962, and his last in 1981, have been packaged in the television viewing archive of the Paley Center for the Media in New York City. A viewer of the two shows side-by-side can see immediately the differences in the format and structure of the *CBS Evening News* that took place over nineteen years.
3. Green's essay in this collection features an interview with Brinckerhoff.
4. In total running time, the excerpts from the 1963 episode represent only eight minutes and twenty-two seconds of screen time in the 1993 anniversary show, or a little more than a fifth of the show. This total does not count the stills from the 1963 episode that appear in the show's closing montage, but it does include the 1963 episode's opening sequence, the scenes from Angie's room, and a scene with Dr. Hardy talking to the night nurse as he leaves the hospital.
5. The 1963 scenes from *GH* discussed in this essay could be found on YouTube at http://www.youtube.com/watch?v=UJRqkTA5ni4 at the time of this writing.

6. A more detailed discussion of the concept of the "thickening moment" appears in Timberg (1981).

7. The 2009 scene could be found on YouTube at http://www.youtube.com/watch?v=3Wh3MtK40_A at the time of this writing.

8. The fourth scene represented in the YouTube clip from the June 17, 2009, episode of *GH* was quite different. It was an action scene shot with longer takes and two-shots of the two characters who were breaking into the office of the mayor of Port Charles. Here, seven shots were used in a sequence lasting one minute and twenty-four seconds, for an average of twelve seconds per shot. Future work to analyze differences in camera and production technique empirically should take different kinds of scenes into account.

9. We are not claiming here that atmosphere does not play a prominent role in soap operas today, but rather that those functions may not be used as prominently to display emotions, when compared to the increasing use of camera work and editing in doing so.

10. The research we have completed here is suggestive and theoretical but could, and we hope will, be followed by empirical and/or historical studies to further assess these claims.

11. The phrase "history, continuity, and intimacy" is the classic formulation of what makes television serial drama distinctive for Newcomb (1974).

IT'S NOT ALL TALK
EDITING AND STORYTELLING IN
AS THE WORLD TURNS

—DEBORAH L. JARAMILLO

> Deborah L. Jaramillo is an assistant professor in the Department of Film and Television at Boston University. Her research focuses on television as a complicated collocation of culture, aesthetics, commerce, and politics. Her soap opera viewing began in elementary school with *The Young and the Restless*, and she has been a loyal third-generation viewer of *As the World Turns* on CBS since her senior year of high school.

INTRODUCTION

On March 1, 2007, two storylines collided on CBS' *As the World Turns* (*ATWT*) to yield a critical representation of romantic and sexual relationships. By exploring only two storylines (instead of the usual three or more), *ATWT* established uncomfortable continuities between a deceptively benign courtship and an overtly dangerous encounter. Just as significant as the textual implications of the episode is the way in which the editing of these two stories departs from the genre's standard style and pace of intercutting—a tradition that differs starkly from episodic primetime dramas that unfold quickly and end with a sense of closure.

Allen (2004) writes that the linchpin of soap opera narratives is the complex set of relationships among characters. Linear trajectories and quick resolutions are less important, he argues, than the ways unfolding events impact character upon character (Allen 2004, 245). Overlapping storylines bind families, friends, and lovers together. In order to ensure cohesive and multilayered narratives, editing is paramount. The interaction of form and content in this particular episode of *ATWT* demonstrates how stylistic interventions can disrupt the routine of soap opera storytelling to yield results that speak to artistic as well as industrial concerns.

CBS' announcement in December 2009 that *ATWT* would conclude its fifty-four year run in 2010 confirmed the expendability of soaps despite the programs' vigorous attempts to attract viewers. Low ratings and the resulting decrease in ad revenue—troubles plaguing even the longtime ratings leader, CBS' *The Young and the Restless*—have prompted a number of stylistic innovations, such as location shooting and hand-held camerawork, to lure and maintain women aged eighteen to forty-nine as viewers (Brodesser-Akner 2008). The genre's underperformance has also affected soap opera casts: The characters of older cast members are killed off, while new hires keep the casts "artificially young" (Brodesser-Akner 2008).[1] On one hand, this particular episode's departure from the norm encapsulates the flexibility and reflexivity of the soap opera form; on the other, it indicates a pragmatic attempt to court a younger demographic disenchanted with (or simply unfamiliar with) the traditional leisurely pace of the genre. In keeping with this mixture of pragmatism and experimentation, the editing decisions in this episode comment on the stories themselves while accelerating the passage of time, thus challenging the paradigm that Allen emphasizes.

SEASON 51, EPISODE #12965

As *ATWT* neared the end of its fifty-first season, Emily Stewart asserted control over her life by becoming a high-class prostitute. Her penalty for this latest misstep was sexual assault.

The attack on Emily at the Lakeview Hotel and the aftermath at her mother's house occur as another heterosexual encounter unfolds across town at Al's Diner. In this second, playful storyline, Craig (a local villain—most of the time) attempts to woo Meg (the nurse-turned-waitress) by following her around the diner as she tries to work. These storylines contain nothing atypical within the logic of *ATWT*, but the way in which they progress alongside each other—at times even colliding—makes the episode particularly relevant to the issues of form and the construction of meaning.

Incorporating twenty-seven scenes, five locations, and six principal characters, Episode #12965 begins at Al's Diner, where Craig sees Meg toiling away and asks her to accept his offer to move into his mansion. Scene two takes us to the Lakeview Lounge, where Dusty (Emily's friend and business partner) remembers her clandestine meeting with Eliot Gerard earlier that evening. In scene three at Eliot's hotel room, Emily assures Eliot, her client, that Dusty will not interfere with their tryst. In the lounge below, Lucinda Walsh tells Dusty a harrowing story of an alleged rape in Eliot's past. Before the opening credits and first commercial break, we return to Eliot's hotel room for the beginning of the assault. In this opening, the episode has introduced us to our twin foci: Emily and Meg. After the commercial break, skillful editing connects the disparate stories through violence.

Back at the diner, Craig accuses Meg of "punishing" herself for losing her nurses' license, and he begs her to stay with him. Knowing Craig's history of criminal acts and emotional manipulation, Meg argues that she is justifiably wary of an intimate connection with him. Back in Eliot's hotel room, Emily experiences the danger that Meg is trying to avoid. We cut from Eliot's sadistic abuse of Emily to the lounge, where Dusty continues to learn of Eliot's past. Before the second commercial break, we return to the hotel room. Eliot and Emily battle each other violently, the former believing that Emily's resistance is foreplay, and the latter processing the events as a repeat of her rape ten years earlier.

The third segment of the episode focuses entirely on Emily, with elliptical editing that quickens the pace. Almost caught by a security guard, Eliot releases Emily from the room, claiming that she is "not worth the hassle." She runs out, and we immediately cut to her room in her mother's house. With simple piano music in the background, Susan (her mother) blames Emily for inviting the assault. Susan then questions Emily's ability to "control" herself and her circumstances—a key inquiry that pinpoints the source of Emily's angst.

Scenes fourteen and fifteen—interrupted by commercials—take place in Emily's room, where Dusty has arrived to comfort her and determine the cause of the assault. Without revealing her new trade, Emily confesses cryptically that her recent behavior protects her from further emotional trauma. The legato piano, oboe, and violin piece ends abruptly, as we cut to Al's Diner. This scene is played for laughs on its own, but, because it follows the aftermath of Emily's assault, the lighthearted music and gaiety barely conceal a sense of foreboding. After Craig buys out the diner for a "private party"—a phrase that recalls Eliot and Emily's appointment in the hotel room—Meg hits Craig on his arm for interfering with her work. Feigning anger, Craig picks up a ketchup bottle and says in a mock-threatening tone, "Don't make me use this." Meg replies, "You wouldn't dare." Though superficially playful, the exchange flirts with the undercurrents of violence endemic to the episode and likewise to any interaction with Craig.

After a commercial break, we return to Al's Diner and the ongoing negotiation between Craig and Meg. Craig hands control over to Meg when he clarifies that any ties between them will be her choice. In return for the transfer of control, Craig asks Meg to dance with him. The figurative and literal dance between Meg and Craig launches a montage that solidifies the connections between the two stories. With the first cut, we see Emily break down in front of Dusty. When Dusty tells Emily that she deserves better than her situation, Emily screams that she has gotten exactly what she deserves. The second cut reveals Meg and Craig dancing for about thirty seconds, and the third cut shows Dusty holding Emily tenderly for twenty seconds. These cuts are followed by dissolves to the dance at the diner and then to a shot of Dusty watching Emily as she sleeps. The pop song that Meg and Craig dance to in the diner plays continuously through the montage, even over the silence of Emily's room. The song's lyrics—"Oh, he's dangerous"—sonically link Craig to Eliot and even to Dusty, whose past is checkered as well.

The final segment pushes the episode's events to a resolution of sorts by opening up on a new day. We begin in Emily's room as she holds the money that Eliot paid her; we then cut to a giddy Meg and Craig discussing their nightlong dance. Still pleading, Craig insists that Meg would not "owe" him anything by accepting his offer. The notion of Meg owing Craig instantly recalls Emily's debt to Eliot for services not rendered. Meg's last line of the episode clarifies her connection to (and distance from) Emily: "If we sleep together again [. . .] I want my life to be under control." The segment concludes with Emily and Dusty confirming their friendship, but, when Dusty leaves, Emily stores her cash with the rest of her earnings. As the episode draws to a close, we can assume that Emily may still equate control with the cold, neutral price of her body, and that Craig will derail any semblance of control that Meg might acquire before they begin their relationship.

CONCLUSIONS

Although it may be an anomaly, this episode of *ATWT* complicates Allen's (2004) primary argument in a compelling way. The two storylines remained separate, and each wrapped up with some semblance of closure. The Meg-Craig relationship and the Dusty-Emily connection would continue, of course, but this episode feels very much like a stand-alone event in which the two casts work parallel to one another and never intersect as they are supposed to in this universe of intimately and almost incestuously related characters. By isolating these stories from each other and from the rest of Oakdale's residents, the episode invites the audience to consider the allegory apart from the characters themselves. Beyond this episode, the characters regain their more traditional significance; however, within this sealed moment, *ATWT* resists its generic trappings in critical fashion.

As ephemeral as even the most experimental of soap opera episodes may seem, one unforgettable remnant of this episode is the larger, almost auteur-driven rumination on the complex and often destructive relationships that drive the genre. An episode like this one demonstrates the ease with which the genre can adjust its form to reflect on the stories it tells and the manner in which it tells them. This is not to say that soap operas are wholeheartedly rejecting the very structure that made them an institution. Allen's thesis, while subject to challenges from anomalous episodes like the one above, is not in danger of obsolescence. Genres are both static and dynamic; they retain certain conventions while transforming others. The uniqueness of episode #12965 depends primarily on the strength of the conventions that remain firmly in place. And, if we step back farther and consider that time was indeed running out for *ATWT*, then we can see how industrial circumstances intermingle with creative ones to produce unexpected scenes and episodes within the established order of genre. In the case of this episode, the editing operates as a kind of shorthand—a shortcut, even—that encourages critical viewing and attempts to attract new viewers without all the talk.

NOTE

1. For more on these phenomena, see contributions from Bibel, Erwin, and Green in this collection, among others.

GUIDING LIGHT
RELEVANCE AND RENEWAL IN A CHANGING GENRE

—PATRICK ERWIN

> Patrick Erwin is a freelance writer and journalist and author of the soap opera blog *A Thousand Other Worlds* (http://1000worlds.wordpress.com/). He has also written about soaps for the Marlena De Lacroix site (http://www.marlenadelacroix.com/). While he watches a variety of daytime serial dramas airing today, he was a lifelong viewer of *Guiding Light* and *As the World Turns*.

U.S. daytime soaps have endured many transitions. As the longest running soap in the genre's history, CBS' *Guiding Light* (*GL*) survived the move from radio to television and witnessed every shift the genre has experienced. In the last decade, daytime has undergone significant changes and budget reductions. These changes had a noticeable impact on *GL*. When the program was renewed in 2005, several actors were taken off-contract and placed on recurring status, paid per appearance. Some veteran actors left the show after being taken off-contract, including Jerry verDorn, who had won multiple Daytime Emmys for his twenty-six year portrayal of Ross Marler. The remaining actors on contract were asked to take a 15 percent pay cut.

GL experienced another transition in 2008 when the show migrated to a dramatically different production model. Because my online writing on soap operas often focused on *GL* and its sister show, CBS's *As The World Turns* (*ATWT*), I was invited (along with four other bloggers)[1] to observe *GL* and speak to actors, writers, and producers during a December 2008 taping.

SIGNIFICANT CHANGES

The new production style began airing on February 29, 2008. As part of this new model, producers TeleNext (funded by Procter & Gamble) started taping the show entirely with digital cameras, made substantial changes to studio sets, and began featuring several scenes a week taped on location. The producers also modified the storytelling approach to match the on-screen changes.

CAMERAS

Prior to the production model change, *GL* had used the traditional stationary three-camera format. Cameras were placed on "runways" and aimed at three-wall sets. They captured the sets and actors just as an audience member would view a theater performance, by facing the proscenium framework of the stage. These cameras were large and had to be mounted on wheels for mobility. Though the camera range was substantial, the agility of the camera itself was limited. In many ways, this method had not changed since the 1950s.

The new model utilized digital cameras for all taping. ABC's *All My Children* began using digital cameras for some of its scenes, but *GL* was the first show to adopt this format exclusively. These handheld cameras can be used in a wider range of settings, and allow for indoor and outdoor taping, as well as taping in spaces that may have been too challenging or too small for a traditional stationary camera. This flexibility was the catalyst for embracing the other major components of the new production model as well.

INDOOR SETS

For decades, *GL*'s primary studios had been in midtown Manhattan. Instead of larger temporary sets, the new approach allowed for smaller permanent sets to be built. According to executive producer Ellen Wheeler (2008), "Before we went to this model, we could play eight sets in a week. We've moved from eight sets a week to what is in our studio space [about forty-five sets]. If you count the sets that are actually built in our studio, the offices, and the other locations around the building that we shot at on a regular basis, we have another forty sort of regular locations here." To maximize the number of available settings, many of the *GL* production offices had a "double life." Writer Jill Lorie Hurst's office doubled as a motel room; Wheeler's office also served as a chapel. Wheeler noted, "We shouldn't be spending money on storing 250 sets in a warehouse when I could be spending that money on the screen."

OUTDOOR SETS

The most visually dramatic change was the migration of a significant number of scenes to an outdoor setting. Traditionally, soaps had taped outdoors only to underscore the culmination of a big storyline; those scenes are generally limited in scope and play out on air for only a few days or weeks. In the mid-1990s, some soaps developed recurring outdoor locations. ABC's *The City* took advantage of a number of New York City locales for outdoor shooting, and NBC's *Days of Our Lives* built an outdoor mall set. The outdoor taping for

GL was notable in its approach—the show essentially set up a second studio in a small New Jersey town—and its scope—approximately 40 percent of the show was taped there.

Wheeler and location producer Lou Geraci looked at several settings before settling on Peapack, New Jersey, located less than fifty miles west of New York City. The production usually coordinated taping all exterior scenes on a single day of the week. *GL* made arrangements with many of the town's businesses and institutions to serve as exteriors in taping. Thus, the local cemetery became the cemetery where Tammy Winslow Randall is buried, the exterior of an ornate stone house doubled as the patio of the grand Spaulding mansion, and the porch of the Gladstone Tavern stood in for the porch attached to Company, Springfield's eatery. TeleNext also leased a house in Peapack that the production staff refers to as the "show house." The house included a kitchen/break room, a makeup/wardrobe area, and several other indoor sets.

STORYTELLING

As the changes in production were being implemented, Wheeler and the writing team also modified storytelling practices. On air, scenes were more streamlined and less complicated. Co-head writer Jill Lorie Hurst (2008) confirmed this observation during my set visit: "It's not that no one will ever have a long speech or that there won't be a big theatrical, dramatic exchange. But, for the most part, we do try to keep it simpler [. . .] It works with the whole look of the show, and for editing purposes." With a trimmed-down cast and enormous visual changes, many of the scenes became "vignettes," short scenes with one or two characters. Rather than move a long-range story arc forward, these scenes were more a slice-of-life. Characters were seen in their kitchens or driving their cars, generally living their lives. Although they shaded the experiences and feelings of a particular character, these scenes seldom fed into a traditional long-range story arc.

TRANSITION TO NEW MODEL

MOTIVATION

For years, *GL* was the lowest rated U.S. daytime drama among the eighteen- to forty-nine-year-old female demographic. Further, after NBC moved *Passions* to DirecTV in 2007 and then cancelled it altogether in 2008, *GL* took on the role of being the consistently lowest-rated soap opera overall. With a downward trend in viewership and thus increasingly smaller licensing fees paid to TeleNext for the show, the production renovation was largely driven by financial necessity. The changes in sets and cameras not only led to the visible on-screen changes mentioned earlier, they also meant a smaller crew could produce the show. Press reports suggested that, in late 2007, approximately fifty members of the show's staff—mostly on the production team—lost their jobs. Wheeler said, "There are people who no longer work for us because this change means we don't have those job functions any more. They don't exist in the new model" (Levinsky 2008). For example, the portable monitors of the new digital cameras led *GL* to terminate its studio-based control room, thus eliminating a number of processes as well as the employees who performed them. After significant budget cuts, staff reductions and the implementation of the new model, *GL* appeared to be more profitable. Manager and agent Michael Bruno indicated that *GL* was, "saving an enormous amount of money. They're *making* money, and believe me, that's what gets looked at" (Levinsky 2008).

When the new model was unveiled, Wheeler attributed the changes to budgetary concerns but also expressed a desire to reexamine old production methods and implement more efficient ones. Thus, a secondary motivation (and the one *GL* and TeleNext strongly focused on in the press) was the need to update and/or bypass some soap opera conventions. In the old model, characters seen at the start of a show could be reasonably expected to appear on the same set and within the same camera range—and visual palette—for the duration of those scenes. With the new model, producers and directors had flexibility in how a scene could be shot. For outdoor shoots, sequential scenes were often positioned at different angles within the same set. A scene might begin with actors inside a gazebo. The following scene might show them at an adjacent bench, or walking on the grounds around the landmark viewed in the scenes before.

PRODUCTION ISSUES

Since soaps are in production continuously, there was no opportunity for substantial testing of the new model. As a result, several problems with the transition had to be resolved in "real time" on air. The most common issue that plagued episodes of the show initially was related to audio quality, a problem that became evident when the show was in post-production or actually airing. Several scenes taped outdoors during adverse weather conditions might have looked more realistic, but the sound of rain, for example, muffled the actor's dialogue and made it impossible to understand. During our visit, Wheeler said that the show had tried lapel "lavalier microphones" (lav mics). When lavs were attached to actors, they rustled against clothing and picked up ambient noise that did not translate well on-screen. The show eventually switched to primarily using boom mics for both indoor and outdoor taping. Wheeler (2008) noted:

If I have to change lavs or the batteries go down, it's tough. We tried that, and it creates more problems than we can deal with and get this much shooting done in a day. While we have to deal with exterior sounds, the clothes rustling on the lavs is difficult. We don't go back and do another take very often. If there is an airplane overhead, we can EQ out the airplane, but we can't EQ out the clothes hitting the mic. ["EQ", or equalization, allows sound technicians to remove or minimize a background sound while retaining the main sound recording.] So, it's mostly a logistical thing.

The change in cameras from stationary to digital also led to some technical issues. Earlier episodes had more of the so-called "shakycam" feel, where the camera's perspective or zoom changed dramatically from one second to the next. The camera crew, who had been learning to tape with new, non-stationary cameras, minimized these issues soon after they became apparent, though they still occurred in later episodes.[2]

One major issue for the launch of the new production model was beyond the show's control: the Writers Guild of America strike. Most daytime soaps write scripts between two and three months in advance and tape shows from four to eight weeks in advance. The strike ran from November 5, 2007, to February 12, 2008. As a result, many of the initial shows that were taped using the new model were produced with scripts that were created by non-staff writers. Scripts written during strikes generally "tread water" and attempt to "hold" the story until staff writers resume their roles. Therefore, at a time where writing was key to support the new model, scripts with little or no forward plot movement aired instead.[3]

REACTIONS TO CHANGES

STAFF REACTIONS

GL actors were, for the most part, publicly supportive of the changes. Several actors commented on the benefits of taping outdoors in Peapack. Elizabeth Keifer (Blake Marler) explained, "I enjoy it. I love the outdoor stuff, and think it looks fantastic. You're no longer confined to a set; you have more sensory stuff to deal with. It's always been a little hurried and crazed, no rehearsal, but this is even more like diving off the deep end. It's always been 'Ready, set, go!' And now it's, 'Ready, set, go . . . . and it's gonna rain!'" (Erwin 2009). But reports surfaced in the press that other actors were not as supportive of the new model. *Soap Opera Weekly* (*SOW*) featured a summer 2008 interview from an unidentified *GL* performer who described a number of negative aspects about the experience, including having to make a wardrobe change in the car en route to Peapack (Hinsey 2008). In July 2009, veteran performer

Maureen Garrett (Holly Norris) returned to the show and was overwhelmingly negative about the new production model (Torchin 2009):

> You do not see the other actors. There are no rehearsals, no monitors on which to watch the action. Actors are led from hair and makeup to a kind of holding pen. Then they're guided through the maze of pieces of sets to their spot. [...] There's no director, no time, no spontaneity. If this is what has to be done to save the form, I think there's room for debate about trying to preserve the process, too. You can't really create connections or foster "chemistry" without the work.

VIEWER REACTIONS

Initial viewer reaction was overwhelmingly negative for a variety of reasons:

- The aforementioned issues with audio and taping.
- The new sets represented an enormous change in how viewers had envisioned familiar places (Company, the Spaulding mansion). Because financial resources were so limited, sets were furnished with materials and props that were inexpensive. Such a quick change in the visual palette was jarring to viewers and interrupted the narrative universe that many viewers were accustomed to.
- Smaller sets and cameras meant that taping captured the actors at a much closer vantage point (and in different proportions) than before. Viewers saw actors in a realistic way—imperfections, pores, and all.
- The new production model received significant press, but, instead of using an attention-getting story point, the first days and weeks featured lengthy, expositional scenes. One such scene showed the character Jonathan wordlessly driving a car. Although it's understandable that the show wanted to underscore its new reality in a vivid way, many scenes did so without any follow-up narrative. As a result, many fans felt *GL* was focusing on the new production model at the expense of the actual story.

Comments in the soap press (particularly *SOW* and Soap *Opera Digest*) and on soap message boards ran heavily negative for the first few months. Even after viewers became accustomed to the new model, the reaction to the "new look" remained mixed.

AFTERMATH AND CONCLUSION

After several months of production, many of the issues cited above were minimized or resolved. Eventually, *GL* also revisited its storytelling approach, making significant thematic changes in late 2008 to return to a more traditional narrative instead of the experimental focus on "vignettes." The most high profile moves included the reintroduction of legacy character Phillip Spaulding (portrayed by Grant Aleksander), as well as characters Olivia Spencer and Natalia Rivera exploring the possibility of a same-sex relationship.

The new production model clearly made *GL* more cost-effective. However, despite these financial and artistic gains, CBS decided against including *GL* in its future programming schedule and announced its cancellation on April 1, 2009. The radical changes to the production model, coupled with the weak ratings performance, led to the ultimate decision that *GL* was no longer financially viable. An attempt was made to find a new network for *GL*, but, on July 24, 2009, TeleNext confirmed no new outlet had materialized (Newcomb 2009a). The final episode aired on September 18, 2009. It was followed in December 2009 by CBS's announcement that sister show *As The World Turns* was also being cancelled. TeleNext has continued to discuss potential new outlets for *GL* and *ATWT* (which last aired on CBS in September 2010) but ultimately was unable to find new homes for the shows.

Many changes, both onscreen and behind-the-scenes, were implemented at *GL* in a short period of time. Ellen Wheeler and her production team made an enormous investment of time and energy in the new model to continue *GL*. Ultimately, *GL* was unable to absorb those changes and survive an intense metamorphosis. Yet the new model represents an important milestone. As with radio and the early days of television, *GL* was at the forefront of new technology. Though no other daytime show has adapted *GL*'s model in full, shrinking audience shares and budgets have encouraged other soaps to take a second look at it. *ATWT* and ABC's *One Life to Live* more consistently incorporated outdoor scenes following *GL*'s new model, while CBS's *The Bold and The Beautiful* (*B&B*) showed interest in the new model. Actor and director Susan Flannery (Stephanie Forrester) visited the *GL* set and directed an episode of the show, and *B&B* subsequently incorporated an outdoor rooftop set. Though the *Light* is now dim, its innovative approach may enable other shows to remain relevant and profitable in a new business environment.

NOTES

1. These bloggers included Sara Bibel and Roger Newcomb, both of whom have an essay elsewhere in this collection.

2. Timberg and Alba provide a more detailed account of camera work in soaps in this book.

3. For a more detailed examination of the strike and its impact, see Metzler's essay in this collection.

THE EVOLUTION OF THE PRODUCTION PROCESS OF SOAP OPERAS TODAY

—ERICK YATES GREEN

> Erick Yates Green is an assistant professor of media production in the School of Communication at East Carolina University and a director and cinematographer. He has fifteen years of production experience on feature films, commercials, documentaries, and network television dramatic series. He was a proud follower of *General Hospital* during the "Luke and Laura" era.

During the early weeks of April 2009, ABC announced that its venerable soap opera *General Hospital* (*GH*) would launch high-definition (HD) broadcast.[1] At the same time, CBS, in a once unthinkable action, killed off *Guiding Light* (*GL*), television's longest-running soap opera, even after *GL* had introduced technological innovations other shows have since considered emulating in an attempt to stay alive. These two developments underscore the impact changing technologies have had and will have on the soap genre at a time when viewers—and budgets—are in a continual decline. While soap opera scholars have made significant contributions to an understanding of the genre's audiences, texts, and production, very little work has explicitly examined the impact of changing technologies on the soap opera form. This essay will focus on recent technological advancements through the lens of modern soap opera directors who are themselves striving to keep abreast of these new developments.

Technological advancements have become a double-edged sword for the soaps industry. There are vast image quality improvements: from standard-definition (SD) to HD images; diminished costs for higher speed HD cameras; less need for high output lighting gear; and more seamless and affordable post-production workflows. But there are dramatically more competitive distribution systems as well. Longtime soap opera dominance over daytime network television audience share is no longer a reality.

The growth of TV outlets beyond three major networks (CBS, ABC, and NBC) to a myriad of distribution outlets—and the success with which these outlets have taken some viewers away from the TV set altogether—has raised the question of how new production conditions and formats may not just change soap opera as a form, but doom its existence altogether. This is a reality of which many practicing soap opera directors and performers are well aware. It doesn't surprise Emmy-award-winning soap director Peter Brinckerhoff (2009): "Broadcast television and the business that we know of that world are perhaps going away." Take, for instance, NBC: Once the home of several daytime dramas, the network today only has the long-running *Days of Our Lives* (*DOOL*), which has endured major cutbacks of its own.[2]

Yet, instead of the traditions of network TV distribution holding firm, or the advent of expanding Internet-based outlets suddenly dominating distribution, the future, Brinckerhoff (2009) believes, lies in the middle—in the world of pay cable: "I just don't see there being enough money in the Web version of individual soap shows to put together a project like [*GH*] or [*DOOL*] on an ongoing basis. Cable will have a defined viewership where they will get their money up front from their subscriptions, and the producers will get a percentage of that to produce a particular show."

Attempts to broaden the appeal of this once-dominant network television genre had been the recent focus of actress-turned-executive-producer Ellen Wheeler, who inspired many innovations in the cancelled *GL*. Purportedly with hopes of garnering a younger audience, her recent innovations, much heralded by CBS in publicity releases in February 2008, included attempts to modernize the "look" of the show with the adaptation of reality TV-like use of hand-held cameras. These innovations failed to regain a consistent audience and ultimately failed to save the program.[3] It is unclear whether and how these kinds of production changes, though improving the bottom line, will affect the survival of the form.

Brinckerhoff started examining the possibilities of new production techniques and distribution outlets when directing *General Hospital: Night Shift*, the spin-off of *GH* that played on SOAPnet.[4] His early visual design for the show included a hand-held approach to camera operating made possible by the new, lighter weight, three-chip HD cameras. Although the network rejected the idea at the time because of its fears that such an approach would be too visually jarring, Brinckerhoff foresees in the long term a soap opera production process that, for the vast majority of soaps (and especially those shows most challenged financially), will necessarily consist of more modest crew sizes and simpler production values.

The consensus of professionals seems to be that change will come. How quickly and how profoundly the new advancements in equipment will impact the genre is yet to be revealed. *DOOL* producer and director Albert Alarr (2009) also said an eventual adaptation is inevitable, though the investment in new equipment may simultaneously discourage change: "We still use big traditional [electronic news-gathering] cameras, and there is little evidence on other shows that the small digital cameras have had much of an impact."

Brinckerhoff (2009) also had a wait-and-see attitude about technology, but he was already concerned about the size of the image that many equipment manufacturers are marketing as a full media experience: "I think that, the smaller the screen is, the more distance you have created between the product and the viewer. God bless iPods, and I am a huge music fan, but why people would want to watch well produced shows on an iPod is absolutely beyond me." The aesthetic concerns expressed by producers like Brinkerhoff have done little to stem the on-going shrinkage of the viewing screen—witness the increasing numbers of viewers who watch television programming on computers (Colker 2009), portable media players, cell phones (Reardon 2009), and even the still-developing wrist watch-fitted 1.5" LCD TV screens.

Like their production cousins in feature film production, TV soap opera producers also have to solve the challenge of changing aspect ratios, as their world of viewers is still transitioning from the Academy Aperture four-to-three TV sets to the new HD standard of sixteen-to-nine. Aspect ratios refer to the length of the horizontal dimension of the viewing screen relative to the vertical dimension. As the producers of the original HD broadcast of CBS's *The Young and the Restless* were well aware in June 2001 (Kaufman 2001, 8), the transition from one aspect ratio to another does affect the compositional frame that the audience is able to view. The airing of an April 2009 episode during the first week of full HD broadcast of *GH* reveals that this change in screen shape issue remains a challenge. It was clear, when viewed upon on an Academy four-to-three aspect ratio monitor, that the camera operators taping the episode chose to favor the wider HD sixteen-to-nine aspect ratio composition. Especially on over-the-shoulder shots and medium-wide two-shots, the foreground characters appeared cramped up very tightly against the right and left frame lines or virtually left out of the frame altogether—save for a nose or a hand just barely visible within the four-to-three frame.

Technical and production issues will undoubtedly affect the struggle of soap operas to stay on the air in the wake of advances in technology, shrinking budgets, and the unprecedented American economic crisis that has been profoundly felt by every producer in the industry. Alarr (2009) said: "The ratings have dropped dramatically for everybody. Most daytime today is pared down, and we are all returning to that earlier version of ourselves because producers don't want to go late and do the big shows with lots of characters."

The *DOOL* director/producer finds this downsizing of budgets—albeit a financially painful process for cast and crew alike—something that might force a return to the roots of the genre. As Alarr (2009) says, "There are certain shows, and [GH] is an example, where the audience will expect high-concept episodes. Yet, I still contend that even these audiences seem most interested in the simple beautiful moments where one character turns to another and says, 'I love you.' I think this is true especially in soap operas."

As in all programming for traditional network television, soap opera production and distribution will be a work-in-progress for some time to come. Though loyal soap opera audiences today can rejoice in the staying power of this time-tested genre, it might well be that determining where their show is playing and in what format will involve a technological and socioeconomic plot as complex as that of the soap opera itself.

NOTES

1. The other two ABC soaps, *All My Children* and *One Life to Live*, announced plans to switch to HD a few months later.
2. See Bibel's essay in this book for more on the current state of the soap opera industry.
3. See Erwin's essay in this collection for more about *GL*'s many innovations in its final days on CBS.
4. For more on this series, see Gonzales's essay in this collection.

FROM DAYTIME TO *NIGHT SHIFT*

EXAMINING THE ABC DAYTIME/SOAPNET PRIMETIME SPIN-OFF EXPERIMENT

—RACQUEL GONZALES

> Racquel Gonzales is a media scholar and graduate student in the University of Texas Radio-Television-Film program. Her research interests in serial television include reception studies of online and offline fan communities and industry history. Her relationship with soap operas began in 1989 with *Santa Barbara* and her lifelong viewing of *General Hospital,* and she has followed more than a dozen soaps at one time or another.

INTRODUCTION

Historically, due to its ability to inspire loyal, long-term viewership and to target a definable audience (Allen 1985, 47), the soap opera has been one of the most stable, dependable, and predictable television genres in the U.S. These characteristics, however, are now open to question, as the formerly stable genre has experienced falling ratings since the 1990s and continues to hit new lows (Janofsky 2007; Levine 2001, 69–70). One crucial contribution to soaps' state in the digital age is the growing disconnect between industry priorities and the desires of the soap audience. A key component for profitable success in television is the ability to identify and retain viewers. The structure of the soap opera, in particular its serial nature and long-standing, complex narrative histories, allow fans to maintain deep connections with the text for years or even decades. The relationship between audiences and their soaps is arguably unparalleled in television (Harrington and Bielby 1995). The soap audience's "intense and lasting loyalty" has defined the medium in scholarly studies and has been regarded as the primary reason for the genre's success and longevity in television (Allen 1985, 47; Baym 2000, 39). Traditionally, the industry has recognized the importance of long-term fan investment, equating "happy fan with long-term viewer" (Harrington and Bielby 1995, 22). Thus, while the industry has historically adapted to meet the desires of its long-term viewers—while appealing at the same time to potential new ones—today's soaps are struggling to understand and connect with the twenty-first century audience.

In 2007, ABC launched a bold experiment. As ABC Daytime President Brian Frons proclaimed, "We are committed to growing the soap genre and the General Hospital franchise by expanding its storyline only on SOAPnet" (*The Futon Critic* 2007). This expansion on ABC's soap-focused cable network came through the creation of *General Hospital: Night Shift* (*GH:NS*), a prime-time spin-off of ABC's *General Hospital* (*GH*) that experimented with the traditional U.S. soap serial format by airing once a week for thirteen episodes in the first season and fourteen for the second. The creation of soap spin-offs as a network attempt to cultivate new audiences is not unique;[1] however, the series was created during a tenuous time in soap history. The ambitious goal for the producers of *GH:NS* was to gain new viewers, specifically the primetime cable audience, while testing the possible migration of daytime broadcast viewership to an original primetime series on SOAPnet. The industry has consistently tried to balance appealing to long-term fans while courting new viewers, not always successfully (Harrington and Bielby 1995, 22).

The two seasons of *GH:NS* differed in their methods. The first struggled to achieve equilibrium between viewers of *GH* and new viewers, resulting in convoluted narratives and confused reception; the second combined the established narrative history of *GH* with new storylines to mediate between these dual audiences. Comparing the two seasons and their respective critical responses provides insight into contemporary viewers and industry negotiations with them. *GH:NS* provides a valuable case study for the soap opera in the digital age, exposing impediments and providing insights into ways the industry can (re)connect with viewers in the new millennium.

The soap opera has a long history of trial and error, as more than one hundred soaps have come and gone from television since 1952 (Schemering 1988, 323–5).[2] Likewise, there are numerous instances of the medium's ability to adapt to changing viewership. For example, to tap into the growing television audience, soaps made the transition from radio to television in the early 1950s, a change few radio genres survived.[3] Currently, networks are making experimental forays into digital film and CGI technology for soaps (Steinberg 2008). Meanwhile, CBS and NBC have for some time made their daytime dramas available online and outside the typical daytime slot (Fitzgerald 2007), while ABC was distributing all three of its shows on ABC.com and SOAPnet.com by December 2009 (Newcomb 2009b). The first season of *GH:NS* was a major step for ABC toward this trend of online distribution, with the show accessible at ABC.com, AOL Video, Verizon's Soapnetic, iTunes, and ABC Mobile (*SOAPnet Medianet* 2007a). Recent soap discourse has examined the industry's contemporary changes and how these transformations have affected long-standing viewers. The "new way to watch soaps," according to its tagline, SOAPnet has tested the daytime format with primetime syndication of soaps. But SOAPnet has also attempted to redefine the ideal soap viewer, to the chagrin of many long-term soap fans (Seiter and Wilson 2005, 141–2). This redefinition is most exemplified by the network's acquisition and broadcast of teen primetime dramas in place of classic soap operas. Investigating the soaps' move into online distribution and transmedia storytelling, Levine (2006) observes: "Fans' interpretations of the changes in the soaps both on-screen and online point out the mismatch between the television industry's conceptions of its audience and that audience's own interests and concerns." In fact, some critics argue that these "gimmicks" actually do more damage to industry-audience relations, because networks sacrifice loyal viewers by employing sensational yet fleeting spectacle to gain new ones (De Lacroix 2008). These experimentations are attempts by soaps to reinvigorate the genre, but such moves may arguably compromise the industry's once secure relationship with long-term viewers.

GH:NS SEASON ONE

The first season of *GH:NS* struggled to address and balance the varying levels of viewing knowledge among new and experienced of audiences. For example, the marketing for *GH:NS* emphasized different facets of the show to attract both audiences. For new viewers, heavy mainstream and primetime ads promoted the series as *Grey's Anatomy*-esque, referring to the popular ABC medical drama.[4] These ads were also featured on ABC Daytime and SOAPnet to attract existing soap viewers. Creators targeted long-term *GH* viewers by focusing the spin-off on the popular couple, Dr. Patrick Drake and Dr. Robin Scorpio, and on mobster Jason Morgan, each with unique and established ties to the longest running ABC soap.[5] The integration of recognizable characters into the spin-off was intended to ease the transition for *GH* viewers.[6]

Former SOAPnet president Deborah Blackwell asserted that *GH:NS* could be understood by both existing viewers and new ones (Umstead 2007). However, this focus on both audiences was not smoothly executed on the show. *GH:NS* focused on establishing newer characters and their backstories in order to appeal to novice viewers with no knowledge of *GH*. Consequently, many long-term viewers were aggravated by these new characters, because the hospital-based spin-off was supposed to serve as a primetime supplement to *GH*. For many *GH* fans, doctors and medicine have been central, though the soap has juggled this focus in the past few decades with storylines devoted to international espionage, corporate takeovers, and mob life. The last decade of *GH* in particular has brought the mob characters and their storylines to the forefront, with comparatively little screen time devoted to the soap's namesake. Thus, *GH:NS* was also a response to longstanding fan complaints about the marginalized medical side of the *GH* canvas (Kubicek 2007).

GH:NS made allusions to celebrated *GH* storylines and characters, acknowledging well-versed viewers and the narrative complexity of the forty-six-year-old daytime drama. Yet new viewers, lacking the context to understand these references to past events and characters, were left puzzled, unable to fully grasp their intended meanings. For example, the significance of Robin naming an orphaned newborn "Anna" revealed her overwhelming attachment to the child and explained the intensity of Patrick's troubled response—but only for viewers aware of popular *GH* heroine Anna Devane, Robin's globe-trotting, super-spy mother.

The series ultimately lacked narrative direction, in part because the show's creators were straddling the fence between appealing to new and old viewers. Writers integrated story arcs that necessitated more than thirteen episodes. Realistic episodic medical stories were accompanied by a hospital takeover arc, a fantastical murder mystery narrative, and individual storylines for the fifteen major characters. Overall, the series was plagued by narrative inconsistency, unexplained character motivations, and stories that were abruptly dropped or clumsily wrapped up before the season's end.

Further, the *GH:NS* creators made the decision to merge the spin-off into *GH* despite previous statements this merger would not happen, ultimately leaving many in the *GH* audience feeling disrespected while alienating new viewers. *GH:NS* was originally designed to stand apart from *GH* as a completely separate "alternate universe," with no continuity between the two shows (*Soap Opera Digest* 2007a). Promotion for the first season emphasized this separation in numerous statements made by Brian Frons and *GH* and *GH:NS* executive producer Jill Faren Phelps. *GH* and *GH:NS* head writer Robert Guza said: "We are not going to directly link it to *GH* [...] The stories will not intersect. That would just be impossible to do" (*Soap Opera Digest* 2007b). The distinction was also understood by the *GH* and *GH:NS* actors. As Jason Thompson (Patrick) explained, "[The spin-off] carried over more than anybody expected. At first we were told nothing was going to carry over" (*Soap Opera Weekly* 2007).[7] After the integration of the two canons at the end of *GH:NS*, Guza explained, "What Brian Frons, JFP [Jill Faren Phelps] and I all wanted to do is, we didn't want to sync the shows up simultaneously because it's literally impossible to do, but we had no problem at all with driving some of the NS stories once the first round ended into GH" (*ABC Soaps In Depth* 2007). However, the merger created significant challenges to narrative coherence and story satisfaction for *GH:NS* viewers, *GH* viewers, and viewers of both. The sync-up also generated negative critical reaction, because many felt the show and its creators had misled the audience.

Narrative coherence and a sense of authenticity are often crucial for audience enjoyment; in particular, soap viewers base their knowledge and pleasure on their understanding of a soap's narrative complexity and history (Allen 1985, 90). However, attempts to recognize *GH* and *GH:NS* as one narrative canon are rendered impossible because of contradictions between their respective storylines and character development lines. For example, on *GH*, Jason was in maximum security prison for the entire duration of *GH:NS*, while he performed community service as a janitor on the spin-off.[8] After the *GH:NS* finale, the subsequent *GH* daytime episode[9] dedicated the majority of airtime to wrapping up *GH:NS* storylines. However, *GH* viewers were not provided prior context for these incoming storylines; neither were *GH:NS* viewers eased into the extensive and ongoing world of *GH*. Further, a cliffhanger "dream" ending for the spin-off was never addressed on the daytime soap, creating a confusing disconnect for those watching during the merger. For *GH* viewers, while some of the crossover *GH:NS* characters had sporadically been seen in recurring roles on the soap, these faces suddenly appeared to take over the canvas. Loyal *GH* fans were resentful when these crossover characters were featured over established *GH* favorites.[10]

Viewers who watched both shows were unable to reconcile the two canons because their logical understanding of *GH*'s narrative and relationship history was compromised (Harrington and Bielby 1995, 14). Nevertheless, viewers permanently had to contend with a muddled timeline, as *GH* continuously referenced the spin-off's characters and storylines as part of the canon for the next year, despite incongruities. While this may not seem like an extraordinary issue in a genre that allows the deceased to come back to life, understanding the canon is crucial for fans because viewers often base their soap "cultural capital" on canonical knowledge. More importantly, the creators' disregard for narrative coherence shows a lack of consideration for their viewers, a troubling issue considering the genre's dependence on fan loyalty.[11]

Despite its pitfalls, the first season of *GH:NS* provided glimpses of the complex reader-text relationship unique to long-standing televised narratives like U.S. soap operas. Longtime *GH* viewers could experience multilayered moments of recall: For example, physician Robin's emotional breakdown with Jason over the death of her HIV+ patient invokes viewer memories of Jason comforting Robin following her own HIV+ diagnosis in 1996/1997. But the backlash to the first season greatly overshadowed its successes, which included a record premiere at SOAPnet, one million total viewers, approximately half from the industry's target demographic (*SOAPnet Medianet* 2007b). According to ABC's statements, SOAPnet gained new subscribers prior to the premiere of *GH:NS*, and the show gained new advertisers as a result of the premiere ratings (Lafayette 2007). Also, the show was cost efficient: scheduling single-day productions; using *GH* sets; shooting digitally; and sharing production crews, writers, and several actors with *GH* (*Soap Opera Digest* 2007c).

Still, critics who were hopeful about the experiment ultimately regarded the first season as a chaotic production. For instance, soap critic Marlena De Lacroix (2007a) said, "I fear because its first episode was so highly rated the entire show will go down in soap history as a hit instead of the incoherently written and produced mess that it was. [...] It was [...] exhausting for us viewers who had [sic] watch and decipher what we were seeing!" The first season showed potential for audience migration but suffered due to its inability to articulate its narrative within the new format, as well as from its creators' cavalier treatment of the relationship between audience satisfaction and narrative coherence.

GH:NS SEASON TWO

The second season of *GH:NS* revealed several changes, undoubtedly stemming from season one failures. First, the second season was fully synchronized with *GH*, and promotional statements emphasized that viewer discomfort owing to the last-minute merger of the first season would not happen again.[12] A new executive producer, head writer, and staff writers were hired to avoid overtaxing the *GH* writing crew and to maintain a more consistent narrative. Importantly, new head writer Sri Rao reached out to *GH* viewers by establishing his fan-turned-soap writer personality: "I want fans to know that I'm one of them—totally one of them. I absolutely come from this world—as a viewer. It's the show that I've watched the longest in my entire life" (West 2008). Rao embodied the former industry-soap viewer relationship and endeared himself to fans. Second, Rao insisted on tapping into the rich history of *GH* by bringing back favorite characters with long *GH* histories and integrating them into the spin-off world (*SOAPnet Medianet* 2008a). In doing so, the second season had a larger backdrop of significant *GH* characters and utilized nostalgia for *GH*'s own past, a technique not readily available to most primetime shows (*Soap Opera Digest* 2008a). The second season once again modeled itself after *Grey's Anatomy* and maintained a similar narrative structure for episodes. However, with less than half the cast of the first season, Rao's *GH:NS* balanced the mix of new characters, *GH*-crossover characters, and returning *GH* characters by building relationships amongst them through rotating A, B, and C storylines. This strategy created a more equitable ensemble, provided episodic storylines for new viewers, and established relationships for long-term viewers.

The second season encouraged long-standing and potentially lapsed *GH* viewers to reestablish their loyalty and viewership with this spin-off, while simultaneously exposing new viewers to the narrative complexity of *GH*. The best example of this dual strategy was the first episode of the two-part season finale, "Past and Present." The episode continued the storylines of *GH:NS* while paying homage to 1980s *GH*, featuring the nostalgic dream sequences of a comatose Robert Scorpio, whose colon cancer storyline had been unfolding since the spin-off's third episode. A fan favorite, Robert was a superhero throughout 1980s *GH*, where he played a spy-turned-police commissioner. The episode featured visits from iconic past and present *GH* characters tied to Robert, as well as allusions to Robert's and *GH*'s past through dialogue, wardrobe, and set design. Even the opening credits tapped into the nostalgia, with the episode featuring *GH*'s old 1980s title sequence rather than its typical flashy, music video-style credits. Uniquely, the dream reenacted actual "memories" of Robert's, like meeting his daughter for the first time in 1985, that long-term *GH* viewers would recognize. This episode rewarded what Allen calls the "diachronic relationship" between soap viewers and soap texts; there is both an established history of soaps and also an established history of reading soaps where "we have a text that might have been begun by a reader in adolescence, but which, thirty years later, is still being read by the same reader" (1985, 73). Moreover, the show accommodated newer viewers as well as those loyal viewers who were not alive or watching *GH* during the 1980s by providing connections to *GH* history. Before each commercial break, SOAPnet aired bumpers featuring old *GH* clips that established the historical connections and complex relationships referenced within the episode. These clips provided both experienced and new viewers with the means to understand and appreciate how *GH:NS* used *GH* history to create a modern storyline.

The season two premiere garnered SOAPnet its second highest debut ratings, second only to the series premiere (*SOAPnet Medianet* 2008b).[13] Overall, it was well received critically, compared to the first season. For instance, one critic called it "a successful soapy hybrid featuring something old, something new, something borrowed and something blue—us, as we wait for season three" (*Soap Opera Digest* 2008b). Soap critic Marlena De Lacroix, who lambasted the previous season, ultimately praised the second: "What I liked the most about [*GH:NS*'s second season] is that it expertly delivered traditional soap opera in a modern form while reinforcing love as the center of the medium, instead of devaluing it as so many soaps do today" (De Lacroix 2008).

The second season was not perfect. Many dubbed it the "poor man's *Grey's Anatomy*" (Casiello 2008), with low budget production values and mixed acting abilities among its cast. However, it was praised for conjuring up the "glory days" of the 1980s, when *GH* was the highest rated show in daytime (Martin 2008). As a result, many have seen the series as an indication of the potential for reinvigorating the whole genre by refocusing audience efforts. Rao outlined the possibility:

> Honestly, I don't know if it's possible to hook a 16 year old or 24 year old into a show at 2 in the afternoon anymore [...] I think networks are trying to win that losing battle. But if you change the fight or the battle, it may turn in their favor. Let's go after the older viewers. That may be what we've demonstrated on [*GH:NS*] [...] [W]e hoped to reinvest in our loyal audience. I think by combining history and legacy characters with [postmodern] new characters, we really invigorated the core characters and brought in some new viewers. That could be a partial solution to daytime's problem" (Branco 2008).[14]

Such considerations as those cited by Branco are important for the industry to reevaluate in their current efforts to gain and stabilize viewership.

IMPLICATIONS

The future longevity of the soap opera depends on experimentation with and investigation of what does and does not work in contemporary times. The digital age has significantly altered how the American public watches television. Soaps have often been seen as a barometer of changing viewing trends, so its current state points to greater shifts in television audiences (Adalian 2007). Subsequently, soaps must identify and adapt to these changes as they have done in the past. *GH:NS* has not drastically transformed the state of the soap opera, but its presence illuminates a shifting, but viable, future for soaps. Broadcast networks should venture out of daytime to explore the potential viewership of those who have migrated to cable. While SOAPnet's seventy million household subscribers and top ranking in viewer loyalty for many years (*SOAPnet Medianet* 2008b) indicated it could possibly be a future platform for the medium, it will go off air in 2012. It could possibly be the primary future platform for the medium. Uniquely, *GH:NS*'s once-a-week primetime airings tap into a more casual viewer who lacks the time or accumulated cultural capital with soaps to start watching five days a week in daytime.

The spin-off also provides additional network revenue through DVD distribution, an area hindered by the narrative complexity and long-running format of daytime soap operas. The reception of *GH:NS* shows there is a contemporary, adaptable audience for soaps, yet reminds the industry that its primary concern should be identifying, understanding, and (most importantly) respecting its audience. New viewers are necessary for soap survival, but no soap can afford to alienate their loyal, existing audience—a lesson particularly important for *GH*, which has experienced significant audience losses and record ratings lows in 2007 and 2008 (Lewis 2008). While changes to distribution and production are important,[15] the industry must remember its major goal: reaching out to current and potential viewers by providing an entertaining text worthy of loyal investment. For television studies, these considerations should open up contemporary avenues for soap discussion, an area with a rich history of analysis, as to the connections between daytime and primetime dramas and the continuing and active negotiation of the definition of the soap opera genre in a digital age.

NOTES

1. There have been numerous soap spin-offs, beginning with CBS's *Our Private World* (1965) from the CBS soap *As the World Turns*. The now-cancelled *Port Charles* (1997) was *GH*'s first spin-off on ABC, premiering in primetime to target new viewers but airing in daytime. ABC also featured a special one-time primetime episode called "General Hospital: Twist of Fate" (1996).

2. Schemering's *The Soap Opera Encyclopedia* is a listing of soap operas aired since the beginning of broadcast television until the mid-1980s. The included soaps ranged from fifteen minutes to one hour, both network and syndicated, with 50 percent of all listed soaps lasting a year or less.

3. Extensive writing has been devoted to the radio life of soaps prior to their transition to television. For instance, see Allen (1985, 122–5).

4. Numerous stylistic comparisons (soft lit/quick-edited ads) and overt allusions to *Grey's Anatomy* appear in the actual show (For example, the first episode of *GH:NS* was entitled "Frayed Anatomies.").

5. Patrick (Jason Thompson) and Robin (Kimberly McCullough) are played by the original actors, a rarity in a medium where characters and actors are consistently recast. Significantly, McCullough originated Robin in 1985, making Robin and McCullough the first and only soap character and actor to date who have aged onscreen from childhood to adulthood in "real time" and without recasting. Steve Burton has played Jason since 1991. The characters Robin and Jason share over fifteen years of narrative history, including a former romantic relationship. All three characters have complex and interwoven ties to popular characters, families, and storylines from *GH* history.

6. "Viewers [...] are generally given a consistency that they can count on. These resemblances also provide a cushion for introducing new shows, easing marketing by assuring new audiences of something with which they are familiar" (Spence 2005, 73).

7. See also Robinson's (2007) interview with actress Minae Noji: "I confessed Soaps.com fans' confusion about the contradiction between the *GH* and [*GH:NS*] story lines with Minae and she shared our frustration. She admitted that even the actors get confused and the writers keep telling

them, it's a different show." Minae Noji plays character Dr. Kelly Lee on *GH* and appeared on the first season of *GH:NS*.

8. Other contradicting examples include: On *GH:NS* into *GH*, Spinelli suffering a gunshot wound to the foot; Patrick and Robin are separated; and Maxie was hospitalized for MRSA and went into cardiac arrest. During the concurrent time frame on *GH*, Spinelli was in perfect health; Robin and Patrick were still together; and Maxie was scheming with her rival's boyfriend.

9. This episode originally aired on October 5, 2007.

10. This case showed an abrupt, dominating "infiltration" of *GH:NS* into *GH*, unlike the ideal described by Allen: "New characters and situations can be introduced in an attempt to attract new audience members, but since the new narrative strands are positioned alongside other, more 'traditional' ones, there is little risk of alienating existing viewers" (1985, 175).

11. For more on issues of narrative coherency on *GH*, see Levine's essay on the online blog for *GH*'s Robin Scorpio in this collection.

12. In contrast to the first season, the creators promoted the second season by stressing the synchronicity of the two shows. Brian Frons said, "One thing we heard from fans was that there were some disconnects. We're working hard to make sure there will be more consistency tying the shows together better" (Reynolds 2008). Second season *GH:NS* head writer Sri Rao said, "[W]e'll be syncing up with *GH*. We heard that message loud and clear from fans of the first season, being confused about the disconnect between daytime and primetime" (Diliberto 2008, 5).

13. The complete ratings for the second season are not publicly accessible, yet it was considered successful, garnering critical commendations, viewer praise in soap magazines, and good online reception.

14. "Postmodern" was an editorial insertion by Branco rather than me.

15. See contributions in this collection from Timberg and Alba and from Green for more on production changes on *GH*.

"WHAT THE HELL DOES TIIC MEAN?"
ONLINE CONTENT AND THE STRUGGLE TO SAVE THE SOAPS

—ELANA LEVINE

> Elana Levine is an associate professor in the Department of Journalism and Mass Communication at the University of Wisconsin-Milwaukee. She has written about soap operas in her book *Wallowing in Sex* and essays "Toward a Paradigm for Media Production Research," "Doing Soap Opera History," and "Like Sands through the Hourglass," as well as various online venues such as *Flow TV*. She is currently writing a book-length history of the U.S. daytime serial drama and has been a faithful *General Hospital* viewer since November 1981.

INTRODUCTION

As overall ratings have declined, as the average age of viewers has risen, and as commercial slots have generated fewer dollars, the U.S. broadcast networks have sought out new ways to attract viewers to daytime soap operas, in particular the young viewers—those eighteen to thirty-four and even younger—most valuable to advertisers. With market research claiming that young people spend an ever-increasing amount of their time online, the soaps' networks and producers have turned to Web-based platforms as a way to retain existing viewers and appeal to new ones. Soap fans have been eager participants in user-generated online communities since at least the mid-1980s (Baym 2000), but the soap industry only began to use the Web as a promotional outlet in the late 1990s and did not expand its online presence very substantially until 2005. Since then, soap producers and networks have sought out means of cross-platform distribution, in which they circulate current soap episodes via download services such as iTunes or as streaming video offerings. They also have engaged in transmedia storytelling as a way of promoting the broadcast episodes and, potentially, of generating new revenue. In these efforts, new content featuring existing soap characters, actors, or backstage personnel extends the soaps' narrative worlds beyond the daily broadcast episodes into podcasts, blogs, Webisodes and other sites.[1] In these respects, soaps have become active participants in contemporary convergent media culture, an environment in which the boundaries between media—in this case, those between television and the Internet—are increasingly loose and porous.

Like all of the soap industry's survival tactics, the soap business's online presence has been fraught with challenges. One such challenge centers on the criticism and derision with which soap fans have responded to some of these online efforts. Indeed, online content has become one of the many sites within which soap viewers and the soap industry battle over the present and future of the genre. In this essay, I present an overview of the soaps' range of online enterprises. More centrally, however, I closely consider the fact that soap audiences have criticized many of these steps, demanding that writers, producers, and networks better serve what fans see as the true nature of soaps. To examine these contentions, I provide a detailed analysis of one instance of online content: the character blogs of ABC Daytime. In particular, I examine the ways in which the blog of *General Hospital* (*GH*) character Robin Scorpio has served as a site of struggle between ABC Daytime executives, *GH* creative staff, and *GH* audiences. Such struggles offer insight into the tensions between today's daytime soaps and the audiences they are so desperate to maintain and grow, suggesting that the use of new media outlets in and of itself cannot save the genre.

SOAPS ON THE WEB: A BRIEF HISTORY

Much as with U.S. network television as a whole, since the late 1990s, daytime soap operas have established an increasingly visible online presence. In fact, daytime soaps have often been at the forefront of network efforts to use the Web as a site of merchandising, distribution, and promotion. In the area of merchandising, one of the soaps' first innovations appeared in 1999, when CBS offered a bracelet worn by a *Guiding Light* (*GL*) character for sale on its Web site (Flint 1999). In a pioneering attempt at Web-based distribution in February 2003, SoapCity.com began to offer downloadable episodes of CBS soaps *The Young and the Restless* (*Y&R*) and *As the World Turns* (*ATWT*). And, in the realm of promotion, ABC Daytime premiered one of the first blogs from a fictional television character, *Robin's Daily Dose*, in October 2005.[2]

The soaps' various realms of innovation within the converging media culture began to take shape in the late 1990s. At that point, the soap industry started to recognize that soap fans had found the Internet to be an ideal spot for collectively communicating about the shows they loved—and loved to hate. The industry's first effort to use the Internet in service of the genre came with Sony's launch of the aforementioned SoapCity.com site in 1999, which was the official Web site for Sony-owned CBS soap *Y&R* and NBC soap *Days of Our Lives* (*DOOL*), as well as Procter & Gamble's soaps, *ATWT* and *GL*. SoapCity's early entry into the business of online soap content positioned it as a groundbreaking force, and its 2003 offering of soap episodes for download was "the first time anyone [had] made a current TV series from a major network available online—at least in a legitimate, non-pirated fashion" (Healey 2003).[3] (In comparison, iTunes did not start offering television series for download until October 2005.) SoapCity was also an early participant in transmedia storytelling efforts—for instance, the emails of *ATWT*'s Abigail appeared on the site in 2001, offering fans a new way of accessing the character's relationships with her boyfriend, her mother, and others (Ford 2007, 147). In each of the soaps' three areas of online activity—merchandising, distribution, and promotion—the industry has built on SoapCity's efforts.

On the merchandising front, CBS's early efforts at using the Internet to sell soap-related products were quickly taken up by others. Also in 1999, NBC launched the first of many of its merchandising campaigns related to its new soap, *Passions*, with the "*Passions* boutique," featuring jewelry and clothing worn or inspired by the soap's characters. In 2006, NBC revised this concept with the debut of Cranecouture.com, a site where viewers could purchase fashions evoking the serial's fashion maven character, Fancy Crane. *Passions* also pioneered special sponsorship deals for online-only content, partnering with cosmetics and hair-care product advertisers to offer viewers clues to mystery stories playing out on the broadcast program (resulting in the show's "Red" and "Vendetta" marketing campaigns of 2005 and 2006, respectively). In 2006, *Passions* offered a mock-tabloid about the on-screen world, also carrying beauty industry sponsorship.

Other than the SoapCity downloads begun in 2003, it took until late 2005 for the three major networks to become especially aggressive in their use of online content, surely at least in part a result of the broader bandwidth and connection speeds that were making audio and video transmission increasingly possible across the Internet. Since then, the soaps have taken advantage of these technological developments to offer new modes of cross-platform distribution, as well as to continue their earlier merchandising efforts and to experiment with transmedia storytelling for promotional ends.[4] Late in 2005, CBS and Procter & Gamble Productions were heralded as great innovators for their decision to make downloadable podcasts of *GL* and, later, *ATWT*, available online (See, for example, Herbert 2005; Vara 2005.). *GL* offered two versions: one was the full audio track of each day's episode, with brief previously-on and scene-setting comments by an announcer; the other (*Guiding Light Lite*) was an abridged version of each episode that included a brief interview segment with a *GL* producer, writer, or actor commenting on that day's material.

In August 2006, *Passions* became the first daytime soap available on iTunes and, in November of that year, the first to be available as streaming video on its network site.[5] As of this writing, NBC's sole remaining soap, *DOOL*, is the only U.S. daytime soap available on iTunes, and NBC also began streaming *DOOL* episodes on the network's site in August 2009. Meanwhile, all of CBS' soaps (*Y&R*, *The Bold and the Beautiful* [*B&B*], and the now-cancelled *ATWT* and *GL*) were streamed on that network's site by 2008. Because ABC owns cable channel SOAPnet, which reruns all three ABC soaps as well as other daytime and primetime soaps (until its January 2012 cancellation), the network was not as aggressive at making episodes of its serials available online, seeking to drive viewing traffic to the cable outlet instead. However, full episodes of *GH* were made available at ABC.com starting at the beginning of 2009, and sister ABC shows *All My Children* (*AMC*) and *One Life to Live* (*OLTL*) made daily episodes available online by the end of the year (Newcomb 2009b). Perhaps not surprisingly, with the arrival of this online video distribution, the *GL* and *ATWT* audio podcasts ceased production as of March 2007. Because these were not sponsored and were downloadable for free, ad- and fee-supported video episodes quickly became the more desirable cross-platform distribution outlets to the networks.

Apart from these attempts at cross-platform distribution, the soaps have also escalated their participation in a number of online promotional campaigns since late 2005. Some have had rather conventional ties to the owners' other outlets or have paired viewing incentives with merchandise promotions. I have already discussed a number of *Passions* campaigns along these lines. In addition, *DOOL*'s production company, Sony, began a program in which *DOOL* viewers could answer questions about each day's episode in exchange for points to be accumulated and applied to Sony purchases. In 2006, Procter & Gamble instituted its Daytime Dollars campaign, offering viewers the opportunity to win money in exchange for attentive viewing (which would net them a code to be entered online for a chance to win).[6]

Even more innovative efforts have promoted the shows by offering some form of added content and thus added value for viewers' involvement in daytime drama. A few of these projects have taken advantage of the Web's audio and video capabilities, such as behind-the-scenes video podcasts offered by *AMC* and *GH* beginning in 2006, or "Ask the Actor" videos on the *ATWT* site. Others have been text-based blogs, some in keeping with the podcast model of providing behind-the-scenes insights. Since 2006, a variety of writers, production staff, publicity teams, and actors from all three networks have launched blogs and Twitter accounts.[7]

More significant as transmedia storytellers, however, are those blogs and Web sites written by fictional characters or presented as part of the fictional on-screen world, which promote the soaps by offering more story, or at least more character, than is accessible through the broadcast program alone. ABC Daytime was first to launch a character blog, with Robin Scorpio's online journal starting in October 2005. This was followed soon thereafter by *Split Reflections*, a blog co-authored by Jessica Buchanan and her alternate personality, Tess, from *OLTL*, as well as *Hart to Heart*, a blog by *AMC* character Kendall Hart.[8] Similarly, in January 2006, *ATWT* character Luke Snyder (played by blogging actor Van Hansis) debuted his blog, which moved to a MySpace page in June 2006 and ceased updates in March 2007. From August to November 2006, *GH*'s Lulu Spencer kept a blog about her unplanned pregnancy and subsequent abortion. From April to November that same year, *ATWT*'s Katie Peretti Kasnoff kept an Amazon.com blog about her anonymously written book, *Oakdale Confidential* (a transmedia outlet in and of itself, in that the book was available for purchase and revealed secret backstories of various Oakdale residents).[9]

The networks have offered a few other sites to access to fictional characters and storytelling alongside these character blogs, such as *Passions*' presentations of Fox Crane's email inbox, a 360-degree tour of Fancy Crane's bedroom, and *Tabloid Truth*, the aforementioned tabloid revealing secrets of Harmony residents. *GL* offered a similar gossip blog, Springfieldburns.com, in Fall 2006. In 2008, *Y&R* made Restlessstyle.com, the Web site accompaniment to the on-screen fashion magazine created by Genoa City characters. Perhaps the most ambitious such effort at transmedia storytelling was *L.A. Diaries*, a five-week series of Webisodes chronicling the adventures of *ATWT*'s Alison Stewart and *Y&R*'s Amber Moore, who meet up in the time period between the two characters' respective departure from and arrival to the broadcast episodes of their shows.[10]

While some of these efforts at promotional transmedia storytelling have been short-lived, others—such as the ABC character blogs—have persisted, suggesting that the network has found them to be beneficial. In July 2006, ABC Daytime claimed that *Robin's Daily Dose* was the second most popular site across ABC.com (Lisotta 2006b). That said, many instances of the online content offered by the soap industry have not lasted long and have not done much to improve daytime ratings. *Passions*, arguably the soap with more instances of online content than any other and a pioneer in terms of cross-platform distribution, was cancelled by NBC in 2007. The network (also the show's owner) licensed the serial to satellite provider DirecTV. However, even with a newly reduced budget, DirecTV found the program to be a detriment, and the final episode aired in August 2008.

As with all of the television industry, soaps are struggling to figure out how best to measure audiences in this converged context and how best to quantify the costs and benefits of the various instances of online content.[11] While such questions remain unanswered, however, it is still possible to assess some dimensions of the soaps' Internet-based efforts, in particular to examine how viewers have responded to them. Ford (2007) has studied the unhappy fan response to Katie Peretti Kasnoff's blog promoting *Oakdale Confidential*. In this case, fans were irritated by the disconnect between the blog and the on-screen story, as well as the credulity-stretching purpose of Katie blogging to promote a book that was anonymously written and that was a source of great mystery and consternation to the Oakdale community as a result. If no one in Oakdale knew that Katie had written the scandalous book, why would she reveal herself as the author on this blog (Ford 2007, 149–53)? Soap fans are well used to suspending their disbelief when it comes to outlandish story developments, but they understandably demand narrative continuity in order to do so. Katie's blog angered fans invested in the seamlessness of soaps' narrative worlds.

Some of the soaps' attempts at cross-platform distribution have also generated fan criticism. As one blogger complained of the *GL* audio podcasts, the podcast announcer's role was limited to the beginning of each episode, which meant that there were few alerts to changes of scene and few-to-no descriptions of visual action (such as fights). This critique saw the podcasts as a cheap and half-hearted effort by *GL* to appear innovative and appealing to youth. As the blogger wrote, "Many people, including yours truly, have made the statement that, by podcasting, GL has come full circle to its radio days. Having now actually listened to it, I take it back. The GL podcast isn't radio. It's the 2005 version of putting your tape recorder up to the TV set" (*Snark Weighs In* 2005). Such criticism speaks to the fundamental disrespect with which many soap fans feel they are being treated as the soaps have moved into online ventures, given the seeming disregard in such ventures for the integrity of the story worlds. Because so many of these efforts have been subject to viewer criticism, even viewer hostility, such controversies reveal the intense stakes for all parties—industry and audience alike—in the soaps' struggles for survival and the particular place of online content within those struggles.

THE BATTLE OF THE BLOG: *ROBIN'S DAILY DOSE*

Robin's Daily Dose began in October 2005, when the character of Dr. Robin Scorpio returned to *GH* after departing the serial in the late 1990s. As Abbie

Schiller (2006) of ABC Daytime explained, the blog was part of the "continued evolution of how people [were] learning about the [ABC Daytime] shows. [The blog was about] strengthening brand awareness to bring it back to the television." Because the network knew that viewers were excited about the return of this long-standing "legacy" character (the daughter of two fan favorites), her return seemed an appropriate launching point for this early experiment in character blogging (Schiller 2006). Since its debut, the blog has been continuously updated, including across the period of the 2007–2008 Writers Guild of America strike. All of the ABC Daytime character blogs are written by writers' assistants at the network's three shows (and, in one case, a producers' assistant), none of whom have Guild membership, a fact that explains the continuation of the blogs during the strike. Because the blogs' writers rank low in the ABC Daytime institutional hierarchy, and because their jobs already consist of a number of different duties, it is relatively easy for the network to continue the blogs at low cost; these blogs are an easy means of generating the online content seen as increasingly crucial in a convergent age.

In the blog, Robin chronicles her daily life in one-paragraph entries that explicitly state what she is feeling. The June 17, 2008, entry, "House Arrest," offers one such example:

> I can't believe Uncle Mac went behind my back and hired Epiphany to take care of me. Last time I checked, at-home nurse was not part of her job description. She doesn't know what's best for me and neither does Uncle Mac. I understand that he's concerned about my health, but I don't need him policing my life. He only did this to punish Patrick and it's totally unacceptable. Patrick is the father of my baby, and I'm not going to let anyone keep him away from me.[12]

Most often, Robin's posts reinforce the meaning of, and sometimes simply summarize, scenes that play out on the broadcast episodes. At other times, however, the blog chronicles events in Robin's life that do not appear on screen. For example, in her April 12, 2006, post, "Protecting Myself," Robin writes of meeting with an attorney to discuss the lawsuit in which she may lose her medical license.[13] This scene did not appear in televised episodes (Robin did not appear at all in that day's show). Whether they are recapping the day's episode or filling in the blanks about what Robin is doing and thinking when she is not on screen, the blog entries are designed to keep viewers up-to-date on plot developments involving Robin. They can also provide useful backstory for viewers who were not watching *GH* during Robin's 1985 to 1998 tenure. For example, viewers new to *GH* or to the character can find out through the blog that Robin's previously presumed-dead parents were government spies or that Robin is HIV positive. In this respect, the blog can perform a sort of initiation function for newer viewers, much as the regular recapping of events in the characters' on-screen dialogue does.

Viewers with a more intense investment in the character and the show have a potentially different kind of relationship with the blog. In theory, the kind of insight into a character's thoughts and feelings offered by a blog could enrich fans' experiences of a show, a soap opera in particular. As numerous scholars have pointed out, much of the pleasure in viewing a daily daytime soap is not primarily in following the progression of the plot. Soap plots move too slowly, are too prone to repetition, and are too resistant to closure to reward viewing with such developments as one's focus. Instead, what is most significant, and most pleasurable, for the regular soap viewer is what Allen (1985) has labeled the paradigmatic aspect of the text. The complex histories of individual characters, as well as the complexities of their relationships to other characters, are the most significant sources of meaning for soap viewers. As Allen puts it, an inexperienced viewer would find the multiple revelations of a character's pregnancy within a single episode redundant. An experienced viewer, however, would understand the differing significance of each individual reveal, watching not for the news of the pregnancy (which the viewer may already know, anyway) but for different characters' reactions to the reveal, reactions that are meaningful only if one understands the characters' histories and their complex relationships with one another (Allen 1985, 69–71).

Viewer responses to Robin's blog evidence their desire for just the sort of paradigmatic meaning Allen identifies. The comments posted under each blog entry reveal these readers' investment in exploring the nuances of Robin's experiences and those of the characters with which she interacts. The following comment, in response to the "House Arrest" post quoted above, makes clear this reader/viewer's attention to the multiple layers of meaning in even the most basic of soap scenes:

> Your Uncle Mac was out of line, but he did it out of love. His little girls are all grown up and he just lost Georgie. It probaby [sic] scared him when he found out that you had passed out on the floor. He doesn't want to lose you, too. He is just trying to protect you and the baby.
>
> He does need to step back, though. Patrick is the father and he deserves to be there to help you and the baby. It scared him a lot to see you lying on the floor and I think it made him realize how impotrant [sic] you and the baby are to him. I think the thought of fatherhood became real to him when he thought something might acually [sic] be wrong with the baby, and that scared him.[14]

Comments posted to Robin's blog have been selected and edited (primarily for grammar) by the blog's writer, but they come from actual readers who seemingly enjoy playing out their imagined relationships with Robin.[15] Many such comments give Robin advice or offer interpretations of the paradigmatic significance of the events she relates. In this respect, the blog can seem a successful effort in promotional transmedia storytelling, expanding fans' involvement with the show and driving them to view the broadcast episodes.

But the carefully selected responses that appear on the blog offer just a partial picture of the ways in which *GH* viewers have responded to this online effort. More typical across the Web are examples of fan *dis*pleasure with the blog, displeasure that ultimately speaks to a broader dissatisfaction with the kinds of stories the contemporary *GH* is telling and with how those stories are told. In this respect, the network's and the show's effort at online content becomes a site within which the breached contract between the soap industry and soap audiences gets exposed. Many viewers feel that contemporary soaps have violated basic tenets of the genre, telling stories in ways that do not fit with what soap audiences believe soaps should be. Responses to Robin's blog reveal two dimensions of this difference of vision: one that concerns the screen time committed to particular kinds of stories and characters, and one that concerns the meanings being encouraged in the blog and in the soap's narrative. Although fans online take issue with aspects of every contemporary soap, in the case of *GH*, much of the viewer hostility centers around questions of gender politics, as many fans see the soap as rejecting women's concerns and interests in favor of a more masculinist, even misogynist, perspective.[16] The controversies around Robin's blog have brought such concerns to the fore.

Even those *GH* and Robin fans that read the blog regularly have found it objectionable at times for the way that it offers access to Robin online without offering as much access to her on-screen as they desire. These fans are especially disgruntled when the blog chronicles events in Robin's life that do not appear during the broadcast episodes.[17] Several readers found the "Happy Halloween" post of October 31, 2006, particularly guilty of this offense, as the post described the costumes Robin and her boyfriend, Patrick (fans refer to the two as "Scrubs," as they are both doctors), wore to the hospital Halloween party—an event not chronicled on screen. One fan described her disappointment at missing the chance to see these events play out and wrote a letter to the ABC and *GH* executives in protest:

> We really enjoy this couple and we will not be silenced by the slap in the face we get from [*GH*] when we read EPISODES on the blog!!! I wanted to see the Hospital Halloween party. I wanted to see Scrubs on a date, having fun,

laughing! [...] EQUALITY on [GH] is a foreign concept. Character driven plots/storylines is a foreign concept [sic]. ROMANCE is a foreign concept.

Such protests not only bespeak fans' desires to see more of their favorites; they also point to the fans' sense that the network is not treating the show, the characters, or the audience respectfully in its use of the cheap blog format to tell a story instead of the more costly airtime.

This kind of frustration with the blog has been so widespread that fans on a Robin and Patrick message board waged an anti-blog campaign, designing postcards for members to print and mail to ABC and GH executives. The postcards featured a photo banner depicting an imagined scene between Robin and her cousin, Georgie, and quoted a blog post describing a talk between the two that did not appear on-screen. The postcards then argued, "The blog, if used properly, should be used as a follow-up to the storyline. It should not be used as a quick fix or filler to scenes we should be seeing onscreen."[18] Underlying these sentiments is not only the fans' feelings of ill treatment, but also their disapproval of the emphases of the serial's stories: namely those stories' move away from conventional soap narratives motivated by character relationships and reactions in favor of more plot- and action-driven tales.

Since the late 1990s, these plot-driven tales have centered primarily on two main characters, mobster Sonny Corinthos and his one-time "enforcer" and later rival mobster Jason Morgan. For fans, the absence of "Scrubs" on-screen (and the consequent relegation of Patrick and Robin's love story to the blog), is a direct result of this narrative emphasis. As one fan complained:

We should do a campaign that demands GH give more love to our Scrubs. More airtime for Scrubs. (which hopefully means less airtime hogging from the Mobster and his ex-wives). It is driving me crazy how certain characters, I don't give a sh*t about, mainly the whole mobster family, gets all the freaking airtime [sic]. It's one of the main reason [sic] why I stopped watching before. It's supposed to be GENERAL HOSPITAL, not MOBSTERS' ALLEY.[19]

Although Sonny and Jason do take up more screen time than other characters, objections such as this are protesting the norms and values highlighted in the soap as a result of its emphasis on violent, criminal, male characters as much as they are protesting screen time or lack thereof. While Sonny's and Jason's "sensitive" and "caring" sides are their most valued characteristics within the narrative, these traits are presented as justification for their violent careers and borderline abusive treatment of women, which many fans find morally and politically offensive.[20] Thus, critiques about the blog serving as the primary site for telling Robin's story are implicitly (and sometimes explicitly) critiques of the broadcast serial's gender politics. Many fans object to what they see as the soap's privileging of "masculine" values of violence, stoicism, and toughness over Robin's more feminized traits. Because Robin is one of the few women characters on *GH* to criticize the violence and machismo that characterize Sonny and Jason, championing Robin over these other characters is a way to reverse the gendered hierarchies inherent to the recent *GH*.

The serial's privileging of masculinist perspectives over more feminized ones—favoring the violence and tragedy of the male (anti-)heroes over the romantic and familial drama of a young, independent career woman—is a source of displeasure for many *GH* viewers.[21] For some of these distraught fans, Robin's blog too often serves as a vehicle to help drive this narrative emphasis. For instance, the blog's support for what was presented on-air was particularly upsetting to fans in the blog's first months. At that point in the story, Robin had returned to Port Charles for the explicit purpose of treating Jason, her former lover, for his long-standing brain injury. Although Jason's injury was the narrative purpose for Robin's return, the story *GH* told was largely about the threat Jason's condition posed to *his* memory and *his* life, especially the way his new memory loss affected his relationship with his girlfriend, Sam, and his willingness to work for the mob. But Robin's stake in the situation received very little airtime. She and Jason had fallen in love ten years earlier after his brain injury resulted in memory loss and a personality change, one that eventually turned him into a mob enforcer. Because Robin's and Jason's break-up had been partially driven by his mafia involvement, the possibility that her treatment could restore Jason to his pre-brain injury, pre-mafia identity was of great import to Robin, as restoration might have removed the primary obstacle to their romance. However, Robin's interests received little attention during the story. Indeed, Robin's experience was never privileged within the broadcast text; she was merely a supporting player in Jason's drama.

One would think that Robin's blog would have, of necessity, privileged her perspective on these events. But her blog did not so much expand the *GH* narrative to take into account its meaning for Robin, as it worked to reinforce the story's current focus on Jason, emphasizing that Robin's romantic feelings for him were long resolved and thereby neglecting to give her a major stake in the story. In the blog, Robin worried about how Jason would respond to the treatment or about whether he would accept the surgery's necessity. But she also insisted: "While I still care about Jason, I have no illusions about a future together. I came back to Port Charles to save Jason's life, not complicate it further. It's evident that Jason's in love with Sam and I have no plans to try and change that. All I can hope is that we'll finally be able to rebuild our friendship."[22] The blog's emphasis on Robin's current role as Jason's doctor and its lack of attention to the paradigmatic complexities that might have made this story as meaningful for Robin as it was for Jason, limited rather than opened up the potential readings viewers were invited to make of the *GH* narrative.

The emphasis of the blog entries as well as the kinds of comments that appeared on the blog during this storyline supported the preferred reading of the broadcast text, one that kept Robin from seeking a reunion with Jason, as well as prevented her from expressing why she might have hoped to reverse Jason's brain condition: to help him leave behind his life of mafia violence. Had Robin's posts or reader comments made this connection, the blog might have offered a critique of the serial's validation of a violent masculinity. Instead, commentators wrote responses such as, "Robin, I am so proud of you. You have saved Jason's life and your motivation is without question."[23] Such comments steered the meaning of the story away from Robin's attempt to "remove" the mafia from Jason's life (and thereby the "life" of the soap itself). Even those comments that were less supportive of Robin's actions affirmed the current direction of the story[24] by supporting the current romantic entanglements while keeping silent about what Robin's "cure" for Jason might mean for his identity—and consequently for the narrative focus of the show.

Even though the comments on the blog come from real viewers, the fact that they are carefully selected adds to the blog's potential to shape narrative meaning. For example, comments never break the "fourth wall" of the story and thus never acknowledge that Robin, being fictional, could not have written the posts (even though viewers do send in such comments questioning the blog's authorship, many of which are also critical of story direction).[25] While the promise of online content may be a greater level of interaction between the mass media text and its audience, tight control of online materials such as Robin's blog not only dismisses all but a particular kind of fan involvement, it also fails to embrace the technological and cultural potential of the Internet as a source of "bottom-up consumer-driven" power (Jenkins 2006b, 18).

That potential is more fully on display in message board discussions and blog posts written by viewers, where one regularly finds oppositional readings of stories, alternate visions of plots, and trenchant critiques. One such post offered up mock entries for Robin's blog. In so doing, the post challenged the blog's fealty to the meanings endorsed in the soap itself, especially those meanings that place the misogynistic mafia and its values at the heart of the show. As this poster has imagined it, Robin might write in her blog:

> In sadder news, Sam may need a hysterectomy. I can tell Jason's just crazy with worry because he called up Alexis and said . . . that she'd best get her butt down to GH and give her uterus to Sam . . . Emily [Jason's sister and

mob boss Sonny's new love interest—a development Jason strongly protests] started jumping up and down saying 'ooh, ooh, let me help! I'm a *med* student!,' but I felt it was my duty to tell Jason there's no such thing as a uterus transplant. Jason told me I was just being judgemental [sic] as usual and that Emily should leave because [as he says very often on the show] she's 'no longer [his] sister and is dead to [him] and is also too stupid to live - and coming from [him] that was really saying a lot.' . . . All the way back to the nurses' station, Emily cried (and cried and cried) and talked about how hard it is to love a man who'd call a hit on your brother and how she was only at the hospital to hide from flying cell phones [Sonny tends to throw things when angry] and avoid the ass-kicking Carly [Sonny's ex-wife] has promised her. I guess I must be as bitchy as Carly claims, because after a while I kind of wanted to slap [Emily]. Hard.[26]

With her mock post, this writer critiques *GH*'s gender representations, pointing out the ways that Robin and Alexis, the other female character who regularly challenges male mobsters, are repeatedly criticized and belittled by characters such as Jason and the women more intimately involved with the mobster men: Sam, Emily, and Carly. In response to this and other mock blog entries, another poster sarcastically remarked that she saw Robin as the writers' "designated career-woman whipping girl. Because let's face it, Robin had a chance to spend her life as a *mob moll* instead of a world-renowned brain researcher and all around brilliant physician who can even do transplants and *she blew it*."[27] This poster affirmed the mock post's oppositional reading, charging that the writers of *GH*, and of Robin's blog, sought to represent strong, independent women as fundamentally flawed, while affirming that these women's opponents, Jason and Sonny, are the moral and emotional protagonists of the show.

Challenges to the dominant meanings offered by *GH* and Robin's blog in online fan discourse have not gone unnoticed by the show's creative staff. In fact, blogging and the hostile responses blogs can generate became part of Robin's on-screen storyline—albeit in ways that lacked continuity with *Robin's Daily Dose*. In a 2008 story, Robin found herself single and pregnant, anxious about her impending motherhood, and at odds with the baby's father, her on-and-off boyfriend, Patrick. While Robin had hoped to become pregnant, seeking out sperm donors after she and Patrick broke up because of his unwillingness to have children, the couple's one-night reunion accidentally resulted in her pregnancy. As they slowly began to rebuild their relationship, Robin launched a video blog (vlog) to share her experience with others. Patrick did the same, and the two engaged in a sort of "blog war."

While broadcast episodes showed each of the characters recording some of their vlog posts, these posts appeared not on *Robin's Daily Dose* but amidst a number of other video clips on the Web site for ABC's SOAPnet cable channel (which included ads before each vlog).[28] On *Robin's Daily Dose*, Robin made reference to her pregnancy vlog, and commenters discussed it as well. It is unclear why these vlogs were not part of the character's existing blog, but the "Robin" that posts on *Robin's Daily Dose* and the readers who comment on it seemed to accept that the pregnancy vlog existed in a parallel narrative universe. Whether other comments that did not appear on the blog challenged the continuity of the two universes is unclear but seems likely, given the displeasure *ATWT* fans expressed at the discontinuity exposed in Katie Peretti Kasnoff's Amazon.com blog.

In this blog war storyline as it played out on-air, Robin began to receive negative comments on her pregnancy vlog, a response that upset her greatly. On the June 5, 2008, broadcast episode, Robin reviewed the comments, exclaiming, "Who are these people? And why do they hate me so much?" Within the same scene, Patrick points out that it's "the same fifteen people" calling her a bitch and that their anonymity is unfair, as it allows them to say anything and escape the consequences of their diatribes.[29] Despite the fact that the pregnancy vlog did not exist in the real-world Internet, viewers who discussed the show online understandably saw this development as a pointed response to their discussions of the show, given the references to "fifteen people," the standard number of comments attached to a *Robin's Daily Dose* post, and to the anonymity of blog posters and participants in message board forums.[30]

But the most pointed part of the dialogue came from Robin, who began the scene reading the nasty comments off her computer screen and muttering, "What the hell does TIIC mean?" TIIC, a commonly used acronym amongst fans online, refers to "The Idiots in Charge"—in other words, the writers, producers, and network executives responsible for creating *GH* and other soaps. As the name suggests, this term is used to criticize choices made by soap industry authorities. Disgruntled fans tend to see those in control of daytime serial dramas as having betrayed the true potential of the soaps.

Online fans debated the intent behind the reference, agreeing there would be no reason for posters to use the term "TIIC" on a pregnancy vlog and that it was undoubtedly a way of responding to the real-world soap fans' discourse. Many took the reference as an insult—especially because of Patrick's hostile and dismissive attitude towards people who make negative comments online. But others saw it as a winking acknowledgment of the fans' typical indignation—or at least as one clever scriptwriter saying, "Yes, we hear that you think we are idiots. We recognize that you are unhappy and think you are actually pretty clever in your critiques."[31] However such a remark slipped into the dialogue, it made clear that the controversies not only around Robin's blog but also around the show more generally (not to mention around all of daytime) are not solely matters of fans bellyaching to one another, only to have their concerns disappear into cyberspace.[32] Instead, online sites are a place for contestation and negotiation, where the soap industry and soap audiences attempt to hammer out whether the soap genre can survive and, if so, in what form.

CONCLUSION

The discourse surrounding Robin's blog makes clear the intense disappointment and concern contemporary soap viewers feel about the changes that have come to daytime. Even while the soap industry has worked hard to bring the genre into our convergent media culture by investing in a host of online efforts, the controversies that have emerged around such efforts suggest that daytime drama faces much bigger problems than whether the shows are distributed or promoted online. The dwindling numbers of soap fans in the U.S. are committed enough to the shows to engage in the sorts of debates and protests waged daily on message boards. That these audience critiques have so often centered on the soaps' online content suggests that this content is not achieving its desired ends, that it is not extending fans' passion for the shows, but rather inflaming their hostility toward the industry that creates them.

By framing the matter in this way, I do not mean to suggest that the broadcast networks and soap producers should get out of the business of generating online content. Such promotional and distributional outlets are now essential for any television program. However, the kinds of complaints that have surfaced around character blogs, for example, point to the real stakes audiences have in such material. Former and current soap viewers express palpable feelings of betrayal, disappointment, and disrespect when they reflect on the way today's soaps tell their stories and the kinds of stories they tell. Such complaints are not about rejecting the genre in and of itself; rather, they are about challenging the ways in which the soap industry, in its desperation over declining ratings, has sought to *change* the genre. Soap fans happily embrace new ways to access the shows and characters to which they feel so strongly connected, but those very connections are increasingly imperiled. Without addressing these concerns by attending to the ways television texts are written and produced, all of the soaps' online efforts may be too little, too late to save the soaps from extinction.

NOTES

1. For more on transmedia storytelling and convergent media, see Jenkins (2006b). For the applicability of these concepts to daytime soaps, see Ford (2007).

2. *Robin's Daily Dose* debuted just a few weeks after the advent of *Schrute-Space*, the blog of character Dwight Schrute from NBC's *The Office* (http://blog.nbc.com/DwightsBlog/). It was the first character blog at ABC, preceding the *Emerald City Bar* blog of *Grey's Anatomy* bartender, Joe, which began in November 2006 (www.emeraldcitybar.com).

3. *DOOL* episodes were added to SoapCity.com within the year. The service was available through March 2005.

4. Procter & Gamble also made episodes of older soaps available online via AOL's PGP Classic Soaps Channel between late summer 2006 and December 2008. Early 1990s episodes of NBC/Procter & Gamble serial *Another World* are also available on online video sites Hulu and Fancast at the time of this writing.

5. *Passions* was removed from iTunes once it ceased broadcasting on NBC, although satellite carrier DirecTV allowed NBC to offer an "All-Access Pass" for downloadable episodes via NBC.com.

6. For more on Daytime Dollars and other Procter & Gamble campaigns, see Ford (2007, 144–7). See also Elliott (2006).

7. During the 2007–2008 Writers Guild of America strike, some striking, and subsequently unemployed, soap writers began blogging as well, including two contributors to this collection: Sara Bibel and Tom Casiello.

8. Neither *Split Reflections* nor *Hart to Heart* have been as regularly maintained as *Robin's Daily Dose*, and neither was up-to-date as of this writing.

9. The same blog was revived briefly in September 2007 with posts from the fictional Aubrey Cross, who had purportedly met *GL* character Jonathan Randall and written *Jonathan's Story* (also available in stores) to chronicle the tales Jonathan had told her. The blog was then revived again in December 2008 with posts from *ATWT*'s Henry Coleman, the purported author of another tie-in book, *The Man From Oakdale*. For more on the history of book publishing around soaps and the specifics of *Oakdale Confidential*, see Ford (2007, 149–53).

10. For more on *L.A. Diaries*, see Ford (2007, 153–6).

11. For more on struggles over audience measurement see Lotz (2007, 193–214).

12. "House Arrest," *Robin's Daily Dose*, June 17, 2008, http://www.drrobinscorpio.com/2008/06/house-arrest.html.

13. "Protecting Myself," *Robin's Daily Dose*, April 12, 2006, http://www.drrobinscorpio.com/2006/04/protecting_myse.html.

14. Lisa S., comment on "House Arrest," *Robin's Daily Dose*, June 17, 2008, http://www.drrobinscorpio.com/2008/06/house-arrest.html.

15. Most of Robin's posts are followed by fifteen comments, none more than a few sentences long, most of which include proper capitalization, punctuation, and spelling. In addition to the fact that little of the writing that appears online as reader responses to content is grammatically correct, fan discussion boards for *GH* and especially for the characters Robin and Patrick reveal that comments are edited before appearing on the blog. Message boards also reveal that fans do indeed submit comments (they are not just fabricated by the blog's writer), despite their awareness that Robin is a fictional character.

16. For examples of this perspective, see De Lacroix (2007b) and Streeter (2008).

17. In contrast, they are often pleased with the blog when it "accurately" reflects Robin's feelings about events fans did get to see. For example, see "robin's blog 5.27.08, first movement," May 27,

2008, *Scrubs: A Patrick and Robin Forum*, http://z14.invisionfree.com/Patrick_Robin/index.php?showtopic=48939&.

18. See the postcard and the planning of the campaign at "robin's blog, December campaign," December 16, 2006, *Scrubs: A Patrick and Robin Forum*, http://z14.invisionfree.com/Patrick_Robin/index.php?showtopic=23910.

19. Kimnjason2154, "Damn the blog!!!!!!, Stuff we should have seen on TV," July 18, 2006, *Scrubs: A Patrick and Robin Forum*, http://z14.invisionfree.com/Patrick_Robin/index.php?showtopic=13013.

20. For example, see Martin (2005), Thomas (2007a), and "What the Hell is WRONG with the writers?" thread, *General Hospital* Board, *TV.com*, January 23, 2006, http://www.tv.com/general-hospital/show/316/what-the-hell-is-wrong-with-the-writers/topic/412-196183/msgs.html.

21. See Aspenson (2003), De Lacroix (2007b), Martin (2005), Streeter (2008), and Thomas (2007a).

22. "The Kiss," *Robin's Daily Dose*, November 28, 2005, http://www.drrobinscorpio.com/2005/11/the_kiss.html.

23. Ula, comment to "The Kiss," *Robin's Daily Dose*.

24. As one reader chastised Robin for being unprofessional in allowing Jason to kiss her, "YOU are not the only hope for Jason. YOU cannot make his life choices. You seem very resentful of the fact that he has moved on with his life and is very, very much in love with Sam.... What I hope you realize is that you don't stand a chance with Jason again." Lisa, comment to "The Kiss," *Robin's Daily Dose*.

25. For a discussion of critical or fourth-wall-breaking comments submitted by fans but not posted see the comments section of Thomas (2007b).

26. SlovakPrincess, Post #49961, *General Hospital* Board, *Television Without Pity*, May 2, 2006, http://forums.televisionwithoutpity.com/index.php?showtopic=2979889&st=49950.

27. Honeybee111, Post #49869, *General Hospital* Board, *Television Without Pity*, May 1, 2006, http://forums.televisionwithoutpity.com/index.php?showtopic=2979889&st=49860.

28. Robin's and Patrick's vlogs appearing on the SOAPnet site are actually clips from *GH* episodes, wherein the characters directly address the camera as they record their posts. The vlogs' broadcast quality—not to mention the multiple camera shots that some include—make them noticeably different from the typical real-world vlog. No vlog posts appear online that have not appeared on broadcast episodes and, while the vlog posts do include comments, they are fewer in number than those on *Robin's Daily Dose* and seem less likely to have been selected and edited. Nor do they contain the hateful comments the Robin of the broadcast episodes and *Robin's Daily Dose* consider so hurtful.

29. *General Hospital*, original airdate June 5, 2008, ABC.

30. "6.5.08 scenes," *Scrubs: A Patrick and Robin Forum*, June 5, 2008, http://z14.invisionfree.com/Patrick_Robin/index.php?showtopic=49211&st=0; Posts #105039–#105068, *General Hospital* Board, *Television Without Pity*, June 5, 2008, http://forums.televisionwithoutpity.com/index.php?showtopic=2979889&st=105030 and http://forums.televisionwithoutpity.com/index.php?showtopic=2979889&st=105060.

31. "6.5.08 scenes," *Scrubs: A Patrick and Robin Forum*, June 5, 2008, http://z14.invisionfree.com/Patrick_Robin/index.php?showtopic=49211&st=0; Posts #105039–#105068, *General Hospital* Board, *Television Without Pity*, June 5, 2008, http://forums.televisionwithoutpity.com/index.php?showtopic=2979889&st=105030 and http://forums.televisionwithoutpity.com/index.php?showtopic=2979889&st=105060.

32. That the characters Robin and Patrick were prominently featured in SOAPnet's thirteen-week series, *General Hospital: Night Shift* (*GH:NS*), and that the characters' *GH* screen time

increased in the latter part of 2008, also suggest some degree of responsiveness to viewers' desire to see more of these characters. In addition, the 2008 season of *GH:NS* included some subtle jabs at the mobster-heavy content of the daytime *GH*. If nothing else, the moments of potential disruption to the more masculinist tendencies *GH* has exhibited since the late 1990s that appeared on-screen in 2008 indicate there is far from universal agreement among ABC creative and network personnel on the narrative direction *GH* should take. For more on *GH:NS*, see Gonzales' essay in this collection.

THE EVOLUTION OF THE FAN VIDEO AND THE INFLUENCE OF YOUTUBE ON THE CREATIVE DECISION-MAKING PROCESS FOR FANS

—EMMA F. WEBB

> Emma F. Webb is a doctoral student at the University of Kansas, with recent work focusing on fan influence and online message boards. Webb has concentrated multiple academic projects on a variety of U.S. daytime serial dramas and their fan communities. She started watching *All My Children*, *One Life to Live*, and *General Hospital* with her grandmother at fourteen and has been an avid viewer since.

INTRODUCTION

I first began participating in an Internet community for ABC's *General Hospital* (*GH*) in 1997, primarily through a text-based Web site called "Port Charles Online." This site, named after the fictional town where *GH* is set, started in 1994 as a free-to-post message board for fans coming online to converse about *GH*. The "Port Charles Online" community focused on discussing and dissecting the soap, both through this earlier incarnation and, starting in 2001, through its transformation into a pay-to-post site called SoapZone.[1]

However, beginning in 2000, I noticed that members of this fan community were creating and posting what they called "slideshows" related to *GH*. These fan slideshows, set to music, consisted of both photographs taken either on-set or for publicity and still shots captured from the soap. These still shots were often referred to as "screen shoots," as they were photographs taken from live episodes. The photographs and/or screen shoots were then set to music (typically popular ballads) to tell some type of story, show a perspective, or celebrate a storyline and/or pairing. These slideshows were then made available to visitors of the Web site for downloading, with the creations used by fans to communicate with one another and with "the powers that be" (or TPTB, as online fans often refer to those who oversee the making of a soap opera). For instance, in 2002, I closely followed a fan group dedicated to the *GH* couple Elizabeth and Jason as it organized a call-to-action on an online forum because of unhappiness that favorite characters were no longer appearing in a story together. These fans posted messages on SoapZone, ABC's message boards, and several other soap Web sites asking other fans of the couple to email their slideshows to the president of ABC and the show's head writer. Observing a group of fans use these slideshows made me wonder how soap opera fan communities incorporate video editing, production, and distribution into their activities, and how soap opera fans' communication with one another and the producers of these shows changed as a result of this new type of expression.

In 2004, as technology evolved, fans at SoapZone also began posting links to Web sites like YouSendIt and MEGAUPLOAD,[2] where a member of the Web site had recorded the soap from his/her television and uploaded it to his/her computer so that any visitor to SoapZone could download clips at their convenience. From that point forward, fans of *GH*, or any daytime drama, didn't need a television to watch their soaps; instead, they could download clips or entire episodes of recent shows, if some viewer were willing to record the episodes, convert them to file formats that could be read by media player software, and make them available for download to other Internet-savvy fans. When these practices were first introduced, soap fan Web sites were not heavily used for purposes of exchanging videos, largely due to the slow upload and download speeds available to a fan base still largely using dial-up connections. The video sharing that did take place seemed to inspire fans to develop the slideshow format into the fan music video (fanvid) format—in other words, to "use music in order to comment on or analyze a set of preexisting visuals, to stage a reading, or occasionally to use the footage to tell new stories" (Coppa 2008).

Then, in 2005, YouTube became the link of choice for online video. Unlike Web sites like YouSendIt and MEGAUPLOAD, which limited the number of downloads for free accounts and were often slow and cumbersome to use, YouTube allowed anyone to view a video with fairly minimal loading times. The platform is Flash-based, and—unlike streaming video or audio content—YouTube doesn't require extensive computer memory to download files. YouTube's rise in popularity also coincided with an increase of broadband Internet in the United States. In 2004, approximately 20 percent of people in the U.S. had either DSL or broadband access in their home (Fadner 2004). By 2007, approximately 50 percent of all households in the United States had broadband (McClure 2008).

As a result of this greater accessibility to broadband and YouTube's ease of use, online video became increasingly popular with members of SoapZone and began to be used as an important part of fan discussion. Links to clips that reinforced a specific interpretation of events that happened on a show or to a piece of evidence about a character's motivation became more commonplace, as did recommendations for particular fan videos. Because of the cumbersome process for accessing video prior to the widespread availability of broadband and platforms like YouTube, videos had generally only been linked to for others to download a fan video or excerpts from a recent show. Because these videos had to be downloaded and watched on one's own time (which could take quite some time for those on dial-up), they could not easily be used to illustrate a point for argument within the fan discussion.

However, YouTube allowed a member of the community to provide one external link that took other members directly to content for immediate viewing. SoapZone had particularly high traffic levels (often as many as 20,000 posts in a day) at the time. Because of this constant stream of content, a stalled conversation could quickly get buried by ongoing discussion around these shows (especially considering that a new episode of each U.S. soap opera airs every weekday). The ability to share video that was only one click away helped drive the incorporation of video into the overall dynamics of the discussions that happen on the site, where the ability to view and comment immediately was vital.

As the popularity of linking other members to content on YouTube increased on SoapZone, so did the number of members who uploaded content for others to view. From browsing other soap opera discussion sites and from conversations with fans who participate in other online soaps discussion boards, I believe that my experiences with YouTube's popularization on SoapZone reflect trends across various soap opera fan communities. As these various fan groups came to use YouTube as a common platform, and as a continually increasing number of soap fans gained home access to broadband Internet (combined with the ease of use mentioned earlier), YouTube itself quickly became an extraordinarily popular Web site among soap opera fans. By June 2008, YouTube hosted approximately 104,000 videos specifically related to *GH*, and the most-viewed *GH* fanvid had been watched over three-and-a-half million times.

When YouTube launched in February 2005, no one knew how the site could and would transform the Internet landscape (Boutin 2006). In 2008, when Google acquired the video-sharing site, 29 percent of the total Internet audience regularly visited YouTube, and the site hosted almost eighty million videos (Wesch 2008). But, even as YouTube has gained a larger audience every month of its operation, soap operas have suffered from a steep ratings decline. As early as 1986, advertising rates for television programming were on the decline (Dugas 1986); twenty years later, Carter (2006a, 40) notes that decreased advertising revenue was an issue for all the major networks. Of course, those falling advertising rates indicate how much revenue each television show brings to the network. Further, since 1994, the soap opera audience has eroded to the point that ratings are at an all-time low for all the soaps currently on the air (Jacobs 2005). *GH* once set the record for the most-watched soap broadcast on November 17, 1981, when more than thirty million viewers tuned in to watch the wedding of supercouple Luke and Laura (Fernandez 2006); by June 2008, *GH*'s audience had eroded to fewer than three million daily viewers (ABC 2008). Despite this severe decrease in the size of its viewing audience, *GH*'s ratings pattern coincides with those of most U.S. soap operas. Of the seven daytime dramas still on the air at the end of 2009 (before the 2010 cancellation of CBS' *As the World Turns*), only two or three had more than three million daily viewers in an average week (Newcomb 2009c).

As soaps' once-solid standing in the U.S. network television line-up continues deteriorating, the popularity of soap opera clips on YouTube and the proliferation of online discussion—and viewing—of soaps across the Web still appears to be expanding. This convergence between traditional and new media platforms has been described by Jenkins (2001, 93) as "an ongoing process, occurring at various intersections of media technologies, industries, content and audiences." While there have been many studies about fan communities and media convergence (Bury 2005; Jenkins 1988, 1992) focusing specifically on textual appropriation and online development of culture, there have been few studies about video streaming and soap operas. How do soap operas converge with a platform like YouTube? More specifically, how soap opera fans use YouTube, how does YouTube change the concept of soap fandom, and how does YouTube change the ways soap opera fans communicate with one other and with the producers of soap operas?

SOAP FANS ONLINE

While the study of fandom began in the late 1980s, it wasn't until the late 1990s that research shifted its focus to fans on the Internet. Since that time, many scholars (Baym 2000; Bielby, Harrington, and Bielby 1999; Bury, 2005; Gray 2005) have examined the role of fans within Internet communities. One common theme in the literature is that Internet fans develop a sense of textual ownership. In their examination of soap fans who attended events, wrote to magazines, and participated in Internet fan groups, Bielby, Harrington, and Bielby (1999) discovered that Internet fans claim significantly more ownership over the television narrative than offline fans or fans who did not participate in online communities. Online fans' sense of ownership led to more evaluative and critical discussions about the narrative (in which fans dissected specific moments of the soap opera and then debated the context and relevance of those moments) than were experienced by offline fans, who interacted primarily at fan meetings.

Baym (2000) also found that Internet soap fans spent much of their time evaluating a given soap text by judging quality, assessing the realism of the actors' portrayals and of the stories told, and criticizing the messages that they believed the show was trying to send. Baym (2000) concluded that the Internet changed what it means to be a fan by more easily enabling members of an audience to establish different types of communities and allowing those communities to proliferate. Prior to the Internet, there were fan communities, but there were a limited number of groups to join, primarily because of physical and communication restrictions. These groups typically only came together for official events, and members were often separated by significant physical distances. The Internet allowed fans who would perhaps not spend money to travel to another part of the country or who could only meet a few times a year to interact on a daily basis at no cost.

One of the first studies exploring how soap opera fans interact with video media was Ng's (2008) examination of fan videos focusing on the same-sex pairing of characters Bianca and Lena from ABC's *All My Children* (*AMC*). Ng observed that, for Bianca/Lena fans, watching romantic YouTube videos made by other fans featuring scenes of the couple's interactions were often more satisfying than watching episodes of *AMC* on television. Ng's is an interesting suggestion because, while many researchers (Jenkins 1992; Lewis 1992) have shown that fans derive pleasure from appropriating and reinterpreting media texts, Ng goes farther in arguing that the romantic videos generated by fans are more satisfying than the source text, specifically in providing alternate endings or reinterpretations. Ng's argument suggests that fan videos, like other expressions of fandom, would become increasingly popular in online soap opera communities, as they provide additional avenues for discussion and interpretation of the primary text. However, Ng focuses exclusively on romantic videos of one pairing, so the implications for the full show or the soap opera genre as a whole are limited.

SOAPS ON YOUTUBE: A STUDY

To expand this body of knowledge, I reviewed *GH*-related fan videos uploaded to YouTube during a two-week period in early 2008. YouTube permits users who sign up for YouTube's free service to upload a maximum of ten minutes of video at a time. Using the search terms "General Hospital" and "GH" on YouTube's internal search engine, I found that 171 video clips were uploaded to the site during this two-week period. Approximately seven were unrelated to the soap opera *GH* (addressing issues such as hospitals under construction), leaving 164 uploaded vids directly related to the soap opera.

The videos fell into four main categories, the first consisting of clips from recent episodes of *GH* (I defined "recent" as clip material which had originally aired within six months of the YouTube upload date). Approximately ninety-three of the YouTube vids were recent clips. The second category consisted of classic clips, which included any *GH* video with an original airdate of at least six months earlier than the upload date. Approximately forty videos in this sample were classic clips, and, of these forty vids, approximately thirty-four were clips from *GH* episodes whose original airdates were at least ten years prior to the upload date. The third category was fanvids; that is, compilations of characters set to music. For example, one fanvid featured a group of female characters and documented each female character's history on the show. Twenty-nine videos were categorized as fanvids, seventeen of which were general (all-cast or featuring multiple characters), eleven of which were couple-specific (i.e., videos that focus on one romantic pairing), and one of which was character-specific. The last category of videos was "other," which included clips of *GH* actors' appearances on talk shows, entertainment shows, or any other type of media. Only two clips fell into the "other" category.

Additionally, I examined the users who uploaded each clip. Forty-one users uploaded the 164 clips to YouTube. I categorized the users as "mega" users, "heavy" users, or "light" users. A mega user is a person who uploads fifteen or more clips; a heavy user uploads more than five clips but fewer than fifteen; and a light user uploads fewer than three clips. Mega users consisted of 7 percent of the sample, but they uploaded approximately 59 percent (n=ninety-six) of the clips studied. Heavy users consisted of approximately 37 percent of the sample and uploaded approximately 27 percent (n=forty-five) of the videos during the timeframe studied. Light users represented approximately 56 percent of all users, and each uploaded no more than one clip during the two-week period under review, or approximately 14 percent (n=twenty-three) of the videos studied.

Finally, I documented the number of views for each clip (indicating popularity) in relation to clip type (i.e., classic, current, etc.). YouTube determines view count by the number of individuals who view each video. As a result, if I decide to watch the same video fifty times, I will count as only one view. The number of views attracted by the 164 videos ranged from a low of five to a high of 10,825.

In addition to examining *GH* clips on YouTube, I also conducted a brief online soap fan survey. The study targeted three *GH*-related Web sites (ABC's official Web site, SoapZone, and Daytime Dish[3]), and participants were invited to complete a ten-question survey about how and why they used YouTube.

My target group was YouTube fans who "define themselves not just in relation to their offline selves or to the medium but also in relation to one another and to the group as a whole" (Baym 2000, 158). I received ninety-seven completed surveys. Ninety-six percent (n=ninety-three) of the survey respondents were women, which is higher than the overall viewer average but generally reflects the female-dominated soap opera viewing audience. Approximately 68 percent of the surveys were completed by subjects who participated on SoapZone. Seven of the questions asked were close-ended queries about YouTube viewing habits, and three questions were open-ended queries about why participants used YouTube as a means for viewing *GH*. Later in this essay, I will present descriptive findings from the review of YouTube videos, with this brief survey providing illustrative examples of how and why fans use YouTube as part of their *GH* viewing experience.

THE DAILY CLIP

The most prevalent type of *GH*-related video posted to YouTube during the period of this study was the daily episode. These episodes were uploaded by two different users during the two-week time frame. One user uploaded each day's episode, condensed into ten-minute timeframes. This user was, by far, the most active uploader studied, contributing one-fourth (n=forty-seven) of all 164 videos uploaded during the study period. The daily episodes were also the most widely viewed category of upload, with more than 3,000 views per video segment. All of the daily TV episodes were uploaded to YouTube within forty-eight hours of original broadcast. Since twenty of the ninety-seven survey respondents (approximately 21 percent) indicated that YouTube was their sole method of watching *GH* (that is, they watched *GH* via YouTube 100 percent of the time), the popularity of these clips is not surprising.

When asked about why they watched *GH* on YouTube instead of on television, survey participants gave two types of responses. The first indicated the ease of viewing on YouTube, as illustrated in the following survey quotes:

Infinitely easier than recording on tape, doesn't take up room on my DVR, and overall ease of access.

The fact that someone clips them all cleanly so there are no commercials and are posted 2–3 hours after the show airs is very convenient, I can skip ahead much, much easier than fast forwarding. I find out online what happened for the day and that dictates which method of watching the show I use, downloadable clips vs. YouTube vs. DVR.

YouTube is love.... Seriously, it had made it easy for me as a college student (now a full time employee) to not miss [*GH*]. It's so easy to just stop watching, but when it's a mouse click away, it becomes so much easier to catch up. You now have devices like the iphone, which has made it possible to watch YouTube clips more readily.

It is important to emphasize that, typically, only the story content of the show is uploaded, with fans editing out advertisements. While we do not know specifically how many people have turned to platforms like YouTube as their sole provider of soap opera content, the trend does not bode well for the declining ratings of soaps under the current business model, since declining television ratings equals lower advertising revenues. While these lower ratings have led the soap industry to provide their shows through a variety of online venues like network Web sites and third-party platforms such as Fancast (on which advertisements air along with online programming), the results of my survey indicate that some soap viewers seek out sites like YouTube specifically because of the lack of commercials in fan video uploads. On the other hand, YouTube allows viewers another venue through which they can stay engaged with *GH*. Viewers no longer have to watch the live broadcast every day, and, if a home recording device fails and/or there is an unexpected pre-emption, they still have an opportunity to keep up with their show(s). Unfortunately, YouTube viewing does not help Nielsen ratings, nor does it provide ancillary income through online advertising.

The second set of survey responses, illustrated in the quotes below, addresses participants' ability to view on YouTube only the stories and/or characters that appeal to them each day:

> I can only tolerate/watch for selected characters/stories. I don't want to waste time watching the show in full on TV, or even DVR the whole show since I either don't care for or simply hate most stories. It's easier to just watch the clips I care about.

> YouTube has allowed me to watch only those scenes of the show that interest me. Members who post daily clips identify segments of the show by character, thus offering a new alternative to the traditional "fast forward" option on a VCR/DVR.

Participants commonly cited the ability to fast forward or only watch select portions of *GH* as a reason for online viewing. Fans indicated that they do not want to "waste" their time on parts of the show they dislike and that watching the live broadcast was too cumbersome. Unlike DVR or VCR technology, YouTube has created an environment where fans can pick and choose scenes/segments, in effect customizing *GH* to suit their particular interests.

For example, if a group of fans only want to see scenes including character Robin Scorpio, one of them may upload sets of clips that only include the Robin Scorpio-related portions of the show. These edited clips allow fans to contribute to the discussion and stay engaged with one storyline or character without feeling obliged to spend time wading through the storylines they are not interested in. As one respondent to the survey said, "I want to know what is going on with my 'fave' and now I can do that in 10 minutes or less, and I have no intention of giving TIIC [The Idiots In Charge] one more viewer."

REMEMBER WHEN . . .

As noted, classic clips are another major category of *GH* clips on YouTube. The vast majority of the clips analyzed in this study were from original broadcasts that were ten or more years old. These videos had fewer views than the daily episode series (fewer than 1,500 each), but they were uploaded by more users. The majority came from the same four users, and each clip was character- and/or storyline-specific. For example, clips featuring Luke and Laura were the most common type of clip uploaded, but other long-term characters like Alan Quartermaine and Robert Scorpio also appeared in uploaded clips. When asked about these classic clips, one survey participant replied:

> For those without the best memories, it provides concrete evidence of current day contradictions of character motivations and storyline history. It also unfortunately reminds me of how balanced and romantic the show used to be versus the current time.

YouTube provides an easy-to-access record of *GH*'s history, though often this record is truncated so that only the highlights of a character's history are available. Viewers who no longer watch *GH* on television or who long for a particular storyline or character from the past can go online and revisit that history.

Classic clips also provide historical context for viewers, giving them the opportunity to evaluate whether the characters' current actions are consistent with their past and to situate a current story thread in a larger narrative context. Armed with this context, fans can participate more robustly in discussions on soap opera message boards. As Baym (2000) notes, the history of a soap opera plays a significant role in online soap fans' connections with one another and the formation of community norms. In addition, the availability of historical clips allows viewers to become familiar with former characters brought back to the soap by producers. For example, since 2005, *GH* has reintroduced at least seven characters (Anna Devane, Robin Scorpio, Robert Scorpio, Noah Drake, Holly Scorpio, Laura Spencer, and Scott Baldwin) who were initially popular on *GH* during the 1980s. From classic clips posted to YouTube, viewers can now learn about these characters' relationships and past storylines and can view the critical turning points of these characters' lives. Access to scenes from a show's history also means that viewers who were not watching during the 1980s have the opportunity, by searching for pertinent clips on YouTube, to become familiar enough with a character's back story that they can detect when he or she is being written out-of-character, even if they never watched episodes during their initial airing.

INTERPRETING THE CHARACTERS

Approximately 18 percent of the *GH* videos uploaded on YouTube were fanvids (again, with clips edited together and set to music), and fanvid producers represented the most diverse set of users. Thirty-four users created and uploaded these videos to the Web site, with 67 percent of these users uploading only this type of video. Fanvids were the least viewed type of upload, often with fewer than 1,000 views per video. These fanvids fell into two major categories: couple-specific and multiple character/cast. The couple-specific fanvids ranged from a summary of the couple's storyline to alternate-reality versions of what the storylines might have been.[4] In some instances, these videos involved pairings not actually seen in the television text of *GH*. The multiple character/cast fan videos featured a particular family (such as the Spencers), friends, or a group of characters the fanvid creator seemed to like.

Unlike the daily and classic clips comments, some survey respondents said that they only visited YouTube to see fanvids:

> I only watch YouTube for the fan videos (unless I'm where they don't have cable). I love to see other people's interpretations of GH and the AR/AU [alternate reality/alternate universe] scenarios they come up with.

Could online fan videos reflect what fans want to see aired on *GH*? It is possible. All of the fanvids uploaded during the review period, including those that were historical in nature, tended to feature highly dramatic and romantic moments in the characters' lives. Given that soap operas air five days a week, thousands of scenes didn't get included by fan producers—however, certain characters and scenes re-appeared in multiple videos. For example, videos focusing on the character Elizabeth, who *GH* fans had been watching grow up on screen since she debuted as a 15-year-old on the show in 1997, included several scenes that fans clearly felt were key to her narrative history: Elizabeth being attacked and saved by her love interest; her first kiss; and the birth of her child. These types of milestone moments were included in the fanvids for almost all characters. While the specifics may have been different, the "big moments" were consistent across videos, and characters who did not experience these types of big moments were rarely featured in fanvids. Additionally, these fanvids seem to provide an opportunity for fans to create their desired ending or storyline for specific characters. For example, one video focused on two characters who rarely interacted on the show and created a scenario that suggested they were romantically linked. This type of video, constructing an alternate reality, did not appear frequently.

CONCLUSION: SAVING SOAP OPERAS WITH YOUTUBE?

Despite their increasing online presence over the past decade, ABC, CBS, and NBC still seem to lack an understanding of television fans who participate in online communities. This study, while limited in scope, suggests that many *GH* viewers enjoy watching and analyzing soaps online, and gravitate toward Web sites like YouTube because of its ease-of-use and the ability to watch clips uploaded by other fans (a feature not available on any of the networks' own sites). Before the creation of YouTube, fan videos were unavailable to the vast majority of online fans. Instead, they were enjoyed by the small number who had the technical savvy and appropriate Internet connection allowing them to upload videos to (often subscription-only) Web sites. With the advent of YouTube, and to a lesser extent video editing software like Windows Movie Maker, almost any fan can create a video and share it with others.

This study also suggests ways that YouTube makes the soap opera watching experience significantly more flexible. For example, viewers who watch classic clips on YouTube can compare the quality of their soap today to its quality in prior years. Viewers can also choose to watch only the characters or storylines that appeal to them. Interestingly, many of the survey participants noted they access YouTube because of their loyalty to certain characters and/or actors on the show. Storylines are secondary. This approach to viewing obviously can contribute to declining viewership over time, since viewers who purposely avoid watching an entire episode are unlikely to become interested in other characters or storylines. However, this type of viewer also represents a potential opportunity for the networks, which could create character-specific clips for their own Web sites.

During the two weeks I studied YouTube, the soap opera clip with the highest number of views (10,825 views) was a twenty-second promotion of what was going to happen during the upcoming week. This one clip had thousands more views than any other clip uploaded in the same period. Even though there was no evidence of video clips taken from ABC's Web site and uploaded to YouTube during my study period, I have observed multiple instances in the past year when a promotion for *GH,* only available on ABC's site, was uploaded to YouTube. Taking advantage of this fan interest on the video sharing site, ABC began uploading its soap opera promos on a YouTube channel in 2009. However, at the same time, the network started being more aggressive in policing those fans who upload entire episodes to YouTube. One fan who had posted "daily clips" on YouTube wrote on SoapZone that her YouTube channel had been disabled due to a copyright violation. Decisions like airing all its soaps daily on ABC.com (Newcomb 2009b) and using an official YouTube channel for distributing videos demonstrate that the network is somewhat aware of how fans are using video online. However, ABC's policing behavior on YouTube and its preference for official distribution of material over fan video sharing demonstrates a lack of understanding and/or an ambivalence of how to capitalize on the ways video content actually spreads in online communities. At the very least, YouTube has provided new avenues for *GH* fans to express their dedication to the show, generating a collective memory of the show's history through archiving historical moments from their show, and analyzing character motivations through both these videos and the online discussions these visual works are embedded within. YouTube offers viewers both new ways to watch soaps like *GH* and an online library of video clips created by other fans. Platforms like YouTube cannot "save" the soap opera genre. However, if networks can become more successful at integrating new media with soap broadcasts,[5] YouTube may provide one venue in which to provide that programming, since soap fans already use it regularly. More importantly, sites like YouTube might be particularly crucial locations for shows to encourage engaged viewer behaviors such as creating fan videos. If shows provided a library of historical clips for fans to view and refer to in their online discussions or even allowed those clips to be edited and remixed for storyline or character videos (even for a small fee), fans may be more likely to stay tuned to what is happening on their favorite shows.

NOTES

1. SoapZone is available at http://www.soapzone.com.
2. YouSendIt is available at http://www.yousendit.com. MEGAUPLOAD is available at http://www.megaupload.com.
3. ABC Daytime's online fan message boards are available at http://abc.go.com/community/. Daytime Dish is located at http://www.thedaytimedish.com/.
4. For more on types of fanvids, see Ng 2008.
5. Levine examines *GH* new media strategies elsewhere in this collection.

SECTION FOUR
LEARNING FROM DIVERSE AUDIENCES

SOAPS FOR TOMORROW
MEDIA FANS MAKING ONLINE DRAMA FROM CELEBRITY GOSSIP

—ABIGAIL DE KOSNIK

> Abigail De Kosnik is an assistant professor at the University of California, Berkeley, in the Berkeley Center for New Media and the Department of Theater, Dance & Performance Studies. She began watching *General Hospital* at six years old and has since been an occasional-to-avid viewer of almost every soap opera airing on NBC, ABC, and CBS.

DIGITAL TECHNOLOGIES FOR "SOAPING OUT"

In his 1986 novel *Count Zero*, a handful of years before the creation of the World Wide Web, noted sci-fi author William Gibson (1986, 38) predicted what soap operas would become in an age of digital technologies:

> [Bobby's mother would] come through the door [and] just go straight over and jack into the Hitachi, soap her brains out good for six solid hours. Her eyes would unfocus, and sometimes, if it was a really good episode, she'd drool a little. [...] She'd always been that way, [....] gradually sliding deeper into her half-dozen synthetic lives, sequential simstim fantasies Bobby had had to hear about all his life. He still harbored creepy feelings that some of the characters she talked about were relatives of his.

Numerous negative stereotypes of soap opera audiences populate this excerpt. Harrington and Bielby (1995, 12–3) summarize the non-soap-viewing public's perception of typical soap watchers as "people desperately seeking mindless fantasy," with the average soap fan being "a bathrobe-clad housewife who has abandoned her domestic duties to sit weeping in front of the television set clutching a half-consumed box of bonbons." In Gibson's story, Bobby's mother, the soap viewer of the digital future, has the same failings as the soap viewer of the analog past: she ignores her son for hours at a time in order to mindlessly experience fictional narratives, to "soap her brains out," and gets so lost in the characters' lives that she lapses into a semi-conscious state, eyes glazed and mouth drooling.

For the purposes of this essay, the most interesting aspects of Gibson's stimsim-addicted character, and her similarity to the decades-old stereotype of the soap-addicted housewife, are: first, her passivity; and second, her conflation of fantasy and reality, as evinced by her having nearly persuaded her son that her favorite soap characters are their family relations. On the first point—the prediction that digital technologies would lull soap audiences into even greater levels of inactivity than television had, allowing them to merely plug in to machines to directly download escapist narratives into their brains—Gibson has been proven wrong. On the second point—the soap viewer's predilection for merging truth and fiction—Gibson has been proven right, though it is important to note that the majority of contemporary media fans elide this difference between the real and the imaginary consciously, in order to construct dramatic narratives that give them pleasure, not because they are confused about what is true and what is not.

The downward trend of daytime dramas' ratings shows that media consumers are turning to other genres for their soap opera "fixes" today, such as serialized melodramatic primetime television programs. This essay will argue that would-be soap fans are also creating their own soap opera narratives, using digital technologies. Rather than losing themselves in floods of fantasy provided by "stimsim" devices, contemporary media audiences use the Internet and basic authoring tools to create fantasies that stimulate them. Since the mid-1990s, mass audiences have adapted soap opera narrative structures and tropes into new media formats.

Two genres of digital production—online celebrity gossip and Internet fan creations—might be regarded as the soap operas of the digital era, since they deliver more of the enjoyments traditionally associated with the soap genre than do most currently airing daytime dramas. Much of the online content offering the pleasures of soap opera consists of "reality" or "real person" discussion, speculation, and fiction. In this sense, Gibson's prophecy that the soap viewer/user of the digital era will fuse reality and fiction has hit somewhat close to the mark. However, Gibson's forecast that new technologies will render the soap fan increasingly passive has been contradicted by the fact that media audiences devote significant amounts of labor to creating their own soap operas online. The contemporary real-life equivalents of Bobby's mother are millions of people who come home from work, open their Web browsers, and, for hours on end, work furiously to entertain themselves and others with celebrity gossip by writing blog entries and posting comments on various Internet communities. Far from mindlessly letting a machine lead them through various fantasy lives, media fans today perform as storytellers for and with each other, using images of—and bits of information about—the rich and famous (gleaned from publicists and paparazzi and circulated on the Web) as the starting points of their tales and conversations.

THE FAMILY RESEMBLANCE BETWEEN SOAPS AND GOSSIP

Scandals involving stars have always piqued public interest, but the Internet has boosted the business of celebrity news to new heights. Samuels (2008) states that "the industrial phase of paparazzi production" began in 2002, when *US Weekly* began running their "Stars—They're Just Like US" photo feature, helping to fuel a massive market for photos of celebrities in everyday situations. According to Samuels:

> [T]he 24/7 BritneyParisNicoleLindsayLohanMaybeDidCokeBradandAngelina SavetheWorldOneChildataTimeBradandJenBenandJenPoorKatieHolmesMarriedTomCruiseWhoIsAnAlien Celebrity Reality Show [...] was a distinctive new kind of participatory media resulting from the cold fusion of the old-school paparazzi business with the new wave of celebrity magazines and Internet sites. [...] Where old media imposed a polite and deadening spectatorial distance between the reader and the medium, [...] the online convergence of instant images and dramatic story lines encouraged the idea that the news was filter-free and that readers were part of the story. Online communities became gladiatorial forums where pseudonymous participants sallied forth to trade insults and shred the toilet-paper-thin reputations built by studio publicists and New York magazine editors.

There are hundreds of Internet celebrity gossip sites, including the Web sites of the major celebrity magazines such as *People* and *US Weekly*,[1] sites dedicated to specific stars, and high-traffic blogs such as Just Jared, Perez Hilton, PopSugar, and Defamer, which is now under the umbrella of gossip aggregator Gawker.[2] On all of these sites, Samuels declares, "[R]eaders [are] part of the story." In other words, readers of celebrity gossip are the writers of celebrity narratives. The various images of celebrities appearing in restaurants, on film or TV sets, at awards ceremonies or other red-carpet events, and in numerous ordinary settings appear on celebrity Web sites with incredible frequency but little exposition. It is the readers' thoughts, interpretations, and opinions about the stars in these paparazzi photos that give the images meaning. In the comments sections of gossip sites, readers deduce and construe the nature of the "reality" that is conveyed by the photographs.

For example, Perez Hilton (2008) posted photos from the film premiere of *The X-Files: I Want to Believe* featuring stars David Duchovny and Gillian Anderson standing together, arms around one another, on the red carpet. In the comments section, several posters claimed to have attended the event and to have personally witnessed Duchovny and (a very pregnant) Anderson sneaking into the cinema's handicapped bathroom together during the event, presumably to have illicit extramarital sex. In response to these supposedly eyewitness accounts of the actors having a scandalous liaison in a public setting, other posters replied with theories that Duchovny and Anderson had carried on a sporadic affair throughout the nine-year run of *The X-Files* television series on Fox.[3]

Speculation that the two stars may have carried out a secret sexual relationship for more than a decade mirrors the "long-arc" romance format that is a keystone of the soap opera genre. On soaps, the most popular romantic entanglements begin and end countless times over many years; marital infidelity, furtive encounters at public events, and a wish to keep the relationship concealed from the eyes of the world are typical features of the long-arc soap couple story. In the soap genre, only a show's viewers are privy to all of the sordid details of any given pairing's history; each time the couple has a clandestine rendezvous, other characters are usually know ignorant of it or are tormented by nagging suspicions, until a dramatic "reveal" occurs. On celebrity gossip sites, the commentators are the omniscient onlookers who can stage their own shocking "reveals" for fellow fans, even if the scandals they disclose are only inventions of their dramatic imaginations.

One of the most often-repeated arguments made in soap opera scholarship is that soaps serve as gossip engines, and that this is one of the genre's most valuable aspects in the eyes of its fans. Soaps make for good talk, both on the shows and amongst the shows' audience members. According to Gillespie (1995, 142–3):

> Soap operas are seen by [...] viewers to be intrinsically based on gossip. Much information is passed between characters, and of course to the viewer, in the form of gossip [...] The reception of soaps is also characterized by the speech forms of gossip. Viewing generates gossip among young people about the characters and their actions. And this soap talk is also fuelled by soap gossip published by the tabloid press, which adds further dimensions by playing with the double existence of the characters within the soap text and the actors outside it.

Comments and conversation on celebrity Web sites collapse the threefold production of gossip on and around soaps that Gillespie delineates. Soap narratives are comprised of gossip, prompt gossip among soap viewers, and suggest to viewers that, somehow, the lives of the actors who portray the gossiping soap characters are themselves leading gossip-worthy lives. Celebrity narratives *are* the gossip that media fans supply, and celebrity gossip *de facto* assumes that real actors and singers are at least as interesting and attention-worthy as their fictional personae.

As constructed by commentators on celebrity sites, the "soap" involving the romantic entanglements of actors Jennifer Aniston, Brad Pitt, and Angelina Jolie is fundamentally a love triangle, perhaps the most common plot device in daytime drama. Posts by Jen-Brad supporters tell a tale of America's sweetheart couple being split apart by a wicked temptress. Posts by "Brangelina" fans narrate a story about how the handsomest man in the world found himself torn between his sense of loyalty to his picture-perfect but boring wife and his unexpected passion for a bad-girl-turned-good; in the end, the hero risked near-universal condemnation by following his heart.

Meanwhile, the "soap" involving actors Tom Cruise and Katie Holmes is about an innocent girl ensnared by a seemingly charming but deeply controlling older man, mirroring another staple of soap plotting: the "naïve ingénue led astray." In the early days of Holmes's and Cruise's relationship, some gossip site regulars frequently expressed concern for Katie's physical and psychological safety and wondered if she will ever manage to "escape" her presumably unhealthy marriage and her spouse's "cult" of Scientology. This group of commentators seemed to interpret Holmes's situation as similar to the predicaments of heroines of nineteenth-century gothic novels (clear precursors to the soap opera genre), who are often trapped in haunted houses and in the clutches of wealthy, physically handsome, and psychologically damaged men. Other gossip fans interpreted Holmes's marriage to Cruise as a strictly financial arrangement and criticized Holmes for trading in her independence for money, a sure sign of an ingénue losing her moral innocence.

Just as soap opera storylines often split soap fans into factions, gossip fans likewise often propose conflicting explanations of celebrities' actions. On soaps, a certain character may be regarded as a protagonist by one set of viewers and an absolute villain by another, or a certain couple may be cheered by one fan group and deplored by another. Likewise, in the comments sections of gossip sites, groups of fans battle and debate each other over which explanation comes closest to the truth (such as the Jen-Brad fans vs. the Brangelina fans, or the Katie-as-victim faction vs. the Katie-as-money-grubber faction).

THE FICTIONAL NEIGHBORHOOD OF REAL PEOPLE

Each ongoing drama concerning a pair or triangle of celebrities is just one thread, however, in the larger soap opera of the celebrity sphere as a whole. The real benefit of the gossipy quality of soap operas—the fact that soaps consist of rumors and prompt endless rumors—is that anyone who participates in the rumor-mongering gets to be part of a certain kind of neighborhood, a luxurious and drama-saturated community where the goings-on, and all the conversations about those goings-on, are far more fascinating than those of any ordinary, actual city or town. Thus, the Jen-Brad-Angelina storyline and the Tom-Katie storyline and the Lindsay Lohan storyline and the Britney Spears storyline all alternately attract the attention and interest of media fans, and are related to one another in the sense that each story is one square in the giant narrative quilt of the soap opera that one could call *The Lives and Loves of Famous People*. Every celebrity is one player in that soap's enormous ensemble cast.

Certain formal aspects of celebrity gossip sites instill this sense of neighborhood participation and ensemble interaction in media audiences. On gossip sites, each post typically contains photos of one star or star couple, a short blurb giving the context in which the photos were taken, and a link to reader comments that greatly elaborate on the scant information provided. The vertical ordering of these posts is such that, as the reader scrolls down her browser page, she encounters one celebrity storyline after another, an experience that creates the impression of different celebrity storylines being broadcast in segments, just as soap opera storylines are segmented. On a soap opera, each mini-narrative is shown for a few minutes in one segment (a segment is each block of the show bracketed by advertising), then a few minutes in the next, and so on; storylines are shown in rotation, and the focus of the soap viewer continually shifts from one character or couple to another. A daytime drama's narrative is made up of the combination of all the smaller stories, and the fictional town in which the soap takes place (Pine Valley or Bay City, for example) is made up of all the characters' individual dramas being played out simultaneously in its hospitals, living rooms, and gala events. Celebrity gossip similarly is delivered to media consumers in a way that permits consumers to rotate among segmented storylines, from the Reese Witherspoon/Jake Gyllenhall-on-Sunset-Boulevard pics to the Zac Efron/Vanessa Hudgens-at-the-Lakers-game pics to the Beyoncé/Jay-Z-at-the-Grammys pics. Short glimpses of stars' lives are composited on celebrity blogs into a grand continuous drama, and the towns in which the ongoing stories play out are Los Angeles or New York, with occasional forays to other major cities.

There are additional ways in which celebrity culture resembles the vast canvasses of television soaps, with their large ensemble casts and interwoven plots. Stars are often sighted together in romantic pairings or friendship groupings, and one assumption that gossip fans often make is that most celebrities know one another—that the social scene of stars is a single network. This network resembles how the social circles of the fictional soap towns of Llanview or Springfield all overlap, and brings to mind the way all the characters on a soap interact frequently with one another, as allies, enemies, or potential romantic partners. Also, stars periodically are relegated to what soap viewers call the "backburner" of the story canvas or "fall off the radar" completely, just as soap characters do, and just like soap characters, these stars can reappear suddenly as they go on film promotional tours or suffer through personal crises.

The fact that the grand celebrity drama actually does take place in neighborhoods where people live only adds to the sense that celebrity blog readers belong to, and occasionally participate in, the physical and psychical world of that narrative. Until it was acquired in early 2009 by blog conglomerate Gawker, Defamer had a regular feature called "Hollywood PrivacyWatch," which now continues under the heading "Gawker Stalker." These posts consist of texts and emails sent in to the site moderators by readers who have sighted celebrities. Here are a few examples of PrivacyWatch sightings:

> 12/4—Paris Hilton at Gil Turner's on Sunset at like 2am last night. Wearing torn stockings, purchasing Red Bull, the New York Post and three slimy packets of Oscar Meyer Bologna. Classy.
>
> 12/30—Saw Meg Ryan at the Brentwood Country Mart today at lunch. She was eating with a nondescript couple at City Bakery. She is too skinny and looked mean.
>
> 2/5—Puff Daddy going through the motions on an elliptical at Equinox West Hollywood. His bodyguard actually approached my friend and said, "Sean would like your number." She didn't give it to him, but she did confess that his *I Am King* commercial makes her laugh her ass off every time she sees it. Grown men riding jet skis in white tuxedos is totally her brand of humor.

Although readers of these celebrity sightings may enjoy their dismissive and cynical attitude toward star culture, they also understand these reports as new "scenes" that expand the ongoing soap-like narratives that they construct out of celebrity lives.

The Paris Hilton sighting adds to her ongoing storyline, that of "rich heartless vixen born into immense wealth but acting out trashy proclivities." The Meg Ryan sighting reads like the latest entry in her drama, that of "formerly good wife and mother, beloved by all, who walked out on her family and now has fallen into a Botox addiction, lost her career, and is a heartless, empty shell of her former self" (incidentally, Ryan initially gained fame playing the popular character Betsy Stewart on CBS' *As the World Turns*, which may further encourage some media fans to read her life's events as soap-like). The Sean Combs sighting augments his character, the "arrogant-but-lonely, successful-but-laughable mogul who has no friends but bodyguards, no dates but girls he spies at the gym." Previous plot twists such as "Paris stars in low-class reality show," "Meg cheats on Dennis Quaid," and "Jennifer Lopez breaks Puff Daddy's heart" inform the readers' reports of the sightings and serve as fodder for the story-building activities of the readers of the PrivacyWatch/Stalker reports. Fresh blog comments weave the new bits of "evidence" together with older events in the stars' lives to continue the vast celebrity drama.

The fact that new sightings come into people's awareness every day via a number of gossip blogs—one might even call them "episodes" of the ongoing celebrity soap opera—also mirrors the daily (Monday-Friday) delivery schedule of daytime dramas. Television soaps have always been an every-weekday experience for viewers, adding to the illusion that its characters are a part of fans' circles of neighbors and friends, or even, in Gibson's formulation, family. According to Bourdon (2000, 550), "[T]he very duration of the series create a sense of real-life temporality, sometimes lasting for years. . . . Some viewers can actually view the whole flow of television as nothing but a life-long serial about the life of television celebrities." Contemporary online celebrity gossip sites have further intensified the sense of "real-life temporality" that media consumers used to get mostly from televised shows, particularly soaps. Because they can access celebrity news on the Internet night and day, every day, and contribute to online discussions that augment celebrity narratives as often as they wish, audiences today can easily conceive of mass media entertainment as "nothing but a life-long serial about the life of celebrities," even more than soap viewers did in the past with their "stories."

Ford (2007) argues that soap operas are "immersive story worlds," referring to the fact that fans immerse themselves in a narrative where the small and large moments in a character's life (all of their decisions, all of their relationships, all of their personal crises) can be interpreted and retold in an endless variety of ways. Since the narrative of a soap opera is so large that no one person can internalize the full history of the characters, the work of fans is often to make connections amongst all these individual moments and to debate how characters can be interpreted based on their history. Likewise, media audiences probably enjoy forming rich narratives from the raw matter of real people's lives because that raw matter is so abundant, allowing audiences to play a never-ending game of meaning-making, tying incidents that occurred yesterday to other incidents from the prior week or the previous year, or—in the case of celebrities with enduring fame—events from ten or twenty years before. For instance, present-day Beatles fans gather online to scrutinize interviews Paul McCartney gave in 2009 for clues about why the band broke up forty years earlier; fans of the late Michael Jackson analyze videos of the superstar from age seven onwards, trying to piece together a legible narrative of how the adorable child star could have become a debt-ridden, scandal-plagued recluse. Besides celebrity culture, soap operas are one of the only sources of such a massive volume of raw material for audience interpretation. Most soap operas, like the public lives of top-tier stars, have continued for decades, while most films and primetime television shows offer only a few hours' or a few years' worth of material for the kind of fun that gossip fans relish.

COLLECTIVE INTELLIGENCE, ANIMOSITY, AND VOYEURISM

Soap audiences have always appreciated the feeling that they somehow belong to the circle of friends and families and enemies and lovers who live in fictional places like Port Charles and Salem. By watching the same handful of people live out their incredibly turbulent-yet-glamorous lives every weekday, often for years on end, viewers have the sense that they live in those towns alongside families such as the Quartermaines and Hortons, and that all of "us," occupying one side of the screen/fence, are neighbors to all of "them" inhabiting the other side of the electronic border.

A number of media scholars attribute soap viewers' sense of neighborly belonging to the fact that soap narratives address real-life issues in heightened but meaningful ways, allowing viewers to use soaps as springboards for thoughts and reflections on their own lives. Some participants on celebrity gossip blogs may derive a similar kind of utility from reading and writing about stars' private lives. Some celebrities have even noted this function. For instance, in 2005, Brad Pitt said he didn't mind fans' vehement expressions of their opinions concerning his separation from Aniston if such expressions aided them with their own personal troubles, stating: "If you wanna use my thing to make your thing a little better, have at it. I'm fine with it" (Kaylin 2005).

However, an even larger benefit that media audiences derive from producing and consuming celebrity gossip, and thereby participating in a fantasy neighborhood inhabited by the social elite, is the sense of collectivity they derive from participating in gossip sites. "Collectivity" is not the same as "community": Comments on celebrity blogs burst into flame wars far too often (in the "gladiatorial" way alluded to by Samuels) for the word "community," and its implications of warmth and togetherness, to be used to characterize what transpires between commentators on star-watching Web sites. But "collectivity" does accurately describe the range of pleasures that audiences receive from coming together in virtual spaces to invent stories about, to mock and praise, and to insult and agree with one another's perspectives on famous people.

The fans who gather on Internet celebrity sites use their collective intelligence to attempt to decode what is "true" and "authentic" about performers. As Holmes and Redmond (2006, 4) state, citing Richard Dyer, "Fandom, and the construction of stars and celebrities, has always involved the 'search' for the 'authentic' person that lies behind the manufactured mask of fame. [. . .] [T]he digital and virtual media technologies have also opened up the number of spaces where the star or celebrity can be found out, re-written, and seen *in the flesh* as they really are." Together, online fans are able to analyze paparazzi photos more closely, weave more intricate explanations for stars' behavior, and test hypotheses about what is "really" going on with celebrities more extensively than they could do in isolation.

Fans also enjoy collectively admiring or despising celebrities—and each other. The bitter in-fighting that fans often claim to detect taking place between stars (Jen vs. Angelina, Mickey Rourke vs. Sean Penn, Paris Hilton vs. Scarlett Johanssen vs. Lindsay Lohan vs. Nicole Ritchie and so on) is mirrored in the bitter in-fighting that takes place among the fans themselves. For example, comments on an April 2009 Just Jared post containing photos of Pitt with his two sons of Asian ancestry questioned whether Pitt's commitment to his adopted children of color equals his commitment to his biological (Caucasian) children. These pictures prompted heated arguments among the commentators, many of whom employed the epithets "racist," "troll," "crazy," and "stupid" numerous times.[4] This level of antagonism is endemic to celebrity gossip sites. Far from decreasing participants' level of interest, such enmity typically generates hundreds of comments. When we recall that fistfights and catfights, control struggles and family wars, and name-calling and shouting matches are all staples of daytime dramas, it becomes obvious that, while media fans watch fantastical neighborhoods on soap operas and create a kind of star-studded neighborhood out of celebrity news and photos on gossip sites, they also make virtual neighborhoods for themselves on Internet gossip sites. Part of the allure of all three neighborhoods is that the neighbors often break into visceral combat.

A third type of collective enjoyment experienced by media fans on celebrity Web sites is the voyeuristic, scopophilic position they can occupy with respect to celebrities, which allows them the fantasy of possessing "insider" knowledge while remaining safely unknown to the people whose lives they gossip about. Spence (2005, 14–5) claims this insider/outsider relation to the famous-and-distant is enjoyable because it allows people to defy the prohibitions put upon everyday talk:

> A soap opera can be an invitation to enter into a specific social world. And upon entering, the viewer is implicitly obliged to respond. [. . . .] Of course, in practical everyday life there are social constraints that inhibit staring at, eavesdropping on, or talking back to the real people who share our physical space. The hazards, liabilities, and risks are fewer with fictional people. Indiscretion doesn't penetrate the iridescent screen, and soap opera characters never gaze back at us.

The virtual neighborhoods constituted by celebrity Web sites serve well as gathering sites for "town gossips" to collect and share what they think they know. Soaps hail us as nosy neighbors; the soap format seems to invite us to engage in the usually frowned-upon act of rumor-mongering. Today, mass audiences who do not watch soap operas are interpellated in this same way by celebrity Web sites, and they revel in chatting excitedly and sometimes maliciously about the residents of Malibu and the Hollywood Hills, just as previous generations delighted in talking about the residents of small fictional towns that existed just on the other side of their TV screens, in neighborly proximity.

REAL-LIFE DRAMAS

The hypothesis that celebrity gossip has begun to replace soap operas—or at least has started to offer mass audiences the pleasures traditionally associated with soap opera consumption— is reinforced by the fact that, on many soap opera Internet sites, participants now gossip about real-life "soap operas" at least as much as they discuss episodes and characters from their favorite daytime dramas. Soap message boards and blogs such as Soap Opera Network, SoapZone, and hundreds of soap-specific or couple-specific fan Web sites now have forums dedicated to "Community" or "Off-Topic Discussion," where members dispute various theories about what really happened between Rhianna and Chris Brown or why Oprah Winfrey's longtime romance has never led to marriage.[5] During the 2008 U.S. presidential campaigns, soap opera sites have hosted much political debate and political gossip as well, with reports of John Edwards' love child and Sarah Palin's teenaged daughter's pregnancy inspiring volumes of rumors and hypotheses (though political blogs generate at least as much political gossip as soap opera sites).

However, the greatest amount of gossip posted to soap opera Web sites has to do with the soap opera industry itself. In soap fan online communities, participants post and read page after page of comments regarding soap actors, writers, producers, soap magazine editors and columnists, network executives, and fans. Based on tidbits of information they glean from industry workers' public appearances and interviews, soap fans effectively write "behind-the-scenes" dramas about the making of soap operas. For instance, the celebrity news site Jossip (2008) was the first to report the firing of Carolyn Hinsey, a longtime editor of *Soap Opera Weekly* and *Soap Opera Digest*, and the item attracted more than 2,000 responses. Many commentators offered reasons for the dismissal, citing specific examples of Hinsey's unprofessionalism. (As with every discussion on a gossip site, there is no way of verifying whether the posters were sharing rumors, guesses, or facts).

The firing of a well-known soap magazine editor yielded an unusual amount of online activity; however, even the quotidian decisions and prejudices and desires of all the makers of soap opera tend to be mulled over and analyzed as closely on soap opera Web sites as the behaviors of the most successful A-list movie stars are on Perez Hilton. Will this head writer or that executive producer be fired soon? Is Actor J refusing to play the romantic pairing with Actress K because he likes working with Actress N too much to allow their characters' storyline to end? Why is this soap columnist or television critic seemingly so biased against this particular pairing, while favoring another couple on the same show? These are the types of topics in which online soap fans show considerable interest. Possibly, the fact that soap opera Web sites currently host at least as much discussion of industry happenings as discussion of the actual soaps indicates that, for soap opera fans, speculating online about real people and events has become comparable to, and in some cases perhaps even more satisfactory than watching the actual daytime dramas on their television screens.

REAL-LIFE FICTIONS

There is one respect in which online celebrity gossip seems to fall far short of delivering the pleasure of soap operas: Gossip does not often come close to being as deeply involving or immersive as fiction. But media audiences have found another way to make daily reportage about stars' activities serve their desires for soap-like narratives. Real person fan fiction, or RPF, is a genre of fan-authored stories that dramatizes moments in famous people's lives and is increasingly popular on fan blogs such as those hosted on online platforms like LiveJournal.[6] Writers typically find the raw material for their RPFs on celebrity gossip sites—in the images and interviews posted by official moderators, as well as the speculation, rumors, opinions, and hypothetical situations submitted by the sites' readers. These writers then elaborate on what has been reported and guessed about stars' private lives, creating scenarios about how and when Brad and Angelina first fell in love, for instance, or how Jen decided to get her revenge on them by starting up torrid affairs with first Vince Vaughn and later John Mayer. The gossip about David Duchovny and Gillian Anderson recounted above prompted several writers on a Duchovny/Anderson RPF Web site to author a number of short stories about the incident, each depicting a different version of how the actors' "brief encounter" at their film's premiere may have unfolded. Many RPF stories about television or movie co-stars, or members of a band, seem to be inspired by a fleeting moment between co-workers captured by press cameras; the fictions turn mere instants of laughter, eye contact, or physical interaction into narratives of deep friendship or intimacy, for which the paparazzi photographs retroactively function as accompanying illustrations.

Much RPF is about celebrities' private relationships, just as most soap opera narratives concern characters' private relationships. Soaps, like romance novels, dwell luxuriously on the formation, evolution, and dissolution of personal entanglements. What might be considered small and insignificant or simply irrelevant and extraneous moments in primetime television or film formats, such as the thoughts of two people meeting for the first time, or the wounds mutually inflicted by a parting couple during their final conversation, are the stuff of which soap operas are made. The ordinary minutiae of real people's lives, the way that glimpses of their mundane activities make their way onto media audiences' computer screens on a daily basis, and the fact that there is no end and no closure to this procession of the goings-on in public people's worlds, all serve to make celebrity gossip resemble soap opera. RPF is an effective means for media audiences who crave soap-like pleasures to turn their rumors and speculations about famous individuals into more extensive and richer narratives. In RPF, the celebrity stories created by the comments and opinions of media fans become more formalized and even more fictionalized; the drama constituted by the various stars' lives takes on the character of "scenes" and "episodes."

Crucially, authors of RPF mark their work explicitly as fiction. Every RPF story is headed with a disclaimer that the author understands the story is fictional, and no resemblance to actual events is intended. Although RPF may seem more potentially slanderous to celebrities than gossip—because RPF stories offer up personal scenarios at greater length and in more detail than speculation and rumor-spreading on celebrity Web sites—it is RPF that is labeled clearly as fiction, while rumors are often posted to celebrity Web sites under the guise of "truth."

REDEFINING TEXT AND PARATEXT

Exchanges between media fans in online communities have been framed by many media scholars as paratextual: external and supplementary communications that circulate around, and are based on, texts produced by the cultural industries. Despite the argument at the core of much reception theory and cultural studies work that audiences actively make meaning from mass media commodities, and that this meaning-making should be viewed as central to media culture and not peripheral, the text/paratext division still informs many descriptions of online media consumers' productions. The text/paratext division results in two ramifications. One is that all paratexts rank lower and lesser in a hierarchy of importance and "official-ness;" paratexts are not regarded as authoritative because they are not produced by the same professionals that created the original or source text. The second is that paratexts are widely presumed to be mere advertising ploys or efforts to boost consumers' dollar commitment to a given franchise; paratexts are considered marketing campaigns or unnecessary commodities which are never required to make sense of the original or source text. Paratexts such as film novelizations, toys and games based on fictional movie or TV characters, and manuals or encyclopedias that compile all the technical and historical details of fictional universes (*Lord of the Rings*' Middle Earth, *Star Wars*' Empire and Rebel Alliance, *Star Trek*'s Federation and enemy alien races), are all typically thought of as lesser than, and extraneous to, the media texts on which they are based.

But, in the last few years, some genres of media-based online activity have developed to a point that they clearly do more than serve as mechanisms for audience entrapment by advertisers. Online gossip, discussion, and speculation—as well as fan fiction, videos, icons, and other forms of creativity—have taken on the volume and importance of primary media for broad publics. Following the efforts of numerous fan studies scholars to legitimize audiences' engagements with media as productions of culture, we might replace this diagram:

Cultural industries' productions		Texts
---	=	---
Audience productions		Paratexts

with this diagram:

$$\frac{\text{Audience productions}}{\text{Cultural industries' productions}} = \frac{\text{Interpretive texts}}{\text{Source texts}}$$

The second diagram indicates that consumers today produce entertainment for themselves, using authoring tools and Internet technologies to make and distribute mass media, and basing their media texts on "source texts" that are supplied by the official producers of media. According to this model, source texts are incomplete without their audiences' augmentation. The interpretation, one might say *activation*, by *some* audience members makes these source texts more deeply interesting to *many* audience members. To make sense of the texts created by fellow consumers requires a familiarity with their source texts, but the source texts do not provide the majority of consumers' pleasure.

Today, many media fans shift between the two models of media entertainment proposed above. People enjoy some movies or songs or television shows without much or any addition of online interactivity; they either do not feel any need to engage with online communities in order to complete their enjoyment of those texts, or they participate in online discussions only to supplement their enjoyment of them. People engage with other mass media productions only as the sources of Internet productions and watch television or movies or other mass media only as a foundation for taking part in, or appreciating the results of, online interactions and creations by fellow media consumers. This second model of entertainment, in which the bulk of media pleasure is generated by audiences' interpretations of media productions and not by the productions alone, prevails mostly in cases where audiences are seeking the pleasures historically derived from viewing soap operas. Soap audiences' meaning-making activities produce the majority of their enjoyment because, according to Laura Stempel Mumford (1994, 177), "[T]he serial form employed by soaps is as a structure whose lack of episodic resolution demands . . . interpretive work from viewers, in contrast to the self-contained (and therefore relatively self-explanatory) episodes we expect in other forms." Soap operas would mean very little *without* viewers' active participation. Soap viewers consistently recall relevant information from previous episodes (sometimes episodes that aired years or even decades prior), archive knowledge of how each character has developed over time and how characters' decisions are informed by their past experiences, and decode characters' gestures and words to arrive at the "truth" of the characters' intentions and feelings.

Celebrity gossip, discussion, and speculation, as well as fan fiction, are clear examples of Internet audience productions that today provide the majority of entertainment for consumers seeking the kind of enjoyment that audiences derived from soap operas in the past. Certainly, other categories of audience productions serve the same purpose, such as discussion and speculation about fictional media properties and other fan productions, including icons, banners, fan videos, and fan music mixes. Harrington and Bielby (1995, 152) highlight the importance of digital technologies in facilitating fans' "pleasure/play" with soap texts:

> One of the booming modes of communication between fans of all sorts of cultural objects (movies, soap operas, comics, etc.) is electronic bulletin boards. Subscribed to by millions, these boards provide forums for fan communication and gossip and put fans in contact with others who share similar interests and pleasures. They give soap fans access to others who find pleasures in the same things and play in the same way.... [T]he emergence and popularity of electronic bulletin boards both affords these fans access to one another and renders their play more publicly visible.

The promise of the early Internet pointed out by Harrington and Bielby has become more completely realized over the last fifteen years, as soap fans and other media fans who enjoy soap-like narratives are joining together to "find pleasure" and "play in the same way" in and on Internet sites to such an extent that they are inverting the terms of fandom itself, making the fan text the focal point for mass audiences rather than the source text. In an interview included in this collection, QueenEve, a moderator of a number of soap fan Web sites, says that what soap fans learned from organizing in digital spaces is that, by conversing and sharing amateur productions with one another online, "we could make our own fun." What's more, fans have begun to learn that they can even make their own soap operas.

NOTES

1. *People's* Web site is available at http://www.people.com/. *Us Weekly's* Web site is at http://www.usmagazine.com/.

2. Just Jared is located at http://justjared.buzznet.com/, Perez Hilton at http://www.perezhilton.com/, PopSugar at http://www.popsugar.com/, Defamer at http://www.defamer.com/, and Gawker at http://www.gawker.com/.

3. Further discussion of the rumors and the Perez Hilton post can be found at http://community.livejournal.com/homeby_five/18558.html.

4. See comments section of Just Jared's April 26, 2009, post "Brad Pitt Visits Niagara Falls," available at http://justjared.buzznet.com/2009/04/26/brad-pitt-niagara-falls/#comments.

5. Soap Opera Network is located at http://www.soapoperanetwork.com/. SoapZone is located at http://www.soapzone.com/.

6. LiveJournal is located at http://www.livejournal.com/.

SOAP OPERA CRITICS AND CRITICISM
INDUSTRY AND AUDIENCE IN AN ERA OF TRANSFORMATION

—DENISE D. BIELBY

> Denise D. Bielby is professor of sociology and affiliated faculty in film and media studies at the University of California, Santa Barbara, where she researches the culture industries of television and film. She is co-author of *Soap Fans* with C. Lee Harrington and has written extensively on the soap opera genre for academic publications. She remembers music from live broadcasts of *Love of Life* and watching *The Edge of Night* with her grandmother, but her enduring interest in soaps comes from seeing the premier episode of *General Hospital*.

INTRODUCTION

At the heart of all relationships between producers and consumers of media texts is a struggle over the meaning of those texts. This struggle is just as likely to occur over popular cultural material, such as soap operas, as it is over canonical literature. Because popular cultural texts specialize in the familiar and are relatively "open" to interpretation, it is not uncommon for audiences to develop expert knowledge and seek participation alongside professional critics in the evaluative process. In the case of daytime serials, dedicated fans have been shown to develop sophisticated insight amassed from long-term viewing and to assert legitimate evaluative claims that significantly challenge the judgments of soaps' producers and professional critics (Bielby, Harrington, and Bielby 1999). In some instances, those claims, which are rendered visible through organized Internet protests, on blogs, or in massive letter writing campaigns, have affected the outcome of a narrative or the direction of a show. Thus, in popular culture, audience-based critical insight competes head-on with professional expertise about the quality or value of cultural products. But what room do such audience contestations leave for professional criticism within popular culture?

Motivated by an overarching interest in the aesthetics of popular culture, including how properties of cultural products are apprehended and valued, the aim of the research reported here is to gain insight into the contribution of soap opera critics and criticism to the field of soap opera studies. Professional critics in music, film, and, especially, television operate within systems of cultural production that are complicated by the overtly commercial nature of these industries (Lang 1958).[1] Despite critics' complex work context, prime-time television critics, for instance, are valued industry members with autonomous professional standing, and their critical practices include codified systems of evaluative criteria (Bielby, Moloney, and Ngo 2005). Although soap operas were one of television's earliest forms of programming and rank—at least for now—as one of its most enduring genres, far less is known about the origin, role, and status of daytime television's critics, other than Harrington and Bielby's (1995) discussion of the emergence of the soap press.

To better understand the work of soap critics, the research reported here focuses on their understanding of the soap genre, the industry's impact on their evaluative practices, and their relationship to television's networks, actors, and audiences. Relying upon interviews with soap critics themselves, this study explores their background and expertise, the criteria they deploy to evaluate soaps, and the effects of the decline of the soap opera audience and the emergence of the Internet upon their practice of soap criticism. A particular goal of this research is to understand better how recent changes to the soap opera landscape have altered the legendary struggle between audiences and critics over claims to the narrative and the implications of these changes to the future of the genre.

THE ORIGINS OF SOAP OPERA CRITICS AND CRITICISM

Soap operas are one of the few forms of television programming that have remained in continuous production in their original form. Despite soaps' early appearance in the industry, from their very beginning they were widely regarded as unworthy of critical appraisal because of their low standing in the cultural hierarchy, a position affirmed by Thurber's (1948, 191–260) well known satirical but derisive critique of radio soaps. When soaps moved to television, they again were largely overlooked by cultural authorities. This situation remained unchanged for decades until, in the 1960s, the television industry began to publicly acknowledge soaps' vast audience and their ability to generate considerable revenue. According to LaGuardia (1974), an important shift occurred in 1968 when industry record *TV Guide* (*TVG*) began publishing features about the daytime medium. This development coincided with a growing set of public acknowledgements by well known celebrities and literary figures of their avid soap opera viewing, including Renata Adler's 1972 piece in the up-market *New Yorker*, which reoriented the critical direction that periodical had taken since its publication of Thurber's critique. Subsequent coverage by this culturally influential magazine grew to include recurring features about the industry (e.g., Hiss 1975) and items by noted scholarly figure Jamaica Kincaid (1978). Evidently, soaps' widespread popularity could no longer be ignored.

While these factors brought newfound credibility, even some respectability, to soaps, what proved pivotal to the emergence of professional soap opera criticism was the launch of magazines dedicated solely to the genre. These publications, a byproduct of the efforts of entertainment reporters in other parts of the industry, recognized that the soap opera audience was an underserved market. The first of the soap fan magazines, *Daytime TV* (*DTV*), was founded by Paul Denis in 1969 and modeled after established publicity magazines with which he was familiar after years of work covering the media for the *New York Post* (Harrington and Bielby 1995). Although the aim of these new magazines was to create interest in the otherwise overlooked, more conventional lives of daytime performers through visits to production studios, photo shoots, and interviews, the magazines also had to cater to an audience that viewed actors and characters every day in their homes and with whom they identified. "Viewers did not want their fantasy destroyed," according to then-*DTV* editor Al Rosenberg (Harrington and Bielby 1995, 67), and so these magazines had to walk a narrow line. Denis's formula proved successful because his foundational magazine was followed in quick succession by many others. At the height of their popularity in the 1970s, 1980s, and 1990s, at least a dozen magazines were in production at any one time,[2] and even *TVG* added a weekly column on the soap industry, helmed by critic Michael Logan. Of these printed periodicals, only a handful remain in existence today, most prominently *Soap Opera Digest* (*SOD*), *Soap Opera Weekly* (*SOW*), and *Soaps In Depth*.

Even as the possibility of professional soap criticism began to emerge, hurdles remained. First, there were practical considerations such as the daunting task of taking on the sheer volume of programming to be reviewed (five shows a week for each of the dozen or more shows). Relatedly, how was one to evaluate a narrative that had no readily apparent beginning, middle, and end, or a format that specialized in multiple and interwoven stories and emphasized characterization over plot? A second factor is that critics in the more respected fields of film and primetime television were, for the most part, not tempted to become experts in this culturally denigrated genre. A third reason is that soap opera was, and still is, a genre to which the television industry has given only cursory attention as an art form in its own right. Instead, soaps are usually regarded strictly in terms of their commercial value—that is, as a useful source of revenue to offset the expense of primetime production (*Time* 1976). To the industry, what was the point?

This lack of industry attention to the creative aspects of daytime dramas was likely the outcome of longstanding beliefs that its audience was not sophisticated enough to appreciate critical assessment; that the genre was not aesthetically complex enough to be subjected to critical appraisal; that there was no need for critics or their criticism since the audience was already firmly attached to the medium and would watch regardless of aesthetic evaluation (reflecting the industry's interest in the commercial bottom line); and that the audience itself was so dedicated and knowledgeable because of avid viewing that it managed to serve ably as its own critic. These factors are consistent with what many cultural analysts assert about popular culture more generally, which is that its forms are too superficial or un-complex for appraisal; that the basis for their entertainment is too emotionally "under-distanced," leaving little room (or necessity) for thoughtful analysis by cultural authorities; and that popular culture is disposable culture not worthy of lasting attention (Bielby and Bielby 2004).

Whatever the reason, the long delayed and ultimately successful introduction of professional soap criticism was not smooth at first, because when professional soap critics finally appeared on scene, they found themselves competing with established and widely accepted audience-based expertise. However, once critics became regular features of the magazines, their inclusion served to underscore the astuteness of fan-based criticism, a development which fans came to appreciate even as critics' presence challenged fans' longstanding position as cultural authorities in their own right. Thus, in a circuitous way, what began as an intense rivalry dissipated as audiences increasingly recognized that critics' practices drew upon criteria already understood and valued by fans.[3]

THE STUDY

A handful of key questions guided this study of soap opera critics and criticism. First, what draws critics to the field of soap opera, describes their standing in the industry, and accounts for criteria they deploy to appraise the genre? Next, how has critics' standing been affected by the declining soap audience and proliferation of alternative sources of critical opinion? Finally, what do critics anticipate their professional role to be in the future? Five critics drew upon their experience to answer these questions. They are: Nelson Branco, eleven-year contributor to *TV Guide Canada* and the *Los Angeles Times' theenvelope.com*; Melissa Scardaville, former staff writer for *SOD* and now PhD student in cultural sociology at Emory University;[4] Linda Susman, professional journalist and retired thirty-year veteran of the soap press who freelanced for many pioneering soap industry magazines, edited *Soap Opera Now*, and served as executive editor of *SOW*; Connie Passalaqua Hayman, adjunct professor of journalism who wrote the weekly column "Marlena" for *SOW* and is currently the Web mistress for the blog *Marlenadelacroix.com*; and Mimi Torchin, twenty-five-year industry veteran who acted on stage and wrote theater criticism before entering the soaps field.

UNDERSTANDING SOAP OPERA CRITICS AND CRITICISM

BACKGROUND AND EXPERIENCE

Respondents were drawn to the specialty of soap opera criticism from a variety of backgrounds, and they entered the field through a variety of routes and in ways not unlike critics in the larger, more established field of primetime television criticism. The critics who were interviewed had done soap criticism for anywhere from six to thirty years. Some were still actively involved; some had transitioned from defunct soap magazines to ongoing participation in online outlet; and others were retired from the occupation altogether after extensive contributions to the field. One was a member of the National Academy of Television Arts & Sciences (NATAS), the New York-based organization that bestows the Daytime Emmy Awards, and another belonged to three professional groups: Women in Film & Television, the American Federation of Television and Radio Artists (AFTRA), and the Dramatists Guild. Thus, respondents included highly experienced veterans as well as the next generation of relative newcomers. Collectively, the experience they brought to the specialty was considerable, along with an abiding appreciation for the genre. One respondent stated: "Most people who worked at the magazine were college educated, went to rather elite schools, and were big consumers of pop culture. We didn't see that soaps should be treated differently than other television shows and we talked academically about them at times."

WORKING IN THE FIELD

Doing soap reviewing is time consuming because of the vast number of hours the shows air every week. All but one respondent wrote exclusively for the soap industry, although this individual devoted the vast majority of her/his work time to it. All were drawn into soap criticism because of their respect for and appreciation of the genre, coupled with opportunities in the field.

The soap opera industry is a specialized one, a "niche industry," and so a focal consideration was how its characteristic insularity from the rest of Hollywood might be consequential to the practices of its members. That insularity was relevant in two ways. The first is in how soap opera journalists maintain professional autonomy within a circumscribed realm and navigate the professional relationships upon which the industry is based; the second is in whether the supplanting of print media by Internet-based professional and fan-based Internet criticism has affected their employment status and journalistic practices.

To understand better the professional relationships soap opera critics navigate, I first sought a clearer sense of how soap critics are embedded within the web of those relationships. To do this, I asked respondents to identify their social networks, in particular the *roles* (but not the names) of up to ten individuals in the soap community with whom they most frequently interact. These could include other critics, fans, actors, producers, writers, and editors. The most frequently mentioned contacts were actors and studio producers or network executives. Close behind with a nearly equal number of mentions was contact with publicists of shows, networks, or actors; knowledgeable fans; colleagues on their magazine staff; and other journalists. Other soap opera critics were less frequently mentioned, as were press photographers (who may generate industry leads through candid photographs or photo shoots). These mentions are aggregated and do not reflect how specific individuals actually prioritized the importance of the contacts they listed. Even so, there were no readily apparent trends in priority (although I did not ask respondents to rank their responses). The prevalence of mentions does, however, indicate the relative importance of particular sources of information pertinent to these critics' work. In that regard, industry sources were most crucial.

Critics' dependence upon the industry for information to do their work creates particular complications, because the overall size of the industry is so small.[5] Even though soaps are in production year round, a schedule that would seem on its face to generate ample opportunities for publicity that could offset the drawbacks of the industry's relatively small size, most soap magazines that were once on bi-weekly or monthly publication schedules are now weeklies. This weekly production schedule increases the need for information, creates a fair amount of competition for news, and leaves a limited amount of time between commentaries published one week and the need for new information from the same (critiqued) industry sources the next.

Critics were forthright about how the industry's insularity complicated their work as reviewers, and all mentioned thin-skinned actors, executive producers, and writers who threatened to withhold access to industry information if they were candidly critiqued. Consequently, most critics—but not all—were faced with the need to accommodate or mollify important sources:

> Print critics need studio access so, oftentimes, their opinions are compromised. Also, access to certain stars and events may be denied if unfavorable reviews are printed.

> In general, the shows had thicker skins [than the actors] for a longer period of time. We forced the issue and made them all accept criticism, but they never liked it, and the shows in particular tried to withhold cooperation if we pushed too far. It was a tricky balance between integrity and conciliation.

> I, however, am fearless and say what I think. I could care less about access, even though I get it very easily.

Although the dynamics associated with access to industry sources as described here are not that different from what primetime critics contend with, they are potentially more intense because of the soap medium's more circumscribed realm. As the following remarks reveal, shows and actors with clout are able to avoid penetrating criticism or to get away with leveling especially intense fire at candid critics:

> Some shows, the ones that supposedly sell the most magazines, like *Y&R* [*The Young and the Restless*, CBS], *GH* [*General Hospital*, ABC], and *DOOL* [*Days of Our Lives*, NBC], are rarely ever criticized in the magazines, in part because of the influence of the EPs [executive producers], certain actors, etc. Other shows, like *ATWT* [*As the World Turns*, CBS], *GL* [*Guiding Light*, CBS], and *OLTL* [*One Life to Live*, ABC] are not "as important," so it doesn't matter as much what the EP says or does. That said, in general—if an EP or a network executive called the magazine (which rarely happened) and said, "How dare you say this?"—you better believe the magazine would jump through hoops to undo what was said. As for the publicists, only a select few really felt committed to the show they covered [and thus were less sensitive to criticism].

> This can be a slippery slope ... There were some execs and actors with less thick skin than others; the degree of "independence" also was determined by a particular critic's associations. Those with larger venues (à la [*TVG*]) or

potentially larger (i.e. mainstream) venues (*People, Entertainment Weekly*) and those who didn't depend on the regular access to storyline information and personnel (execs as well as actors) were able to function with the greatest degree of "independence."

Given these constraints, critics' power and independence from the publishers and their editors or managers was a matter of interest; in that regard, their perceptions fell along a continuum from "no interference from publishers or editors" to more complicated work arrangements that varied by standing in the field.

On the one hand, we're free to write [criticism] any way we see fit. No one says you have to say this or that. Very rarely was anything of mine edited for content. The effect is more subtle and structural. For example, it's predetermined what weeks you write a "thumbs" [referring to the title of a regular critical column] and, often, it is already decided whether you are doing an up or down. Theoretically, you could write one at any time, but you're so overworked and stressed, you're not going to write one unless it's your turn. Also, the magazine is big on balance, but how to enact that balance doesn't feel very balanced. The magazine wants to be a critical leader, but some shows have to be praised, so that leaves the negative critiques for other shows. *ATWT* and *GL* often get much more negative press because negative press is needed, yet other shows are immune to it.

DOING SOAP CRITICISM

As the above findings suggest, the matter of structural location of critics and the navigating they must do to manage competing interests provide an important context for this analysis and represent potentially significant constraints upon the ultimate product of the critics: their reviews. Critics in all fields evaluate cultural objects relative to aesthetic systems to arrive at judgments of value or worth. When reviewing television programs, "[T]he television critic is on the lookout for novelty, quality, controversy, and the new and different (as, to a degree, is the Broadway playgoer or art gallery habitué)" (Littlejohn 1976, 152). But how does one perceive these qualities in soaps?

Genre is one of the central organizing conventions of television production, and, as a result, it informs "standards for evaluating and appreciating cultural objects" (Crane 1992, 112). Primetime television critics are responsible for covering shows regardless of genre, and they readily understand how type of genre prescribes the kind of entertainment to be derived from a particular show and how its potential should be evaluated. Alongside genre, critics are responsible for studying the structure of both individual episodes and, likewise, the overall arc of a series: Did it accomplish its goals as a narrative or its intent to entertain? Unlike most primetime television shows, soap operas are arguably infinitely complicated to evaluate. The distinctive and unique properties of the genre include content that specializes in narratives about emotional life, often told in real time, with stories that foreground character over plot. Other characteristic properties include open-ended narratives with storylines that never achieve closure, complicating the boundaries of what should be subjected to appraisal.[6] Critics who specialize in the soap medium recognize and understand the uniqueness of soaps and regard their work as considerably more specialized than that of their colleagues elsewhere in the industry:

> [Soap reviewing is] very specialized! You must be well versed in soap history and understand how the industry works because it's unlike primetime and film.

> I think it's very specific, and I think a critic has to understand and also have an affinity for soap storytelling. I don't think a theater critic is necessarily going to be the best person to write a soap critique, nor would I presume to write a review of a Broadway play.

> The genre has its own special rules and conventions that other forms of entertainment don't possess. Also, its continuing nature offers both special freedom and problems for the writers. On the plus side, stories can actually develop in real time, which is possible only in novels. On the negative side, writers have to take into consideration often years of history when moving forward, though they can find creative (and often terrible) ways to deal with inconvenient facts of the past.

> On the one hand, I don't think it's specialized at all; the same evaluative criteria applies [sic]. On the other, you "read" a soap differently than music or movies or even other television shows, and people outside the soap world do not have that lens.

Because the portrayal of genuine emotion is so central to the genre, respondents were also very clear about what comprises the core focus of their appraisals. One critic stated:

> Writing, acting, direction, and *authenticity/soul* . . . quality of writing, acting and *believability* within the parameters of the conventions that govern the form [emphasis added].

Another concurred by clarifying the particularly important role of authenticity and believability to her assessments of quality and the importance of those attributes to audience pleasure:

> If the show hits all the right beats; if the particular storyline is true to the soap genre; if the writing, directing, and acting all come together in some magical way; *if it touched your soul*; if you felt different after watching it [emphasis added].

One respondent stated her bottom line this way:

> The litmus test for me is this: does it resonate; has *a truth* been shared from artist to viewer? Has that transcended the medium? [emphasis added].

Scholarly work on soap operas underscores the fundamental contribution of emotional authenticity in soap opera storytelling to fans' engagement of the genre, so it is not surprising that critics, too, would validate such authenticity as being at the crux of the genre. Perhaps the pivotal reason for this aspect being so fundamental is that character development is the centerpiece of soap narratives, and viewers expect characters to resonate "true to their conceptualization," as one critic said, especially in how they are written and in the genuineness of the emotions they express.

Certainty about and consensus on standards for appraisal notwithstanding, the profound and dramatic decline in the soap opera audience has had an equally profound and dramatic effect on the scope of the work of soap opera critics. Declines in the audience base have eroded the market for soap magazines. This eventuality, in turn, has dramatically affected employment opportunities for soap critics. Some reported that, although still employed, they have lost their columns altogether or have had to modify how they go about the critical analyses they do. A particular effect of the dwindling audience was ruefully observed by one respondent, who stated that, in order to continue to do her job, she felt compelled to adjust her criticism so as not to further erode the industry. She made this adjustment by "accent[ing] what was working more than what was not working in hopes of maintaining a balancing act between valid critique and not hurting the shows more." As if such concerns were not enough, these declines also intensify the insularity of the industry itself, noted earlier. Speaking of adjusting their criticism, respondents said the following:

> I found an increased sensitivity among soap execs about anything they perceived might have a negative impact on their shows. The Internet, which provides a forum that is much quicker and easier than writing letters to

the editor, made the dissemination of soap information and criticism a cottage industry for viewers; the daytime press—unable to hide behind cutesy pseudonyms and having to adhere to journalistic standards—became a more benign presence. However, the press always was an easy target (not only where soaps are concerned, of course), and, in this new, free-for-all environment and with ratings on a steady and alarming decline, the magazines were on the receiving end of closer scrutiny and wrist-slapping, simply because they knew where to find us.

It is such an insular industry that you [now] have to be very careful what you write. Actors and some EPs can be so sensitive and then withhold interviews, etc. from the magazines. For example, our "best and worst" issue is not bylined. Inevitably, one of my actors would be furious over something, and the publicist and I would tell him/her that I didn't write it. I did, of course, but it went a long way to soothing hurt feelings while not damaging my relationship with the show.

CRITICS AND THE AUDIENCE

As noted earlier, contestations between the soap audience and the shows over who is better able to ascertain the quality of the narrative and of the emotional authenticity it offers are legion (Bielby, Harrington, and Bielby 1999; Harrington and Bielby 1995). While fans of all genres pass judgment on the quality of the narratives they consume, they differ in their tactics and in the legitimacy with which their claims are received by executive producers. All forms of popular culture, soaps among them, strive to achieve broad appeal based upon cultural knowledge widely shared, while simultaneously incorporating sufficient novelty to perpetuate interest. But, in light of soap critics' delayed entrance into the industry, how do they maintain their place relative to the audience, and how has the rise of the Internet affected critics' relationship with fans? To examine these aspects of soap critics' work, I focused on the critics' perceptions of their influence on the audience and the emergence of fan-based Internet-located criticism as a major competitor.

Because the soap audience's cultural authority is so well developed, soap critics perceive they do not hold much sway over viewers (nor, they say, do fans have much sway on them). So, when asked how much influence they have on fans, respondents said, for instance:

> None; soap fans know what they like and hate—and that's the way it should be. However, fans crave a dialogue with critics, and LOVE a debate, I think, more than they love their stories.

More than [critics] did before the field expanded around the time of [SOW]'s inception. I'd say [fans] pay attention to what we say to see if it mirrors their own opinions. No one stops watching a soap because critics don't like it, but [fans] might choose one over another if they are wavering and critics laud one over the other.

I don't think soap critics influence viewers to turn off or turn on a show; however, I think they do fuel the fires for the "lay" critics, and that criticism of the critics' views has become a subset all its own.

While I worked at the magazine, I would have said "Not much." Now, talking to fans, I realize that they paid much more attention to critiques than I thought they did.

But, even as most critics expressed skepticism about the extent of their influence on viewers, one also observed that her work was guided by an abiding respect for fans' judgment. A key element in her work was maintaining "a soap's respect for the viewers' intelligence and loyalty." Elaborating further, she stated, "There is a responsibility to fans who watch a show daily, for years, to maintain the integrity of characters and to respect the show's history."

THE INTERNET

The rise of the Internet has appreciably complicated soap critics' established relationship with the audience and the industry, creating both drawbacks and benefits related to how they practice their craft. For some critics, the Internet has been circuitously beneficial to their cultural authority, and fan postings have even proven useful on occasion:

> The Internet hurt the magazines but gave me a whole other area of employment. I think the magazines have much more impact than the Internet because, online, everyone is a critic! I could be a little more open in certain kinds of criticism online, but not much more. I was working for ABC/SOAPnet online, and, in the beginning, I had a lot of battles with "management" over what I could say and not say. At [SOW], I was the boss, and only my own discretion and need to keep good relations kept me from saying anything I pleased. Later, I had more freedom online than I had in the beginning.
>
> The Internet can give me a flavor of what is out there, what the fans like and don't like. Often, if I was writing an "Editors' Choice," I would read what

certain fans had written because they often had great insights on why a particular scene worked.

But, for others, the Internet has brought unforeseen and clearly unwelcome effects because of redefined access to cultural authority. Some critics expressed considerable dismay about the future of professional soap criticism in light of the erosion they perceive the Internet has driven:

> The future of soap critics? I have a (sadly) gloomy opinion about this, as I see the genre in a state of decline, both in quality and in number of viewers. With the additional impact of the Internet, and the immediacy and broad reach that it provides, I fear that the traditional avenues open to soap critics are decreasing, along with their status.
>
> Bleak. Soaps are continuing to fall off the radar, and "in general" people think that because they don't "get" soaps, they're bad. While directing has auteur theory, soaps do not yet have a unified theoretical lens.

One critic explained and summarized the complications the Internet has introduced this way:

> Soap criticism was almost unheard of, other than a few lone voices, until [SOW] made it a standard and accepted practice. We dragged the soap industry kicking and screaming into the real world, rather than the la la land of adoring fans and fluff press. I think they'd all rather we'd go back to the old ways, but the cat is well out of the bag and will never be crammed back in. The problem is that, because of the Internet, every fan has the opportunity to be a critic. Some of them have incredible insights; others are just cranky, unhappy people with an ax to grind. But all have equal opportunity. When I lost my column at SOAPnet, I heard that Brian Frons [president, Daytime, Disney-ABC Television Group] wanted to replace me with free, fan-based critiques. It's cheap, but not as effective, in my opinion, as most fans watch one or two soaps (except for the real fanatics), while the critic watches everything and has much more information by which to judge each soap. We also are somewhat objective, though all criticism has some measure of subjectivity, as that is human nature. Fans are totally subjective and biased toward their favorites or against some actors, writers, and producers. [SOW] was the first publication to cover the behind-the-scenes workings of the industry, and it made people eager for this insider viewpoint. They also learned the names and functions of people who had remained anonymous. Although we very much legitimized the genre in the eyes of the outside world, we might have also hurt it by making fans too demanding. Sometimes, ignorance IS bliss.

Clearly, the field is in transition. There was no agreement on where soap opera criticism is headed, with some asserting that, as long as soaps are around, there will be critics, while others are in despair about their future.[7] But, in the context of these comments, it is worth considering that what some may actually be revealing as notable about this transforming field is that the sheer *volume* of Internet activity among soap fans is astonishing, if not overwhelming, in the context of the soaps' declining audience. That is, even as the genre has become a niche industry and its audience has eroded, fans have not lost their intense love of the genre. During preparation of this essay, a search conducted for Web sites devoted to soap commentary found at least seven such readily available—indeed, major—locations.[8] These sites provide just about anything a critic could offer, including insider scoops, hints about the future of story lines, opinions about plot developments, news about what's happening to the genre, and so on, all of which is being *very* well documented by fans on these sites in ways that rival even sites devoted to the most popular primetime shows.

These findings raise interesting considerations about the future of the soap genre and the place for critics and criticism within it. If critics have traditionally played an intermediary role between the industry and its fans, these sites may be signaling that the critic's role is no longer needed in a formal way, or that this aspect of critics' function has become an informal one that can be easily sustained by sources with access to Web site outlets. On the other hand, the sheer volume of Internet activity *about* soaps suggests something is being overlooked by attention directed solely to traditional, objective measures of audience activity. In particular, the extremely high level of public interest in the pleasures of the genre itself *continues* to be generated and sustained in the midst of the precipitous decline of the soap opera audience. Evidently, for connoisseurs of emotional authenticity about everyday life, there is no substitute for what the soap opera genre offers.

CONCLUSION

By exploring soap opera critics' standing, practices, and relationship to the audience, this study seeks to advance insight into the complex relationship of critical authority to popular culture. Although professional soap critics provide an important function for the industry and its audience, they face an uncertain future due to the "fall" of soap operas, combined with the "rise" of the Internet and user-generated content. Both trends have hit the contribution of soap critics particularly hard, from different sides. These factors suggest that it is not possible to understand fully what has happened to soap criticism—or what will happen to the future of the soap genre or its industry—without first understanding both of these trends as important contributors to the larger world of popular culture.

NOTES

Acknowledgement: Research assistance on this project was provided by Clayton Childress.

1. Although elite art is often considered to be "pure" and free of commerce, it is just as motivated and influenced by market considerations. See Becker (1982), for example.

2. Paul Denis founded a dozen magazines for Sterling's Magazines Inc. in the 1970s alone (Denis 1985).

3. Eventually, soap fans' claims to critical authority evolved into a mechanism that enabled the magazines to provide additional evaluation while at the same time remaining on the good terms that were necessary to stay in business within the insular daytime industry. According to established soap opera journalist Michael Logan, the magazines' increased reliance on fan opinion "is the magazines letting the fans [do much of the criticism] so it doesn't reflect badly on the magazines" (Harrington and Bielby 1995, 73). But, above all, incorporation of professional critics by the magazines, and the parallel inclusion of fan opinion columns and features, reflected what had long been the case about the soap opera audience: It is made up of a demographically complex viewership that draws from all social strata, educational backgrounds, occupational experience, ages, and genders, and it is astute at appraising the complex formulas, codes, and conventions of a specialized narrative form.

4. Scardaville has an essay elsewhere in this collection.

5. The dynamics of this situation are not altogether different from what primetime critics must contend with. Early on, dependence on the industry became a cause for concern among television journalists who feared industry-sponsored access to publicity compromised their professional autonomy and objectivity (Lang 1958). As a result, critics took matters into their own hands in an effort to effectively monitor and maintain journalistic integrity and standards. In 1978, they established the Television Critics Association to assure their independence when covering the networks' biannual rollout of the primetime schedule (http://www.tvcriticsassociation.com). None of my respondents reported belonging to this organization.

6. One critic stated in that regard, "As Agnes Nixon said, 'If it ain't on the page, it ain't on the stage.' Character and story were key for me. One of the reasons I loved *Edge of Night* was that the characters were rich and layered, and that they and their stories 'meshed'—this was one of creator Henry Slesar's favorite words. The best soap writers do this so that it appears seamless and effortless, but that's a tall order (as we see in recent years)." *Edge of Night* aired on CBS and later ABC, from 1956 to 1984.

7. In fact, one established critic declined participation in this study because the Internet has rendered soap opera critics "officially irrelevant."

8. The sites include "Marlena De Lacroix" (http://www.marlenadelacroix.com), "SoapZone" (http://www.soapzone.com), "Media Domain" (http://www.mediadomain.com), "Soap Central" (http://www.soapcentral.com), "Soaps.com" (http://www.soaps.com), "Soapdom" (http://www.soapdom.com), "The Soap Dispenser" (http://www.thesoapdispenser.com), and "Soap Opera Network" (http://www.soaperanetwork.com).

HANGING ON BY A COMMON THREAD

—JULIE PORTER

> Julie Porter is a longtime newspaper editor and reporter. She is Webmaster of *talk!talk!* (http://ghtalktalk.homestead.com/talkpage.html/), a site that aims to collect links to all *General Hospital* sites online, as well as Webmaster of the forthcoming site *Soaps of Our Lives*, an aggregation of the Web sites of hundreds of soaps worldwide. She has watched *All My Children* since its debut, seen the entire run of *As the World Turns* on CBS, and followed every current CBS and ABC soap at some point.

Soap operas are the stories of our lives—women's lives. Watching CBS's *As the World Turns* with my mom in 1960, both of us sipping iced coffee, was my entrée, as a six-year-old, into the world of adults. As my mom took a break from housework and we perched in front of our tiny black-and-white TV screen, our mother-daughter roles were temporarily suspended. I gained a view of my mom as an individual as she stepped outside her usual role as a caretaker and fixed herself on the goings-on in Oakdale. My values and my views on adults' roles in the world were formed as Mom enthusiastically passed judgment on characters good and bad, their errors and their relationships. I bobbed my head and took note of it all. The murder trials, the affairs, the lives of the doctors and lawyers and executives and families all gave me a window into Mom's thoughts and opened a discussion that might otherwise not have come so easily. The demands of life faded a bit as we sat together in front of the TV screen, riveted by the drama of the day.

When she turned eighty-four, we still shared two lifelong threads: our love of soaps and our reasons for watching them. Near the end of her life, I realized that I knew her values, and I knew her as a woman and as a person—not just as a mom—because sharing our soaps had opened a dialogue between us that we wouldn't have had otherwise. From their inception, soaps stepped in and addressed the issues that were on many women's minds: What were a woman's priorities? What endeavors were important to them? Was it wrong to have interests or concerns outside the family? As soaps encouraged my mom to question things a little more than she might have otherwise, she passed on to me the inclination to do the same: question and share. That's what I did with my mom in our living room, and that's what women do now online.

Mom became hooked on soaps during the Golden Age of radio, when two out of three families welcomed radio into their homes (Schoenherr 2007). Radio became America's voice, and the nation's focus shifted away from newspapers. When television came into our homes on a big scale in the late 1940s, as both the number of broadcast stations and the sales of televisions soared (Federal Communications Commission 2005), it was appealing but just a little mysterious. Everyone wanted television, even if it was a little off-putting to have an image beamed into your home from thin air. But soaps had already established a strong bond with women via the radio dramas that already were immensely popular. TV was a strange new medium, lighting up our living rooms and generating talk—but soaps gave it a friendly face. In 1952, housewives enthusiastically welcomed a new incarnation of *Guiding Light*, the first radio soap to make the move to television, into their homes.

Similarly, soaps paved the way for Internet users. In the mid-to-late 1990s, when computers flew off the shelves and into our homes, many new users were at a loss as to where to point their mouse. ABC jumped in to promote its official Web site at the end of its shows, and the most interesting aspect by far of the late-1990s/early-2000s version of the ABC Daytime site was its message boards. In particular, the *General Hospital* (*GH*) boards were constantly awash with battling Sonny and Carly fans versus their opponents, and the two sides consistently feuded into the night. Online fans gathered there, faithfully and daily, to share predictions, opinions, song lyrics, encouragement, thoughts, recipes, birthdays.

Many of the *GH* fans who were looking for more extensive discussions than the ABC format allowed spilled over to SoapZone,[1] a site launched in 1994 as Port Charles Online that eventually became the home base for *GH* onliners. Once again, women responded, and they picked up SoapZone as a daily habit, providing a waiting pool of potential friends to anyone who signed up and logged in. Today, when we watch soaps, we're sitting in a larger "living room," and there are a lot more of us "talking" at once. But soaps still bring women together. On Internet message boards, these shows steer us toward other women who share our values. There are plenty of people in our lives who would say that both of those activities—viewing soaps and posting at online soap message boards—are unproductive. They would be wrong, in my view. I believe that just about any bonding activity between women is a good activity, and soap message boards are a bonding venue like no other. Thanks to hundreds of forums, ours is the first generation of online relationships, and soap discussions still help us to evaluate a lot of the tough questions. What's our role in society? What's important to women today? Whose side do we take when the lines are blurred between good and bad?

Internet message boards and social networking sites aren't the only ways that soap fans bond online, but they're among the easiest to access and use. If you're a soap viewer, once you have a password, you could be only five minutes away from a dialogue over a passionately shared interest. And, while it isn't a girls-only clubhouse—there are certainly men in the trenches of online soap fandom, posting their hearts out—the number of men is relatively low, and any who flirt, scoff, ridicule, or veer outside the ladies' comfort zone in other ways aren't welcome for long. The soap forums are still a woman's world. There are likely no statistics to tell us how many friendships using soap talk as a launching point are formed per minute, or how many words are posted before an online friendship becomes a real one, but I would bet that these relationships are being formed more frequently than we could imagine—probably at this very moment. Within the enormous Web that connects us all, there are niches where friendships are quietly formed. Pre-Internet, we could only wonder what others were thinking in their living rooms when Erica Kane hid those birth-control pills, or CBS/ABC's *The Edge of Night* district attorney Mike Karr prosecuted another suspect guilty. Now, we can all sit together and watch a soap scene "together," tapping our thoughts out on our laptop keyboards.

Often, fierce bonds can form among soap viewers who are irritated about the same issues. Onliners can join together in their outrage over unpunished characters and open-ended storylines. Soaps appeal to women's desire for resolution, even though soaps are never-ending; when resolution doesn't occur, frustration ensues. Discussion about a story thread that a given soap has seemingly dropped can give viewers the feeling that some kind of resolution has been achieved even when a soap's writers fail to provide it. Likewise, factions of fans may bond over a common perception about a show's overall direction, as with the *GH* fan wars mentioned earlier.

Posting through screen names, we soap fans disclose a lot about ourselves as well. The talk veers from soaps and characters toward the details of our daily lives: schedules, children, husbands, careers, money, plans, aspirations, regrets. Ultimately, we've opened discussions online that otherwise might not have come so easily. Disclosure comes easily because we share strong bonds, our mutual years- (or decades-)long viewing of daytime dramas. And, even if one or more of us stops watching "our" soap and deprograms the TiVo, the friendships based on that easy online discussion remain intact. Just let someone criticize our soap viewing as being a waste of time; today, the first aspect of being a soap fan that many of us will defend is the strong online friendships we've forged with one another.

By allowing women to bring deep emotions to the surface and providing a space for women to make their private thoughts public, some Internet soap message boards become interesting soap operas in their own right. Just as soap viewing coaxed my mother into spilling a few secrets here and there, message boards make it easier for us to let down our guard. Miles apart, posting as Feather1982 or TuffGurls or OhYesImaPrincess, women feel free to divulge more than they might otherwise. Friendships form. Cliques form. Truths are told. Lies are told. As one of my online pals observed, "I guess it isn't called 'the Web' for nothing." It's an apt way to summarize the tangle of relationships and emotions, the lines that become woven together and the connections we make—and break—online.

There's a lot of fear that the Internet makes it easier for us to keep to ourselves and live solitary lives. The Web makes it possible for us to shop alone, at home. We watch sports at home. We can take classes or see a movie, all at home—alone. We can accomplish a lot without ever leaving the house, and without making any one-on-one, personal contacts. But soap fan boards bring us together daily. When we post that we "flove" the bad girl or "lobster-hate" the goody-two-shoes heroine "with the white-hot heat of a thousand suns," we're hoping that someone else, sitting at a screen ten miles away or a thousand miles away, relates to our comment, leans into the keyboard and taps out a connection. Like other message boards, online soap boards can be a refuge from real-life events. For women seeking friendships or advice or a place to retreat from problems, online forums ease the rough spots. These online conversations can easily become a part of daily life—a "happy place" to go when life becomes frustrating—and a place to meet good friends. The joy of forming friendships based on genuine fondness for an online buddy is rewarding, despite the challenges of physical distance when there's no substitute for "being there." Thus, online friendships can lead to phone meetings, or "IRL" (in real life) meetings (honest-to-gosh, face-to-face get-togethers). Conversely, online friends can be there for each other in ways that "real friends" can't, being more available and sometimes more dependable. A comment on the MSN group General Hospital Nite Owl Style, a group of late-night posters, describes the group this way: "We were born on the boards, our wings grew and we became great friends."[2]

Many soap viewers are selective about sharing their habit with non-soap friends for fear being judged, making conversations with online buddies even more appealing. Since soap watching is so widely sneered at, participation in soap opera social networking sites may be particularly derided. It can be dicey or even somewhat embarrassing to admit that a "friend" with whom you have a deep connection is in fact an online friend. "Oh, that isn't a real friend," is often the response. Your IRL friends may also look down their noses at online posting in general. "I don't chat," said one of my colleagues. "I don't waste my time." Never mind that the same person probably wouldn't consider a face-to-face personal conversation to be a waste of time. Soap fans may at times feel apologetic or even make excuses for their online attachments, but they have an understanding community online.

It's not all a bed of roses once one gets online, of course. Unlike other sites where members bond over a common interest—for example, breast cancer or parenting—on soap sites, there is likely to be acrimony. Lots of acrimony. As referenced earlier with regard to *GH* fans, if a soap's producers or writing staff seem to favor a particular character or a couple, the fans who dislike that character or pairing will launch an online backlash, which is certain to erupt into a fans-vs.-haters feud. In environments like Internet soap communities where contributors are invited to make their opinions public, emotions run high. It's not unusual to see a lot of frustration expressed when disagreements break out online. "Seriously, I need for people who keep harping on [their beloved character] to just STFU [shut the fuck up] and die," is a common, nightly vent on at least one board.

When the argument hits its peak, a soap onliner can end up banned from his or her favorite site—but there are still alternatives, such as the aptly named The Banned Club, which advertises "sarcasm, served fresh daily." For a decade, thebannedclub.yuku.com has been a welcome port in a storm for folks who were ejected from their regular message boards because they were too feisty for their own good. If a discussion participant ends up at this board or another like it, she can cool her heels before trying to return to the mothership (i.e., the board from which she was excommunicated), or just take up permanent residence in her new location. For a time, places like the now-defunct *GH* and *AMC* [*All My Children*, ABC] Fan Fussin' Forum or Bash Board welcomed onliners who were too confrontational to be happy at a sedate message board. The likewise defunct SoapZone After Dark was a place where regulars at the uber-popular SoapZone message boards could go and post comments that were too overtly sexual or obscenity-laden to make it past the eagle-eyed SoapZone censors.

Smaller boards seem to lend themselves to more intimacy; when six or twelve or fifteen members post together, the bonding comes easily, and posters tend to be friendly and accommodating, in a leaning-over-the-fence-to-chitchat sort of way. At the larger sites, however, turf defending can develop. On any given day, an innocent comment can explode into a "board war," a situation in which dozens of posters jump into the fray at once to argue and bully and huff and puff and hit and run. Interestingly, though, even the skirmishes can pave the way for bonding, as posters jump in to defend their pals, or send thank yous or consolations via private messages. Exhibitionism can take over the larger boards, too, of course, as posters fight to be more than a face in the crowd, and bad behavior is multiplied by the sheer numbers of others who see it.

We may work carefully to construct the image we want to convey, but our words can quickly give our real selves away. Rosen (2007) sees our online portrayal of ourselves as a kind of self-portrait, carefully constructed in pixels and digital bits rather than oils, adding: "But when one's darker side finds expression in a virtual space, privacy becomes more difficult and true compartmentalization nearly impossible; on the Internet, private misbehavior becomes public exhibitionism." Certainly, these types of bonds formed before the Internet became an everyday fact of life, but opportunities to make a connection weren't as numerous. The Internet brings another factor into play: anonymity. We are much braver, unseen and miles away, more willing to take the risk of connecting with someone, or of offending someone. We jump in headfirst, and we tend to tell all.

As soaps have cliffhangers, so too do online friendships. As happens in face-to-face friendships, we may "say" something in a moment of anger onine we later regret. But mending the rift is somehow easier in cyberspace: Shooting off a friendly message, hoping to smooth things over, with time to compose your words and rehearse your spiel, is immensely easier than making the gesture in-person. I've learned, too, that, when your career takes you from city to city and you have repeatedly to pack up and leave your "real-life" friends behind, your soap life and online life make the move along with you. Turn on the TV in a new town, and your friends in Pine Valley are still waiting for you, welcoming and unchanged. Likewise, your online friends are never farther away than the keyboard. Like a soap storyline that never ends, these online friendships mirror soaps' longevity, lasting over the years through many changes.

The open-ended nature of soaps lends itself to bonding as well. As people move away, and their relationships fade, many good things in their lives may come to an end, but soaps—which have been broadcast for as long as many of us can remember—will be here forever, we hope and believe. There's comfort there: Our shows aren't going anywhere, we hope, and neither are our connected friendships. As the shows span episodes, then years, then decades, so can our connections. And even if the soaps do eventually leave the air (anything's possible, right?), it's likely that those friendships will have more staying power than the shows themselves, thanks to the strong bonds that we've created. We love our soaps, but don't good friendships endure longer than love affairs? Our soaps may be in jeopardy, but our online friendships are strong—irrevocably held together by a common thread.

NOTES

1. SoapZone is located at http://www.soapzone.com.
2. Comment made on MSN Groups' General Hospital Nite Owl Style site at http://groups.msn.com/GeneralHospitalNiteOwlStyle (accessed January 7, 2009).

PERSPECTIVE
FAN SITE MODERATOR QUEENEVE ON FAN ACTIVITY AROUND AND AGAINST SOAPS

—BASED ON AN INTERVIEW BY ABIGAIL DE KOSNIK

| QueenEve is the pseudonym of a career professional and soap opera fan who has moderated and/or founded several popular soap communities online. |

My involvement in online fan Web sites was a progression. In 2000, I got involved on a fan site dedicated to a particular romantic pairing on ABC's *General Hospital* (*GH*) and posted regularly. I developed a persona; I had never done that online. Then I started reading fan fiction, and I asked, "What is this?" I had no idea it existed. I thought, "This is much better than the show. There isn't that extraneous stuff you don't care about: just the stuff you like." I thought I would try writing a story, and I did. I got all sorts of positive feedback. So I thought, "Applause; I must do more!" I ended up doing more fan fiction, and I was asked to be one of the moderators at the site.

What was wonderful about that site was that, for the most part, when it started out, really anything went. Maybe the stories got racy; nobody cared. We tried to keep things within reason, but it was a PG-13 board, and you could say what you wanted; it wasn't overly regulated. Then, there came a point when the site began getting attention. I think one of the soap magazines mentioned it. And, as soon as that happened, as soon as there was an awareness, then the people running it became concerned about what people thought: "Oh, take that down; we don't want people seeing that." They began curtailing people's posting. At that point, the site was hopping. It was getting nearly as much traffic as SoapZone. People went to the site and stayed there and posted all day, not just about soaps but about political stuff as well. And then the site became self-aware and that was the end of it. Other moderators said, "Don't post that here; I'm going to close this discussion thread." But hosting a Web site is like hosting a good party. You put the food and drink out, and you leave people alone. You don't say, "Don't put the cups there!" You don't over-regulate, or you immediately cut down on your traffic. People don't want to be treated like children.

I think there can be a shelf life for online fan communities. A problem comes into play when a Web site starts thinking it has an effect on what happens onscreen. As soon as that happens, the moderators say to the community members, "Don't say this; don't say that." I know for a fact that soap writers from various shows do troll these Web sites—some have admitted it in print. But that knowledge doesn't make for a good online community.

The relationship between soaps and Internet fan sites has evolved over time. The network executives and show producers and writers used to take online fandom a lot more seriously. I think the Sonny/Carly fandom really influenced *GH* in particular. But the ratings told another story. Now, they've gone another way, where they really think we're all obsessed nuts. I think we're used, though; if it feeds what they want anyway, they'll use an online community to push something. If they want to say, "Something's not working on the show," they'll justify what they want to do by citing online fan reactions, fan communities.

I was one of several organizers of a fan campaign called Target GH, protesting the violence and the misogyny that were rampant on *GH* and are still rampant on the show. We did everything right: we were smart, we were organized, we had pictures, we had example after example of offensive incidents, and we wrote to sponsors religiously. One of the people participating in the campaign was actually a big mucky muck with an ad agency, so she was kind of checking around to see what the response was to this kind of thing, and the response of the advertisers was that soap fans are nuts. So, we learned that we weren't taken seriously. *GH is* a horribly misogynistic show. Women are victims on the show; women are not allowed to win— even the heroine wins only because a man comes in and saves her. It's like stereotypical 1950s depictions of women—it's terrible, and, unfortunately, nobody cares.

The networks and soap producers do dumb things. They're not willing to try to adjust—they rearrange the deck chairs on a sinking ship. The writers just shift shows, just shuffle—God forbid they should bring someone new in to try something different. They're just going to ride it down to the bottom of the sea.

It's also weird who's got the power over soaps. There are too many straight men trying to run the shows. Soaps are not meant to be run by straight men. You've got someone who's valuing young nubiles who can't act, over women who are over forty who *can* act. The people watching the shows are primarily women who would prefer to see the women over forty who can act. There are exceptions—men who are enlightened. It's so frustrating; it doesn't have to be this bad. I just wish they'd get more women involved.

However, I do think online communities empowered the fans to make their own fun. The shows are not doing it for them. Soaps are going to die as a genre unless something changes. Online communities are the way we make our own fun outside the show. People make incredibly creative music videos where they make their own stories up—stories that never happened on the shows. And the critique and the hate are fun.

THE ROLE OF "THE AUDIENCE" IN THE WRITING PROCESS

—TOM CASIELLO

Tom Casiello is a current member of the writing team for CBS' *The Young and the Restless* and has served as an associate head writer for ABC's *One Life to Live* and NBC's *Days of Our Lives*. He won two Daytime Emmy awards as a breakdown writer for CBS's *As the World Turns*. He has written extensively on the soap opera genre at his blog *Damn the Man! Save the Empire* (http://casiello.blogspot.com/). His career path as The Writer began at *Another World* in 1999, but his membership in The Audience started in 1981 in his grandma's lap, and he has watched all six soaps currently airing on U.S. network television.

Walking into NBC's *Another World*'s writers' room for the first time in 1999 was a terrifying experience. Prior to getting that job, I was part of The Audience, one of those passionate fans who drank every last drop of history, of nuance, of the daily struggles of soaps' fictional characters. Now, here I was, entering the sacred ground of the writers' room of what had long been my favorite soap. What I found was both fascinating and completely the opposite of what I thought it would be. There wasn't a lot of focus on The Audience. That is not to say that nobody cared about The Audience. Far from it; there was much talk of focus group results and ratings fluctuations throughout an episode, and we discussed how our storytelling related to audience response. But the main focus was always on the story: what works for the story and what scenes are needed for the story to progress. The assumption was that, if the story worked, The Audience would follow. It wasn't so much about what The Audience wanted but how we could tell The Audience what they would want.

However, an interesting thing happened over the course of the next decade. Ratings started to plummet, and online fans became more outraged. As head writers and executive producers rotated from soap to soap, a general malaise washed over The Audience. Viewer comments were more often, "How are they going to screw up my show?" than "How are they going to save it?"

It wasn't until late 2007/early 2008 that I truly changed my way of thinking, and I did so because of one of the scariest events to hit daytime: the Writers Strike. Once we all got the order, "Pencils Down!," we were no longer firmly entrenched in just getting from one story point to the next. We could now step back from the trees and look at the forest.

My own journey back into The Audience actually started quite by accident in 2008. I had launched a blog on MySpace in mid-2007, soon after starting my new job at NBC's *Days of Our Lives* (*DOOL*). For the next seven months, I would blog once or twice a month about what it was like, writing words for Stefano and Marlena, for Sami and Lucas, for Tony and Anna—these wonderful characters I felt I had known my whole life. A few fans found me, but they seemed content not to share with the Internet at large that they had discovered a "soap writer's blog." Other fans looking for gossip quickly moved on, once they realized there was none to be found.

Everything changed in February 2008. After being part of an entire writing team whose jobs were terminated in the last weeks of the strike, my "Goodbye to Salem" blog entry (named after the town *DOOL* is set in) suddenly, and quite unexpectedly, thrust my blog into the forefront online. The average audience member still had no idea who I was, but those who haunted the threads of many a soap message board were suddenly more interested in what a writer had to say—especially in the light of the strike that had, in some cases, saved and, in others, decimated fans' shows.

My journey into the online world of fandom taught me so much more than I could ever have imagined. My detractors in writers' rooms still sneered that "The Online Audience" was nothing more than "five or six angry wannabes with nothing better to do than post snide comments about writers over and over again under different screen names." It is true that there are many who just enjoy the process of venting on a message board. But what I soon discovered was that many fans online were not only literate and eloquent, but also understood how the marriage of plot and character should produce quality family and romantic drama. They could also see quite well how many soaps are failing miserably at reaching a balance. This is an audience that is not only well read and theatrically literate, but also educated in the ways of great daytime writers like Agnes Nixon and Doug Marland.

And speaking of Marland . . .

In the midst of my study of what The Audience is truly looking for, I made a discovery: the original long story that Marland wrote in 1985 for CBS' *As the World Turns*. Monumentally iconic, Marland's original story included passages I had never seen in any long story I read at any other show I had worked for or under any other head writer. Marland's long story included The Audience as part of the document. In numerous places, Marland "takes time out" from his storyline to explain what he wants The Audience to be thinking and feeling at any given moment. He discusses when The Audience should "be ahead" of the characters in the information they're given, and when The Audience shouldn't be privy to certain pieces of the story. He talks about when The Audience will turn on a character and begin to hate them, and at what point in the redemption story The Audience should finally be back on that character's side. It's no longer a story told from the perspectives of a handful of characters, but a storyline rendered from the perspective of a handful of characters *and* The Audience.

I shared my thoughts—and portions of the long story—with fans on my blog (Casiello 2009a; Casiello 2009b; Casiello 2009c; Casiello 2009d). I also talked about Marland's document with fellow writers and discovered something amazing—that the distance they got from their shows during the strike helped them see what was right and what had gone horribly wrong. I've never felt that The Audience should dictate storyline, but that doesn't mean viewers shouldn't be a factor in the storytelling process. For years, I had been part of writing teams that struggled with what to do halfway through a storyline that The Audience was not connecting to, not identifying with. But, in recent months, I have witnessed conversations among writers about whether or not The Audience will emotionally connect with a story *before* it's written.

Thirty years ago, writers were offered the luxury in daytime of being auteurs—of guiding their own journey and allowing fans to follow. But in this new daytime landscape of falling ratings and desperate scrambles to discover "what works" and "what doesn't," I find more and more of my fellow writers turning to The Audience. Not for direction, so much as . . . but for inspiration. Gone are the days when we can craft our own progressions without fear of seeming like we're turning our backs on viewers who have given so much of their lives to our shows. At some point, the train went way off the rails, and if ever there was a time to band together, producers and actors and writers and network executives— . . . and yes, The Audience— . . . it is now.

At the time of this writing, I find myself on a team of writers at CBS's *The Young and the Restless* who all not only have had experience in the field for many years, but who all are fans of the genre. Many of the breakdown and script writers on my current team grew up on soaps and are part of the Internet generation. And being part of a team where we discuss our favorite stories, our favorite characters from years gone by, without a trace of cynicism and fatalism is a pure joy. I also talk to writers at other shows who continue to pay closer attention to how The Audience should partake in the story process than they ever have before. Who knows how long this will last? After all, we writers still live in a world where we're constantly looking over our shoulders, wondering when the beast called Cancellation will come nipping at our heels.

In the meantime, though, this writer is enjoying the give-and-take many of us have discovered since the Writers' Strike. Walking into a writers' room now, ten years after my first writers' room, is starting to feel as if that empty chair in the corner is finally meant for those who should always have been a part of the equation—: The Audience.

THE "MISSING YEARS"
HOW LOCAL PROGRAMMING RUPTURED
DAYS OF OUR LIVES IN AUSTRALIA

—RADHA O'MEARA

> Radha O'Meara is completing her PhD in screen studies at the University of Melbourne, Australia, where she is also a lecturer. She has published on soap operas in the Australian journal *Metro* and on *Gilmore Girls* in *Screwball Television*: Gilmore Girls. She began watching soap operas with the Australian primetime soap *Neighbours* and first took interest in U.S. soaps with the after-school hours screening of *The Bold and the Beautiful*, later moving on to *Days of Our Lives* and *Passions*.

INTRODUCTION

After screening episodes of the NBC daytime drama *Days of Our Lives* (*DOOL*) in a continuous sequence for over thirty years, Australia's Nine Network skipped approximately one thousand episodes in 2004.[1] The Nine Network continued to broadcast *DOOL* daily, but most Australian viewers missed four years' worth of episodes. Dubbed the "missing years" by fan magazines, the disjunction caused by this programming decision undermines common conceptions about how viewers engage with soap operas, raising a number of questions. How does this fissure change the viewers' relationship with the text and the network? How do viewers understand their own position in the power structures of global cultural trade? How do viewers make sense of the abridged story, within the context of serial narrative? The repercussions of this textual rupture give us unique insight into dynamic broadcasting relationships and the unfolding soap narrative.

Analysis of this event demonstrates some of the implications for the transnationalization of U.S. soap operas. U.S. soaps are regularly distributed around the world, and these global flows present a new range of viewing experiences for international audiences. In international markets, U.S. soap operas often begin broadcast midway through their life and are sold in blocks of one or two hundred episodes. The serial form of soap operas must be adapted for the programs to become global products, contingent on international trading practices and myriad local broadcasting practices. Under these conditions, the fragmentary experience of serial narratives is amplified. Further, this format adaptation alters conditions of reception for global soap audiences and changes modes of engagement. Soap operas have traditionally been associated with long-term viewers hooked on long-term storylines, but this scenario is uncommon in the export market. In the sea of global media flows, U.S. soap operas become fractured texts, offering different kinds of viewing pleasures for international audiences. The case of *DOOL*'s "missing years" in Australia offers an opportunity to explore the implications of these global forces.

The incomplete, fragmentary nature of soap opera narrative is highlighted by *DOOL*'s "missing years." Viewers understand that, in soap operas, there is no such thing as a whole, definable text. Even within this context, the Nine Network's decision to "jump ahead" in the *DOOL* diegetic chronology created an extraordinary gap in viewers' knowledge and experience. One thousand episodes was simply too much to catch up on via alternative media or ancillary texts. The rupture thus caused a loss of narrative threads, events, and connections. Australian viewers missed out on events and connections in the characters' stories, and they also missed connecting with each other through sequentially unfolding events in Salem. The rupture created a new and expurgated sequence of episodes for the Australian viewing community, and, with that change, new narrative meanings unavailable to viewing communities in other countries. The fissure produced an innovative amnesic effect, reconfiguring the complex textual histories of story, memory, and viewing experience. First, let us consider the context of this rupture, as part of a history of local programming practice, with established discourses structuring responses to the event.

SITUATING THE TEXT: LOCAL PROGRAMMING, LOCAL AUDIENCES

The Nine Network's programming decision to skip hundreds of episodes of *DOOL* reminds us that television texts and audiences are always socially and culturally situated. The event mobilized popular discourses about how relationships of power between viewers and networks function, but also underscored the dynamic nature of these relationships. Focusing on both industry and fan discourse, we can see how the relationships are strained and negotiated over this event. Analysis of industry discourse here is based on media representation of the event, which included some quotes from key network players. Analysis of audience response here is based largely on anecdotal evidence and fan reaction recorded in online forums, such as eBroadcast.com[2]. Audiences and broadcasters are inextricably bound in shifting relationships of power, just like the tightly knit network of characters in Salem.

The jump was announced by Nine about six weeks before it took place and reported in a range of media and sources, from specialist soap magazines to nightly network news and broadsheet dailies. The jump was mentioned in broader public conversations sporadically for a few days after the announcement, but fans debated the move hotly for the following months, through real and virtual social interactions. Nine continued to broadcast *DOOL* daily in the established sequence until Friday, September 10, 2004. On Monday, September 13, Nine screened a special "catch-up" episode and then commenced daily screening of the new episodes. Between that Friday and the following Monday, most Australian viewers missed approximately four and a half years' worth of episodes of the daytime soap opera; viewers in Western Australia missed six years' worth. These episodes of *DOOL* that were broadcast in the U.S. and other territories have never been aired in Australia.

The discontinuity between broadcasts of *DOOL* in the U.S. and those in Australia developed through a specific history of the intersection between international program trade and local scheduling practice. *DOOL* began broadcasting in the U.S. on NBC in 1965; it began broadcasting in 1968 on then Channel Nine in most major Australian cities. The Nine Network has broadcast *DOOL* more than two hundred days per year since 1968. Regular scheduling such as this represents a commitment between broadcaster and audience: The broadcaster achieves a proven rating to sell to advertisers, and the audience gains reliable viewing pleasure. Soap operas are based on the principle of returning regular audiences, and daytime television schedules in both Australia and the U.S. are traditionally less mercurial than their prime-time counterparts. The exceptional longevity and intensity of these commitments is part of the historical commercial success of the soap opera genre. This symbiotic relationship between broadcaster and audience has developed to the extent that soap fans exhibit extraordinary loyalty and often report feeling a sense of ownership over "their" shows (Harrington and Bielby 1995; Spence 2005), and networks use daytime soaps as scheduling and branding anchors. Television scheduling regulates the relationship between audience and broadcaster, building shared expectations and understanding of their imbricated roles. The discourse of a common commitment between networks and viewers frames popular discussion of television scheduling.

Yet this commitment was shown to be malleable even before the "missing years." The daytime screening of cricket during the summer months means that episodes of *DOOL* are not broadcast in their regular weekday slot around twenty to thirty days per year in Australia. Even after Nine's decision to skip four years' worth of episodes, *DOOL*'s broadcasts in Australia fell further "behind" *DOOL*'s broadcasts in the U.S. owing to sports delays. In a wider media context, where viewers felt increasing control over the television text as a result of technological and cultural changes, the network's practice of regularly pre-empting a program with a loyal viewing audience appeared to be an unusual exercise of programming power. Like audience engagement, network programming is a contingent, fluid, and active practice.

Seasonal interruption of *DOOL* by cricket showed the commitment between network and viewers to be mutable, but the jump of four years' worth of episodes was much more radical. At the time, the rupture was popularly framed as a schism of apparently stable broadcaster and audience expectations. Both the network and the audience tried to articulate their understanding of the situation in terms of a mutual commitment represented by *DOOL*, but they failed to see eye-to-eye. Anecdotal evidence suggests that some viewers accepted the jump as part of the already pliable nature of the commitment between broadcasters and audiences, while other viewers objected. Diverse audience responses are reflected in magazine and newspaper letters to the editor written in response to the event and printed in Australian publications such as *TV Week*, *Soap World*, *TV Soap*, and *Daytime TV*. Viewers active in online fan forums expressed dismay prolifically. Fans often angrily lamented their powerlessness to affect the decisions of TV networks, which they seemed to perceive as monolithic corporations.[3] It appears that casual, occasional viewers were more likely to bear the intervention, while fans were more likely to object. In effect, those who had invested more in the show, and felt more committed to it, were the most likely to feel aggrieved. The network pitched the change as something desired by viewers; fans felt that their long-time loyalty had been betrayed. The implicit understanding that existed before the jump, that the network and audience members were mutually invested in *DOOL*, was shattered.

The network's rationale for the move was never made entirely clear and is the subject of some speculation. In 2004, the director of daytime programming at the Nine Network, Lauren Sidler, defended the decision to skip years' worth of story time in a statement she made to Melbourne's *The Age* newspaper. She emphasized the viewer's putative desire to watch "more up-to-date" episodes (Moran 2004). Sidler's statement suggests that the soaps should be broadcast contemporaneously in Australia and the U.S., and that viewers in the digital age will not tolerate delayed broadcasts in different regions. By implication, the Nine Network rearticulated the broken relationship of scheduling expectations, arguing that the daytime soap audience betrayed them first by seeking their entertainment pleasures online. It is quite possible, indeed likely, that the Nine Network had experienced declining ratings for daytime soaps as a result of online entertainment; online media was—and is—a serious competitor for viewers, ratings, and advertising revenue. Many Australian viewers read online about upcoming storylines in various foreign television shows, though, at the time, it is probable that few watched newer episodes of daytime soaps via the Internet. If one of the goals of the jump was to dissuade Australian audiences from turning to the Internet for information and entertainment, it likely failed. The rupture may have actually caused more viewers to engage with online media. If, however, in the era of media convergence, luring more people to engage in online discussions of the show was one of the network's aims, then Nine can be said to have succeeded. Moreover, the "new" episodes of *DOOL* were still several months behind American broadcasts. Significantly, this delay nullifies the network's published argument that viewers demand contemporaneous broadcasting.

Nine's bizarre "fast forward" may have been primarily motivated by ratings. It is curious that the Nine Network never publicly discussed ratings, because ratings are the most common way that the scheduling commitments and conflicts between broadcasters and audiences are publicly articulated. Audience ratings are still a crucial element in the complex relationship between networks and viewers, despite changes to television distribution and reception in the global, post-network era. In the U.S., daytime soap operas have been steadily losing ratings for more than a decade now, due to a range of factors, including competition from talk and reality shows, cable television and the Internet, and social changes including increasing numbers of working women.

Daytime soaps have most likely suffered declining ratings in Australia, too.[4] In an effort to attract new viewers, or just to tempt old viewers to come back to *DOOL* in the U.S., NBC arranged for James E. Reilly to return as head writer in 2003. One of the things Reilly did to shake things up was to kill off half the cast, and the Salem serial killings did cause a temporary increase in ratings for *DOOL* in the U.S. After the jump, new episodes of *DOOL* in Australia began in the midst of the sensational Salem serial killer storyline. This storyline, which may have been a bid by the Nine Network to jump straight to episodes that were earning higher ratings in the U.S. and to skip the years of mediocre ratings experienced in the U.S. Not only could the network skip straight to episodes that were likely to achieve higher ratings, but it could manufacture an "event" from the decision to leap ahead four years as well, thereby gaining press from its own programming decision. In this way, the scheduling rupture might be considered an example of what Caldwell (2000) terms "stunt television," meaning, "producers and broadcasters refocusing interest on a series (and inflating its ratings) through a highly promoted deviation from the series profile" (Sconce 2004, 107). A short-term ratings boost may seem incongruous with the long-term nature of the soap opera genre. If the case of the "missing years" does suggest a daytime ratings "quick fix," then this effort may be indicative of the internationally precarious state of a genre based on long-term loyalties and pleasures.

The textual rupture created a localized, communal experience for Australian viewers, which is significant for what it reveals about the popular discourse surrounding this event. The network and audience responses were discordant in the ways they positioned themselves within global media flows. The Nine Network's rhetoric suggested that Australian soap audiences wanted to be part of a larger global audience, or indeed already felt a part of it. This suggestion implicitly supports cultural and social homogenization, with the assumption that synchronous broadcasting across territories would produce a single, trans-cultural audience and experience. The fan response reasserted a sense of cultural specificity, based on shared regional viewing experiences. Australian fans had always been behind in the episode count, and fans on the East Coast of Australia had been watching for decades, with full knowledge that they were asynchronous with the U.S. viewers (and viewers in many other territories). Fans were more concerned with their own histories of viewing *DOOL* than with acknowledging the source of the product or other viewing experiences.

Although Nine endorsed notions of cultural globalization, the network's actions actually reinscribed national and regional identities in the viewing audiences. Australian soap audiences experienced something international audiences did not. Viewers from Western Australia bonded over their even greater misfortune, as they lost two more years' worth of episodes than the East Coast had lost. The technology and commerce underpinning the rupture are part of the intensification of global mobility, but the effects are felt on a local scale. While Australian fans commonly surf American soap sites for information, they tend to stick to Australian forums, reinforcing the importance of scheduling and contemporaneity for building a communal audience bond. This focus on Australian forums illustrates Robertson's (1995) thesis that globalization always takes place in local contexts. The rupture demonstrates how global flows of media into Australia are localized through programming practice, whereas scholarly attention tends to focus more on the success of Australian television program exports (e.g. Craven 2008; Crofts 1995). In 2009, Australian audiences for *DOOL* saw episodes twelve months behind the US and shared a deeper bond through their common loss of the "missing years."

RUPTURING THE TEXT: THE NARRATIVE GAP *PAR EXCELLENCE*

Like the complex cultural and social histories shaping local programming and reception, audience understanding of the text is also a dynamic process, which was significantly affected by the rupture of *DOOL*. Everybody has missed episodes of their favorite soap, and regular viewers know that their knowledge of the text is always only partial. According to Hans Borcher in his essay with Seiter, Kreutzner and Warth (1989, 233), "Because of the vicissitudes of their personal circumstances, working careers, and everyday lives, even the most loyal fans are perfectly aware that at best they only have a very sketchy notion of the text in its totality." The incompleteness of the text is a familiar idea for soap audiences and scholars alike. In the case of the "missing years," the textual incompleteness is based on institutional practices of distribution rather than personal habits. The sheer size and scale of the rupture of *Days* in Australia makes this an interesting case study for understanding how inherently incomplete texts produce narrative meanings.

A typical response to the "missing years" in everyday discussion and popular media was to suggest that there was nothing to miss. For example, Garry Williams's (2004) *Herald Sun* article said, "Some cynics would say the shows move so slowly that not much will be missed except thousands of shots of actors staring into the middle distance." So it was commonly implied that the rupture would have no noticeable effect on how *DOOL* is understood and enjoyed. However, the "missing years" had significant effects on how the text is understood by Australian audiences, particularly how audiences comprehend the narrative. First, the jump in episodes created a narrative gap *par excellence* and thereby stimulated viewing which was newly productive in the creation of meaning. Second, the rupture engaged viewers in a complex revision of layers of history and memory in the soap text.

Viewers struggling to come to terms with the loss of textual experience from *DOOL* could understand their position as akin to that of an individual suffering amnesia—a plot device common to soap opera narrative. Indeed, amnesia forms a leitmotif in the life history of *DOOL* character John Black. Like John, Australian audiences of *DOOL* have suffered some kind of accident and trauma, thereby losing the memory of several years' worth of events and experiences. Although other characters in Salem naturally remember this period and continually refer to it, Australian viewers struggle to make sense of the "missing years." For Australian audiences, the flashbacks of absent experiences and dialogue explaining lost events help to piece together what transpired in the "missing years."

In order to examine the amnesic effects of the "missing years" on the text of *DOOL* in Australia, I will expand the concept of "narrative gaps." Put simply, narrative gaps invite the reader or viewer to engage actively in the production of meaning. Bordwell has outlined this concept in detail, tracing how narrative gaps suppress story (fabula) information in the plot (syuzhet). For example, mysteries conventionally suppress a murder by beginning the plot with the discovery of a body, engaging the viewer in a whodunit enigma. Central to the idea of the narrative gap is the gap-filling reader activity it implies: "Gaps are among the clearest clues for the viewer to act upon, since they evoke the entire process of schema formation and hypothesis testing" (Bordwell 1985, 55). Iser (1978, 169) also stresses the activity of the reader in his theorization of gaps, arguing more broadly that gaps are all those asymmetries between the reader and the text that constitute the very act of reading. Scholars such as Iser and Bordwell agree that gaps are both necessary and productive parts of narrative, engaging readers and viewers in the creation of meaning.

In the context of television, the idea of narrative gaps has been expanded to include gaps in the text. Allen (1985, 78) notes that serial narrative television regularly relies on the textual gaps between segments and episodes for viewers to produce narrative meanings. Fiske (1987, 103) describes the gaps between episodes as spaces in which "the viewer 'enters the text' in the imaginative and creative way," going on to explain that, "These gaps quite literally make the soap opera a producerly text, for they invite the reader to 'write in' their absences, and the invitation is readily accepted by many viewers." The regular textual gaps of narrative television are therefore understood as larger versions of narrative gaps. Furthermore, Fiske suggests that the producerly activity of viewers in filling gaps delivers significant viewing pleasure and offers ideologically progressive potential. Television viewers use the gaps provided for commercials and between episodes to engage in schema formation and hypothesis testing, often articulated as speculation and gossip.

The rupture of *DOOL* created a bounty of narrative gaps for Australian fans of the show, dislocating the seams of sequence, causality, and character association. Australian *DOOL* viewers were left to wonder: What happened to Chloe, Philip, and Eric?; What happened to Brandon, who apparently married Angela, then disappeared?; How did Nicole convince Victor to marry her?; Why did Marlena remarry John after he reportedly had an affair with Hope?; How did Stefano DiMera die (this time), and how did his brother Tony return from the dead? Viewers actively engaged with the ongoing text to answer these questions and make sense of the ongoing narrative.

Bordwell provides the descriptive tools for analyzing narrative gaps, which can be extended to the textual gaps of television programs. Bordwell argues that narrative gaps can take the form of temporal, causal, and spatial gaps (1985, 54). He further describes their characteristics: Gaps can be temporary or permanent, diffuse or focused, and a plot (syuzhet) can flaunt or suppress gaps (1985, 55). The development of narrative television has seen texts become increasingly aware of their gaps. Textual gaps in soap operas tend to represent focused temporal as well as causal gaps, and viewer attention is drawn to these gaps by the frequent use of cliffhangers. Soap opera, due to its excessively serial nature, particularly embraces the productive power of narrative and textual gaps. The massive textual gap created by the "missing years" was a focused, temporal, causal, and spatial gap. While not flaunted by the text itself, the gap was highlighted by the network through promotion and advertising.

The "missing years" might appear to be a permanent textual and narrative gap in *DOOL* for Australian viewers. Bordwell suggests that gaps can be filled in two ways: through narrative experience and through information. Gap filling through narrative experience has been limited but significant in the case of the "missing years." Small parts of the gap have already been filled by a special episode, trailers, and regular textual redundancy in the forms of recaps and flashbacks. Viewers' endeavors to fill in the "missing years" made them particularly active and productive during this period, and for some time afterwards. Further, this activity generated new meanings, anxieties, and pleasures for audiences.

The Nine Network simultaneously flaunted the gap and offered to fill it with a "catch-up" episode. Nine announced the leap about six weeks before it happened. The network advertised the jump in fan magazines and the popular press while continuing to broadcast episodes in regular sequence until Friday, September 10, 2004. On Monday, September 13, Nine broadcast a one-hour catch-up episode, a locally produced program offered to appease the fans of *DOOL*. This special episode was immediately followed by a new episode dating from four years later in the U.S. chronology. The *DOOL* catch-up episode was called "A New Day" and included approximately ten minutes of material to tie up the loose ends of story lines which had been current for Australian viewers, such as Sami's eventual success in her bitter custody battle against Lucas after Sami's lover Brandon married Angela. Most of "A New Day" concentrated on the later parts of the gap, showing several murders in Salem's serial killings. In the original text, each murder in the spree would have served as the climax of sustained story development, but the special episode dismantled the regular sequence to present a new one. In "A New Day," seven murders were shown together in a montage, and further recontextualized by extra-diegetic graphics, the host's introduction, and voice over explanation. With cast interviews, behind the scenes material, and advertising, less than half an hour of the "missing years" of *DOOL* was actually broadcast in this special episode.

The new sequence of episodes created by the gap created new meanings, both within the special episode and beyond it. The position of retrospection alters the interpretation of some events. For example, the first killings are not presented as individual points of dramatic focus or narrative mystery, but as the prelude to a killing spree. Before we even see Abe Carver's murder, the host tells us that Marlena is the killer, and Abe is introduced to the sequence in freeze frame with a graphic labelling him "Victim #1." Australian viewers missed months of speculation over who the killer might be and jumped directly to conjecture over why Marlena has apparently become a serial killer: Is the murderer her twin sister Samantha, her doppelganger Hattie, or is she possessed (again)?

The four-year fissure means that some moments and events will be read "together" by Australian audiences, but not by audiences elsewhere. On Friday, Sami and Lucas were in a bitter custody battle over their son, Will. Kate and Lucas had drugged Sami and she felt she was going insane. The following week, Sami and Lucas were kissing. The gradual and clearly sign-posted changes in character relationships that constitute the substance of a program like *DOOL* are elided by the new sequence of events. Sami and Lucas' relationship has been tumultuous for years but appears particularly so for Australian viewers. In soap operas like *DOOL*, the motives of friends and lovers are often called into question, as many relationships are based on jealousy, deception, money, power, and manipulation. Changing the narrative sequence alters the ways these relationships are interpreted. Kate was reportedly in love with Roman during the "missing years." However, viewers who didn't see them together might not accept the relationship, reading it simply as one of Kate's manipulations. This effect also has long-term consequences: Years later, when Austin returns and reunites with Sami, Australian viewers are less suspicious of the returning lover because they never saw him callously dump her at the altar. These are some of the ways Australian viewers configure a localized reading of the abridged text.

Scholars and fans acknowledge that American daytime soaps rely heavily on repetition and redundancy (Allen 1983, 100). In *DOOL*, this redundancy commonly takes the form of repeating actual parts of previous episodes through recaps and flashbacks. In addition, plot information is regularly summarized by characters' soliloquies and dialogue and used frequently in the opening scenes of episodes. This summary assists viewers in filling gaps and also recontextualizes events. For example, in the first new episode of *DOOL* screened after the jump, John summarized Tony's death in his conversation with Marlena in this way:

> It's been a hell of a day, I mean ... the tiger getting loose, the rampage, Tony getting mauled, and you ... you took a hell of a shot on your head. And you never officially got that looked at either, I mean, all the medical attention was

hovering around Tony, wasn't it? Armies of doctors, police everywhere, and still that murderous son of a . . . got right inside Tony's cubicle, gave him a lethal injection before we could get the killer's name.

This informational recap through dialogue is intended to assist viewers who may have missed parts of the story, and such snippets became particularly important for Australian viewers. Flashbacks serve a similar function. During John's speech, quoted above, Marlena and the audience experienced visual flashbacks of Tony's lethal injection. The flashback re-presents the footage for the audience, but it also subtly suggests that the killer's fragile mental state is decomposing through Marlena's unsteady performance before and after the flashback. This recontextualizing of prior events is underscored by the prominence of flashbacks within regular *DOOL* recaps. This strategy demonstrates that earlier events are integral to current narrative developments, but, more pertinently, that it is important for viewers to understand how characters are currently reacting to and integrating their pasts into the present. These processes of narrative recontextualization are complicated by the textual rupture.

Audiences fill permanent textual gaps with information gathered in two key ways: from the text itself, and from extratextual sources. As in the quote from John above, soap opera dialogue is littered with summaries, descriptions, and reinterpretations of past events. Gap filling through extratextual sources is so common that it has spawned economies parasitic of soap operas. Fan magazines benefited from renewed interest and confusion caused by the "missing years." Australian television magazines such as *TV Week* and the local soap opera magazines mentioned previously exploited their privileged position of authority by publishing a slew of special articles, sections and issues.[5] Moreover, there is a legion of official and unofficial Web sites providing access to written summaries of episodes, video highlights, and communication forums. In addition to the fan rancor registered above, many discussion boards indulged fans in their communal search for information and speculation on what happened to the characters of Salem during the "missing years."

Most viewers who experienced the "missing years" probably combined these sources to plug the amnesic gap. Like the process of filling ordinary narrative gaps through reconsideration of previous scenes, lines of dialogue, and actions in the light of new episodes, the gap-filling caused by the "missing years" never stops. According to Bordwell (1985, 55), "Temporary gaps point us forward and build up surprise; a permanent gap invites us to apply a 'scanning' strategy, sorting back through single episodes looking for information we might have missed." Viewers are constantly revising their understanding of the narrative, filling the gaps in new ways, following this snippet from a new episode or that tidbit from a Web site. In the wake of the "missing years," Australians viewed *DOOL* more attentively, more purposefully, and more self-consciously.

Allen (1985, 78) suggests that enforced textual gaps heighten the protensive indeterminacy of soap operas. The larger narrative gap caused by permanent loss of textual material probably caused viewers to increase speculation about what would happen in the future: In such circumstances, the text became more open and full of possibilities. This enforced gap means that viewer engagement changes in intensity, as well as in how it produces meaning. Significantly, this greater openness might also be part of the reason the Nine Network anticipated higher ratings. After the shift, regular viewers in Australia might be stimulated to view more often and more attentively to gain footholds in the new narrative terrain.

RUPTURING THE TEXT: LAYERING HISTORY

Bordwell's description of retrospective scanning is particularly pertinent for this case study, because it raises the issue of textual temporality. Soap operas are notorious for their complex chronologies, and an event like the "missing years" further complicates soap chronologies. The idea of history is very important to serial form, noted for the longevity of the programs and for viewer commitment. Many of the narrative events in a soap opera actually refer implicitly and often explicitly to previous narrative events and make little sense without knowledge of the diegetic history. For example, the slayings committed by the Salem serial killer were highly individuated, referring to significant events associated with the murdered characters. The serial killer bludgeoned recovering alcoholic Maggie to death with a bottle; this method would carry less meaning for a viewer unfamiliar with the character's past. White (1994, 339) suggests that there are two meanings of history in soap operas: First, she discusses the history of the diegetic world, which roughly equates to the history of story time; and, second, she discusses the history of the viewers' experience of the program. Temporality, chronology, memory and causality are closely related in soap opera. Since story time and viewing time are often linked and conflated, these different senses of history have a complex relationship.

The "missing years" complicates this history further. Most obviously, four years' worth of diegetic history effectively disappear. Viewers make meaning from the text in different ways in an attempt to account for this change in diegetic history. In addition, the rupture and the special recap episode have the effect of pulling viewers out of the diegetic history and making them very aware of their temporal relationship to the text. The rupture acts as a significant milestone, punctuating viewers' recollections of diegetic stories and of their own narratives of engagement. The temporality of soap opera is often said to encourage the impression of realism, as it coincides with the viewer's own experience of ongoing time. For Australian viewers of *DOOL*, the two senses of history are no longer so easily conflated. This could be another reason that many fans resisted the leap: It represented interference in what is commonly understood as a cohesive, diegetic world.

Soap fans know that these programs are at least as much about the past as the future. The narrative drive emphasizes changes in perspective, knowledge, and understanding of past and present events as much as future narrative actions. So, when it comes to filling the gap exposed by the "missing years," viewers want to know how the missing events also affect past events. Australian viewers watching *DOOL* after the leap speculated about how to fill the subtext of pregnant pauses without historical knowledge. When journalist Williams dismissively suggests that what all viewers will miss are "thousands of shots of actors staring into the middle distance," he hones in exactly on the gaps in narrative meaning, which viewers most want to fill. When a close-up lingers on Hope's pregnant pause, do viewers make connections to last week, last month, or last year? As police officers Bo and Hope pursue killer Marlena over the coming months, are we deprived of levels of subtext as a result of missing the affair between Marlena's husband John and Hope?

The complex textual histories of soap opera are always complicated by forgetting. As Coward (1986, 172) explains in her exploration of recasting practices in soap operas, "If these programmes require such feats of memory, how is it that they also require equally spectacular acts of forgetfulness?" (1986, 172). The "missing years" must be one of the most spectacular acts of forgetfulness required in soap opera. The tension identified by Coward highlights how the new text presents apparently contradictory expectations of its viewers: We should easily skip four years' worth of story, but also recall intricate character genealogies and significant events from decades past. Complex issues of memory and forgetfulness are conventionally addressed explicitly in the soap text, through stories of amnesia and changing past events with retroactive continuity. As referenced earlier, *DOOL* characters including John Black, Marlena Evans, Roman Brady, Mickey Horton, Kate Roberts, and Steven "Patch" Johnson have all experienced amnesia. To this list of *DOOL* amnesiacs, we might add the Australian television audience. Like John Black, for some viewers, their "missing years" represent involuntary brainwashing and ardent power struggles. Like Mickey Horton, other viewers accept the loss of their "missing years" and choose to reinvent their relationships and histories. The history of each soap opera character, and the history of each soap viewing experience, is always incomplete. Amnesia has long been considered a storyline with the ability to reenergize a soap opera, and the "missing years" has yielded an extraordinarily productive relationship with the text for Australian viewers.

This case study demonstrates the precarious position of U.S. soap operas as they increasingly seek global markets. Soaps rely on a uniquely serial narrative form to attract and sustain audiences but cannot deliver as much universally. In order to be rendered successful global products, U.S soap operas are subjected to the vagaries of international trade and local programming practices. Their migratory paths necessarily transform the nature of the programs and the viewing pleasures experienced by international audiences. Fans and theorists of U.S. soap opera will become increasingly adapted to the fractured texts and amnesic effects created in different ways around the world.

NOTES

Acknowledgement: Thanks to Dr Wendy Haslem for her archival and editorial assistance with this essay.

1. The Nine Network broadcast CBS daytime drama *The Young and the Restless* (*Y&R*) in Australia at that time and also skipped four years' worth of episodes of that program in 2004. This discussion will focus on *DOOL*, but would be equally applicable to *Y&R*.

2. For instance, see the *DOOL* fan forums at eBroadcast.com from 2004 at http://www.ebroadcast.com.au/cgi-bin/blah/Blah.pl?b=TVS,v=display,m=1068155552,s=120.

3. For instance, see the *DOOL* fan forums at eBroadcast.com from 2004 at http://www.ebroadcast.com.au/cgi-bin/blah/Blah.pl?b=TVS,v=display,m=1068155552,s=120.

4. However, Australian television ratings are not routinely published like their American counterparts. The only program-specific ratings published regularly in Australia list the top primetime programs and always overlook daytime.

5. Note that Australia's most popular television magazine, *TV Week*, is owned by ACP/PBL Media, part of the same corporation as the Nine Network.

AS THE WORLD TURNS' LUKE AND NOAH AND FAN ACTIVISM

—ROGER NEWCOMB

> Roger Newcomb is the Editor-in-Chief of *We Love Soaps* (http://www.welovesoaps.net/), a soap opera blog aggregating and commenting on the latest news in the genre several times per day. He has produced two Internet radio soap operas and co-wrote and executive produced a full-length feature film, *Manhattanites*, featuring talent from the soap opera industry. He started watching soap operas in early childhood with his family and has been a lifelong supporter—and critic—of the industry.

When Luke Snyder (played by Van Hansis) came out of the closet in May 2006, a number of new fans tuned in to CBS's *As the World Turns* (*ATWT*) (Fitzgerald 2008). However, it wasn't until August 2007, when Luke shared his first kiss with future boyfriend Noah Mayer (played by Jake Silbermann), that a strong fandom was formed. Soaps' history of short-term same sex relationships made some potential fans of a gay storyline hesitant to check out *ATWT* at first, believing it might be another example of a daytime drama exploiting the issue without ever intending to follow the story to its next logical place: a full-blown romance. However, despite multiple viewer frustrations at aspects of the story's progression, Luke and Noah remained a regular fixture on the show during its final three years on CBS and have broken several barriers for soaps in the process.

Two daytime soap operas first flirted with the idea of gay storylines in the 1970s, but only one made it to the air before the story was quickly written off. In 1974, NBC's *Another World* proposed a "coming out" story that was killed at the last minute by Procter & Gamble, the show's producer (which, significantly, was *ATWT*'s producer as well). In 1978, *The Young and the Restless* brought in a character named Joann, who became the object of Katherine Chancellor's affections. The show's ratings quickly declined, and Joanne was written off the canvas after telling Katherine she did not share her feelings. However, in the 1980s, daytime serials told their first coming out stories. Lynn Carson was the first openly gay character on soaps in a 1983 storyline on ABC's *All My Children* (*AMC*). Hank Elliot became the first gay male character in daytime, coming out on *ATWT* in 1988. Both characters lasted only about a year before disappearing.

In the 1990s, a few stories focused on teen characters struggling with their sexual identity, most notably Billy Douglas on ABC's *One Life to Live* in 1993 and Bianca Montgomery on *AMC* in 2000. Billy's story was integrated into a larger, powerful narrative addressing sexual identity, homophobia, and the HIV/AIDS crisis, while Bianca's coming out narrative was situated in the context of a close mother-daughter relationship with iconic soap opera character Erica Kane. Of these various coming out storylines, only Bianca remained on her soap's canvas for an extended run, including a same sex wedding in February 2009 to her partner, Reese.

ATWT head writer Jean Passanante, who penned the coming out story for Luke in 2006, also wrote the coming out story for Bianca Montgomery on *AMC* in 2000. Both stories featured a disapproving parent and a teen's determination to be accepted for who he or she is (Harrington 2003, 225). The difference between these two stories and past coming out arcs was the fact that Luke and Bianca were children of core characters which fans watched grow up on their screens with the help of "Soap Opera Rapid Aging Syndrome," or "SORASing" (the characters age faster than "real time" through various actor changes).[1] The Bianca storyline was well received by fans, and this character continued on the show after coming out. The success of using a core character in this type of storyline for the first time was an obvious influence on future stories. For example, ABC's *General Hospital* began a coming out story of its own with Lucas Jones (son of long-term character Bobbie Jones) during the time *ATWT*'s Luke narrative unfolded. Unfortunately, Lucas, a victim of gay bashing, was written out after a few months.

Due in part to its creation in a media environment that included a wide range of online soap opera fan sites and discussion boards, as well as a diverse range of online communities for gay audiences, Luke's coming out and the development of Noah and Luke as a couple attracted particularly intense fan response, at a magnitude not publicly visible for previous coming out storylines. The most prominent clip of the first kiss between Noah and Luke (known to fans as "Nuke"), uploaded by YouTube user LukeVanFan, has received more than two million hits on YouTube as of July 2009,[2] a record for any soap opera video. Several other versions of the kiss were also posted on YouTube and other video sharing sites. These videos were posted on numerous blogs and Web sites, including many online gay forums and blogs that did not normally feature material on soaps. Further, fan site VanHansis.net became a gathering place for fans, and the membership doubled from 250 to

500 in the month after the first kiss, and had more than 2,200 members by May 2009. Meanwhile, LukeVanFan's YouTube channel, which features every *ATWT* segment Nuke is involved in as its own series, gained hundreds of new subscribers in the days after the kiss and has grown to more than 15,000 subscribers as of July 2009. In the months following the first kiss, the influx of new fans appeared to lead to an increase in viewership for *ATWT*. Viewership among women eighteen to forty-nine and women twenty-five to fifty-four were both up 9 percent, to an average 868,000 and 979,000 viewers, respectively. The median age of the viewer also decreased the most for any daytime program in 2007, down 4 percent from 57.7 years old to 55.6 (Fitzgerald 2008).

Nuke's active fandom was fundamentally a mix of heterosexual females and homosexual males. A late 2007 online survey of more than three hundred fans of the couple on a Van Hansis fan Web site showed that 64 percent were female (the majority of whom were straight), while 36 percent were male (mostly gay). Further, many of the female fans were in the coveted eighteen to forty-nine demographic that *ATWT* needed to attract the most advertising dollars.[3]

The influx of these new fans, including large numbers of gay men, constituted a surplus audience that *ATWT* had not seen before (Ford 2007, 103–6). The majority of these new viewers began keeping up with the story through online clips relating solely to the Nuke story. The Nuke story became its own entity and, essentially, its own television show through series like LukeVanFan's edited version of Nuke's scenes. Gay Web sites like AfterElton.com and Out in Hollywood began to feature regular coverage of the storyline. The show did not immediately know how to take advantage of the new audience or how to grow it, despite the fact that discussion in many fan forums began to incorporate a variety of characters whose stories intersected with Nuke, indicating some interest among those who initially started following just Nuke's story, only to invest more heavily later in the larger narrative.

However, just as fan energy was building around this storyline, and *ATWT*, something odd began happening. After the couple shared another on-screen kiss a month later, Luke and Noah started having loving or romantic scenes where a kiss would be logical but which resulted in a hug instead. For example, a 2007 Christmas episode in which the couple declared their love for each other ended with the camera panning up to mistletoe instead of the soap staple for such events—a long, passionate kiss. Fans began to suspect something was going on behind the scenes, preventing another kiss from airing. Soon, fans launched the "Kiss Campaign," in which Hershey's Kisses were sent to executives at CBS and TeleNext Media, Inc., the show's production agency. Kisses were also sent to the owner of the show, Procter & Gamble, at its Cincinnati headquarters. Nuke fans also started an online petition that quickly collected more than 3,000 signatures.

Then, on Valentine's Day 2008, *ATWT* aired a special fantasy romance episode highlighting many of the show's most prominent couples. Every romance featured included a kiss except Luke's and Noah's; these characters again shared only a hug on screen. This was the last straw for many in the online fan communities surrounding the couple, who decided the show must have had a deliberate kissing ban in place for the couple despite those two initial kisses (Jensen 2008). Fans were not sure who was behind the ban, and the show would not respond to their queries (Fitzgerald 2008). Although neither the network nor the show ever admitted a kissing ban existed, fans speculated that conservative organizations that regularly boycott shows with gay content may have exacerbated the situation. This assumption was later fueled by press outreach from the American Family Association (AFA) in opposition to the storyline (Weiss 2008). According to AFA spokesperson Randy Sharp, the Tupelo-based conservative group received hundreds of complaints from *ATWT* viewers about the Luke and Noah romance and particularly the kisses: "It was a big turnoff for them.... The word 'repulsive' was used once or twice. 'Offensive' was used more than once.... It was overtly gratuitous. It's not necessary to the story line itself" (Weiss 2008).

Fans at VanHansis.net organized a "Media Blitz" in late February 2008, following the Christmas and Valentine's Day episodes. Prominent members of the press were contacted with information about the differences between how the gay couple was being handled, versus the treatment given *ATWT*'s heterosexual couples. In their outreach, fans praised the soap for being willing to tell the couple's story, but likewise asked for answers about why kisses between Luke and Noah had suddenly disappeared. As a fan of the storyline and an out gay man, I felt it was important to speak out on this subject. When the press started responding to e-mails from fans, I was quickly elected as a spokesperson for the cause, in part because of the prominence of my soap opera news site, We Love Soaps, in the online soap opera fan community.[4]

This is not the first time gay or lesbian daytime characters have been the focus of major media activist efforts. For example, the Bianca storyline on *AMC* was targeted for a 2001 audience reception study conducted by C. Lee Harrington and commissioned by the Center for the Study of Media & Society at the Gay and Lesbian Alliance Against Defamation (GLAAD). In contrast to GLAAD's institutionally instigated activist campaign, however, the Nuke campaign was entirely generated and driven by fans. Soap opera fans regularly campaign in an effort to protest a decision or advocate for a character or couple with "the powers that be" (or "TPTB," as they are often referred to online). Most often, these campaigns have involved emailing or writing letters to a particular show or the soap opera press and making phone calls to comment lines. Taking a cause directly to the mainstream media, much less gaining substantial interest from publications that rarely if ever covered soaps, was a new experience for soap opera fandom.

This Media Blitz led to articles published in *The Los Angeles Times* (Smith 2008), *The Boston Globe* (Weiss 2008), and *The New York Daily News* (Hinckley 2008). An Associated Press story for which I was personally interviewed was picked up by hundreds of newspapers in the United States and internationally (Bauder 2008). Further, the fan campaign was mentioned in *Time* (2008), *Entertainment Weekly* (Brown 2008), and numerous other mainstream publications, and CNN did a feature on the story (Wynter 2008). As a spokesperson for the campaign, I was also a guest on the National Public Radio show *The Bryant Park Project*, as well as *The Agenda with Joe Solmonese* on XM Radio.

The campaign was not only successful in driving media coverage, but also in soliciting comments from TPTB. Jeannie Tharrington, Director of Communications for *ATWT* producers TeleNext Media, said, "We're trying to make a show that appeals to our entire audience." The recent changes, she said, were due to "some of the feedback that we've gotten, and because of what we thought was best for the show creatively" (Weiss 2008). In the same article, CBS Senior Vice President for Daytime Barbara Bloom said, "In the soap-opera business, you walk a very fine line between love stories, happily-ever-after, yearning, and obstacles. The drama comes from the quest." In multiple articles, show representatives talked further about taking their time with this love story, but fans still saw a gap in their explanation. From a story perspective, no obstacles prevented the couple from kissing. Not only had the couple already kissed in the past, but Noah and Luke were still very much in love and had even moved in together. Thus, Nuke fans continued their campaign, including further promotion of the "Kiss Clock" many fan sites had incorporated as a means of counting the days since Luke and Noah's last kiss (Weiss 2008).

Then, unexpectedly, the apparent ban was lifted. On April 23, 2008, Luke and Noah kissed, stopping the "Kiss Clock" at 211 days. Fan sites pointed out that this scene was taped only weeks after their "Media Blitz" campaign and further rejoiced when kissing between the couple became a routine occurrence. The show even promoted a special Gay Pride episode featuring Cyndi Lauper on July 3, 2008. Fans continued to campaign and lobby the soap for additional airtime and better stories for Nuke. Leaders of the Nuke fan community routinely organized days of calling into the show, sending postcards, and writing letters, as fans watched the story unfold. In particular, Nuke fans lobbied *ATWT* to better capitalize on the success of the couple and the new viewers they had drawn to the show—especially since the couple was not typically among the show's "front burner" characters who receive the majority of the airtime. Noah, in particular, had several periods of being off-screen for weeks at a time.

Nuke fandom has not, however, limited its focus to the storyline itself. When Van Hansis made a comment in November 2007 about not being in the holiday spirit, fans quickly launched Project Holiday Spirit to raise money for the Broadway Cares/Equity Fights AIDS charity, in honor of Hansis and Silbermann.[5] Subsequently, Nuke fans raised more than $50,000 for various charities dedicated to AIDS research, cancer, and other worthy causes.[6] For many of these fans, Luke and Noah were not just a soap storyline, but a call to action. The overall Nuke fan community dedicated its energy to rallying around the couple and these social causes. The group and even talked about how to use its organization and success to tackle relevant political issues as well.

Aside from its impact on fans' public activism, the Luke and Noah storyline also fundamentally affected some fans personally. Not only did local and even international friendships developed within the Nuke fan community, but many said the storyline and community helped change their perspective and, in some instances, their behavior. For instance, Sandy, a grandmother from Colorado, said that previously she had not known any gay people and was not comfortable with homosexuality; however, seeing the first Luke and Noah kiss captured her attention. Through following the couple, she not only changed her perspective but became a vocal contributor to fan campaigns.[7] Other fans, typically teenagers and young adults, spoke about being able to come out to their friends and families after watching Luke and Noah and seeing them become more comfortable with their sexuality.[8]

As with all soap opera fandom, Nuke fans developed a complicated relationship with TPTB of their show. Once the couple was kissing again, Nuke fans began to wonder if the show would move the couple to the next step and feature a sexual relationship between Noah and Luke. The show had employed a number of roadblocks which delayed the sexual consummation of the relationship, including Noah's fake green card marriage, Noah's unbalanced homophobic father, and sexual advances from Luke's closeted step-grandfather. Then, unexpectedly, on the January 12, 2009, episode of *ATWT*, Luke and Noah kissed passionately in one scene and were shown coming out of the shower in the next, discussing their first time together. The event was not promoted in the soap press, known for spoiling most stories in advance.

The reaction to Luke and Noah having sex was mixed. Many fans rejoiced because the long wait was finally over. On the other hand, some fans and journalists complained that, after such an extended build up, the actual lovemaking scene should have been longer, or the couple should have been shown in bed together. Others felt the show should have promoted the sex scenes, as they do quite often for heterosexual couples, and accused *ATWT* of wanting to appease supporters while trying not to alienate conservative viewers. Balancing these two opposing views was an ongoing struggle for *ATWT*, which continued to feature the couple but never as prominently as some of its other leading couples.

Despite the sporadic airtime devoted to the couple and missteps in the stories, *ATWT* deserves credit for being a trailblazer in daytime television. No other soap has ever had a continuing gay male couple or portrayed gay men positively on the air for an extended period of time. Nuke has laid crucial groundwork for gay future stories on soap operas and television in general. As this book goes to press, *ATWT* is airing its final months' episodes on CBS. Whether the Nuke story plays a prominent role in the show's final episodes—or whether there is a continuing life for the characters or Nuke fans—remains to be seen. However, the impact Nuke and its fandom has had on the soap opera world, and the lessons that can be learned from the couple, are unmistakable.

NOTES

1. For more on the SORAS phenomenon, see Ford's essay elsewhere in this collection.
2. The video is available at http://www.youtube.com/watch?v=DVsX9RnoHGk as "Part 41" of the Nuke story in LukeVanFan's ongoing series of clips from the couple's appearances on the show.
3. See http://forums.vanhansis.net/topic/229624/1/.
4. We Love Soaps is located at http://www.welovesoaps.net/.
5. See more about Project Holiday Spirit at http://vanhansis.net/projectholidayspirit2007.html.
6. See more about these charity drives at http://forums.vanhansis.net/pages/charity/.
7. For more on Sandy's experience, listen to the initial version of *The Nuke Fancast* at http://www.blogtalkradio.com/nukefancast/2008/05/31/Episode-1.
8. Several fan stories are featured in subsequent episodes of *The Nuke Fancast* at http://www.blogtalkradio.com/nukefancast/.

CONSTRUCTING THE OLDER AUDIENCE
AGE AND AGING IN SOAPS

—C. LEE HARRINGTON AND DENISE BROTHERS

> C. Lee Harrington is professor of sociology and a Women's Studies Program Affiliate at Miami University. She has been conducting research on the daytime industry and soap fans since the late 1980s and is author of many published academic works on soaps, including *Soap Fans*, with Denise D. Bielby. She has watched soaps since the late 1970s, but the only one she currently time-shifts is *General Hospital*.
>
> Denise Brothers is a doctoral student in social gerontology at Miami University. Her research focuses on gender and the life course, specifically on how changes along the life course can challenge and alter traditional gender relations and roles. She started watching soap operas in her teens with ABC soaps, later shifted to NBC's *Days of Our Lives* and *Another World*, and has been watching *General Hospital* faithfully since 2000.

INTRODUCTION

As in many other media markets, the target audience for daytime soap opera is viewers eighteen to forty-nine years old, with those younger than eighteen considered desirable as future long-term viewers, but those fifty and over considered (by implication) undesirable. The industry's efforts over the past several decades to court younger viewers as a strategy for combating declining ratings have occurred, somewhat ironically, in the context of a rapidly aging U.S. population, and thus a rapidly aging television audience. We explore the possibilities of re-constructing soaps' target audience to include older viewers and suggest alternative economic models and new narrative directions that soaps might follow. Drawing on original interview/survey data with long-term viewers, veteran soap actors, and other industry experts, we examine age and aging in the soaps.

AGE AND AUDIENCE CONSTRUCTION

The ways in which audiences get constructed are important to understanding the entertainment content that ultimately gets delivered to them (Turow 1997). Analyzing the shift that the U.S. has undergone since the 1970s from "society-making media" to "segment-making media," Turow (1997, 3) focuses on the "dark side" of target marketing. As media firms mount strategies to attract certain groups of consumers, they simultaneously mount different strategies to drive others away. For Turow (1997, 7), a troubling aspect of audience fragmentation is the disruption of larger social unity that can result: "If you are told over and over again that different kinds of people are not part of your world, you will be less and less likely to want to deal with those people." For example, the network era of television constructed mainstream audiences as "replications of the idealized, middle-class nuclear family, defined as monogamous heterosexual couples with children" (Buxton 1997, 1477), thus marginalizing a wide range of family forms and sexual identities. In contrast to the network era, current marketing practices can lead individuals "annoyed by... shows or what they read about them to feel alienated from groups that appear to enjoy them" (Turow 1997, 7). As Turow argues, industry struggles about audience constructions are ultimately struggles about which people in society matter, and why. How those struggles are resolved is reflected in the types of characters and narratives depicted on-screen.

From an industry perspective, the most important age bracket in the post-WWII era has been eighteen to forty-nine year-olds (Turow 1997), with CBS known for practicing more broad-based audience construction than ABC, NBC, or Fox. More recent marketing efforts have resulted in narrower age clusters based on presumptions about common cultural characteristics and consumption habits shared by people of a given cohort. By the early 1990s, "the notion of connections across generations had little meaning" among marketers who used media to separate desirable age groups from undesirable ones (Turow 1997, 74). For example, in advertising, eighteen to twenty-four year-olds are preferable to eighteen to thirty-four year-olds who are preferable to eighteen to forty-nine year-olds, and so on, and advertisers routinely pay more for programs that deliver younger audiences (Ahrens 2004). Older adults are undesirable because they are believed to be too brand loyal (not easily influenced by advertising), to have less disposable income than younger viewers, to be low-volume consumers (buying for small households), to be easier to target than younger viewers (less elusive translates to less desirable), and to be un-hip by definition, thus negatively impacting a company's reputation (Lee 1997; Riggs 1998; Russell 1997; Tedeschi 2006; Turow 1997; Wolfe

2003). While there is evidence that the focus on eighteen to forty-nine year-olds is beginning to break down (Collins 2009), adults fifty and over have long been marginalized by most media marketers.

This marginalization is echoed in media studies, where older adults are under-theorized and under-studied compared to other age groups. Despite the late 1980s emergence in cultural studies of the active audience perspective, media consumption among older persons is still presumed to mimic the long-rebuked hypodermic model of information transmission—that is, TV watching as a "monolithically unimaginative" experience for elders, a "catatonic absorption of meaningless fog" (Riggs 1998, 5, 172). Research finds that older adults watch more television than any other age group due to increased leisure time, declining entertainment options, and/or the continuation of life-long leisure patterns (Mares and Woodard 2006). Scholars tend to regard these viewing habits as de facto problematic, signifying loneliness, potential vulnerability in contributing to TV-based perceptions of reality, and leading to "deadly passivity" (Riggs 1998, 171; see also Mares and Woodard 2006; Tulloch 1989). In short, scholars have presumed that heavy TV viewing among elders represents their disengagement from (not active engagement with) meaningful social life. More recent research has moved past these assumptions to question the presumed homogeneity of the "older audience" in media studies and to examine empirically the meanings and usages of television in older adults' lives, such as its role in the maintenance of social-psychological well being (e.g., Goodwin, Intrieri, and Papini 2005) and engagement with formal political culture (Riggs 1998). However, this research trajectory is nascent at best.

A NEW DEMOGRAPHIC LANDSCAPE

The world's population aged sixty-five and over is growing at an unprecedented rate, which has begun to alter expectations regarding age-based consumption patterns. Of particular interest to media marketers in the U.S. is the aging of Baby Boomers, the first generation to be raised with television and whose first cohort became eligible for Social Security benefits in 2008. Beginning in the mid-1990s, trade publications noted that Boomers deviate from the stereotypical age-based purchasing patterns summarized above, having greater discretionary income and devotion to active leisure pursuits than prior cohorts of elders (Lee 1997; Russell 1997). More recent research confirms that older adults engage in more consumer spending than any other age group and have become major players in the Web economy (Tedeschi 2006). In addition, a survey commissioned by AARP found that brand loyalty varies more by product category than by age, thus disputing one of the major stereotypes of older consumers: that they are unwilling to try new things (Chura, Fine, and Friedman 2002). Finally, research confirms that the majority of the coveted eighteen-to-thirty-four year-old demographic living in households earning $75,000 or more per year is dependent on someone else's money—often, Boomers' money (Ahrens 2004). Indeed, Boomers have been characterized as "chameleon consumers," shopping for more people (parents, children, themselves) than anyone else and adjusting their purchasing needs accordingly (Waldman 2008). In short, the youth bias held by media firms seems increasingly irrational from an economic perspective, as even marketers admit.

As the U.S. population ages, so too do media audiences. In the context of television, Nielsen Media Research reports that the number of viewers fifty-five and older is increasing twice as fast as the general TV audience (Ibarra 2008). The median age of primetime network viewers reached forty or more for the first time in 2007—even that of Fox television, known for its youth orientation since the network's inception in 1986, crept upward from a median age of thirty-five in 2002–2003 to a median age of forty-two in 2006–2007 (Levin 2008). This shift is occurring in daytime as well, with the median age of soap viewers increasing an average of seven years between 1991 and 2001 (Bauder 2001). The median age for *The Young and the Restless* (*Y&R*), for example, is currently close to sixty (Lidz 2009). Since the soap industry's ideal viewer is a child introduced to soaps by a parent or grandparent, who continues to engage with the genre as s/he matures—an ideal that directly contradicts marketers' dismissal of intergenerational connections, as noted above—the aging of the audience contributes to widespread concerns about profitability (Wittebols 2004). Growing recognition among advertising executives that adults fifty and older are not a homogeneous group has led to marketing distinctions between "mature adults" aged fifty-five to sixty-four and "senior citizens" aged sixty-five and older (Turow 1997, 74). However, eighteen to forty-nine remains the "money" demographic in most television contexts (Ahrens 2004), soaps included, despite evidence to the contrary of elders' growing buying power.

AGE AND AGING IN SOAPS

As noted earlier, Turow (2005, 106) reminds us that "the ways media organizations search for and describe their audiences have important implications for the texts that viewers and readers receive in the first place." In the case of soap opera, this means that constructing older audiences as less desirable than younger ones has a direct bearing on narrative content. One marker of this impact is soaps' on-screen marginalization of older actors/characters, though there has been surprisingly little research on age/aging in soap opera compared to the body of research focusing on other identity markers (especially gender and sexuality).[1] Cassata and Irwin (1997, 227) established more than a decade ago that, while 13 percent of soap characters are in the fifty-one to sixty-four demographic, generally matching that of the U.S. population, "the 65 and older demographic is grossly under-represented," at just 3 percent of the soap opera landscape. Through content analysis of a week's worth of soap narratives, they found that older characters are generally portrayed positively but are rarely central figures, acting instead as sounding boards for younger characters. Strong intergenerational ties are a hallmark of the genre, where they persist (at least on some shows), in contrast to the neglect of such ties by media marketers, noted above (Turow 1997, 74). Overall, the marginalization of older actors/characters continues to be accompanied by a dearth of soap narratives addressing later life issues such as retirement, widowhood, and long-term care (Cassata and Irwin 1997, 218), though recent narrative emphases offer promising redress.

Mindful of the different institutional logics with which professionals and viewers assess soap narratives and characterizations,[2] we update and extend prior research by exploring how long-term viewers, veteran actors, and other industry experts evaluate representations of age and aging in the soaps. We are interested in exploring implications of industry re-construction of the audience, with our focus on prioritizing those who *are* and *will be* watching—those who represent current and future U.S. age demographics—rather than the young(er) audience the industry still hopes to attract. What would soap storytelling look like, in other words, if soaps were written for its existing older audience? How might narrative priorities change?

PROJECT DESIGN

We draw on three original data sources: (a) telephone interviews with eleven veteran actors who have played the same roles for at least fifteen years;[3] (b) telephone interviews with five outside soap opera experts;[4] and (c) email surveys with thirty-four soap viewers who have watched the same soap opera for at least twenty years. Following IRB guidelines, informed consent was obtained from all participants. The actor sample includes five males and six females, all white, ranging in age from thirty-five to seventy-seven at the time of the interview, with a median age of sixty-two.[5] They represent shows (current and defunct) on all three networks (ABC, NBC, and CBS) and both coasts (East/West). Per agreement, we identify actors/outside experts by name in this manuscript.

Viewer participants were identified through snowball sampling and invited to complete a survey designed by the co-authors. Participants' ages ranged from twenty-four to seventy-three with a median age of fifty-four (age listings refer to age at time of survey completion). Thirteen viewers (38 percent) are between eighteen and forty-nine, and the remaining viewers (62 percent) are fifty years of age and older. Viewers of each of the eight network soaps airing in the U.S. when the study was conducted are included in the sample, though the majority of participants (71 percent) watch CBS's now-cancelled *As the World Turns* (*ATWT*). Given their length of viewing investment and knowledge of industry trends and practices, these viewers are not representative of the overall soap audience. Following IRB protocol, returned, completed surveys indicated consent to participate. Viewers are identified by participant number rather than by name in accordance with confidentiality agreements.

All quotes included in this essay are verbatim, though they have been edited for clarity. Transcribed interview quotes (from actors and outside experts) have also been edited for minor grammatical errors; quotes from viewer surveys are presented as written. Both interview and survey data were analyzed using a grounded theory approach (Glaser and Strauss 1967) and a constant comparative method.[6]

DISCUSSION

REPRESENTATIONS OF AGE AND AGING: INDUSTRY PERSPECTIVES[7]

The professionals we spoke with agree that storylines focusing on later life concerns (such as changes in physical appearance, menopause, retirement, and parental caregiving) hold value in the world of soap storytelling, both in their dramatic potential and because they reflect the realities of older adults' lives:

> Most of us who hit fifties and sixties, you start to [gain weight]. I think it would be something millions of women, especially in the Baby Boomer generation, could relate to . . . Maybe the sex drive [decline] that happens to both older women and older men . . . Maybe talk about having to wear glasses . . . Just facing the fact that you aren't what you once were, and how do you deal with that? . . . I think it would be very interesting . . . Soaps are very good for giving people life lessons (Mimi Torchin).

> I do believe the audience wants to see that, because ... soap operas are about life. They can't [say] OK, we're only going to show what happens ... from fifteen to thirty ... I think we need to respect and honor [aging] and bring [viewers] something that is a true picture (Jacklyn Zeman).

> The difficulty from an industry perspective is how to script such stories in a way that does not put viewers "in a state of depression" (Suzanne Rogers).

Explains Kay Alden, head writer of CBS' *The Bold and the Beautiful* (*B&B*):

> Here is where one runs into the inevitable "business versus art" [question]. I think it's hard to figure out how to appeal to that [eighteen to forty-nine] age group by telling stories that focus so much on the other end of life ... There are wonderful, moving tales to be told. They are very difficult to tell in daytime [because] you're asking people to tune in ... day after day, and some of these stories are just too painful. People ... don't want to tune in to see somebody having to care for their parent, especially when they've watched that person [who] is the parent from the time they were in their vital adulthood.

Given these challenges, veteran actors are all-too-aware of the risk of "aging out" of a central place on their show's canvas:

> When the actors really get old on a soap, they don't use them that much[... If I got to be seventy-five and eighty years old and was still on [air], I wouldn't be working three days a week ... I would be working ... two shows a month (John McCook).

> Those of us who have been there for many, many years are not used as much as we used to be. The show had different age groups when Irna Phillips wrote the show, when Doug Marland wrote the show, you always had all of the different age groups reacting [and] you had different points of view. And now, they don't seem to be as interested in that. They go from one story involving the young people to another story, and we [veteran actors] are really kind of in the background now to give the show some texture (Don Hastings).

In a pre-emptive strategy to avoid aging out of the core narrative, actress Eileen Fulton secured a now-infamous contract clause in the 1970s which prohibited her character Lisa from becoming a grandmother. The clause, explains Fulton (1995, 192–3), was designed to maintain Lisa's sexual vitality and counter narrative marginalization:

I wasn't refusing to become a grandmother out of pure vanity. My story line was running hot and heavy.... Would the audience accept a seductive, conniving, sexy vixen who had to baby-sit for her new grandchild? More important, would the story line be dropped because of an impending grandchild? I felt at the time that the writers had a preconceived notion of how grandparents behaved. They didn't know how to create interesting storylines for them.

Not surprisingly, most of our industry participants spoke with ambivalence about the realities of aging in the public eye in a medium that adds visual pounds and in a genre whose production routines include harsh, unflattering lighting:

You go through life as this character, as this personality, as you're aging. I aged twenty-some odd years in front of millions of people who were aging with me ... I think I got out [of soaps] just in time ... [If] I'm at the gym now and look up at the TV at different friends of mine ... everybody's looking pretty rough. I mean, we're all fighting a good fight but time has really had an impact ... It's a very vulnerable and unique position to be in because ... you're putting it right [out] there from day-to-day. Week-to-week. Month-to-month. Year-to-year (Stephen Schnetzer).

Actors agree on the need to maintain a youthful appearance to survive in the industry, maintainence accomplished through exercise, dieting, Botox, and surgical intervention when necessary. Veteran actresses Linda Dano, Kim Zimmer, and Jeanne Cooper have openly discussed face-lift procedures; rumors fly regarding which male actors have had cosmetic surgery. As Peggy McCay, a veteran actor of NBC's *Days of Our Lives* (*DOOL*), remarks, "Whatever you can do to keep it looking good, I think you should." Interestingly, culturally based age norms contribute to different types of audience reception around the world. John McCook, whose show is wildly popular on the global market, explains:

What's nice about it for me is that European women find [my character] Eric—the older man—more interesting than the younger man. And so I'm very happy with that, of course. They give the older women and the older men a lot more weight, especially in the Mediterranean ... [They] love those older characters and they give them a lot of credit for being older and wiser.

Narratives focusing on later life issues entail economic risks in the domestic context beyond ensuring that the stories are "not depressing." One such risk is the salary costs of veteran actors. Soaps cost about $1 million per week to produce; a veteran actor earns about $350,000 per year, with several legendary stars pulling in annual salaries of $1 million and higher as of 2008 (Lenhart 2008a; Flynn 2008). Production cuts underway at all three networks are hitting veterans hardest, as once-established contract perks disappear, guarantees shrink, and contract negotiations become more hardball for each renewal. Recent casualties include superstars Drake Hogestyn and Deidre Hall, who were fired as part of *DOOL*'s contract restructuring with NBC in 2008, and the highly publicized salary cut of daytime legend Susan Lucci from ABC's *All My Children* (*AMC*).[8] Increasingly veterans are demoted from contract to recurring status, as in the recent practice of killing off veteran characters but bringing them back as ghosts (e.g. Stuart Damon on ABC's *General Hospital*) as a way to keep fan favorites on-screen but on a recurrent basis only.

Another area of economic risk includes the presumed relationship between age-of-viewer and age-of-actor/character. A key reason for the marginalization of veteran actors is the industry assumption that viewers prefer to watch characters in their own age bracket; if young viewers are targeted, young characters must be featured. None of our interview participants have seen market data confirming this point, and they personally find it hard to believe:

> Kids like to watch older people.... Kids are interested in what's going to happen when they get to be older. Yes, they're interested in some degree in their own [peer group] but it's much more exotic to see what the grown-ups are doing because [they] do things that are so much more interesting than what you're doing (Mimi Torchin).

> Children will often hear things from their grandparents or from their aunts and uncles that they can't discuss with their parents [or] or peers.... There's plenty of dramatic conflict that could be mined in there (Lynn Liccardo).

> It's such a weird thing to always think that young people don't care about where they're headed, you know, what's life going to be like for me?... [Viewers] want to see [older characters] living full, amazing lives..., a validation that being older is fascinating (Michael Logan).

Academic research generates mixed results. Scholars who examine links between media choices and social group identity find that, in general, child, young adult, and older adult TV viewers *do* prefer to watch characters their own age, though viewing patterns of older versus younger adults are fairly similar. Older women in particular have a distinct preference for characters sharing their gender and age bracket (Mundorf and Brownell 1990, 690). Although younger adults prefer not to watch older characters in sexual/romantic contexts, they do enjoy them in non-romantic settings and stories (Harwood 1997, 210; see also Scodari 2004).

There are several points to take from this research in the context of soap operas. First, as the above quotes suggest, compelling story is compelling story regardless of age, as long as depictions of physical intimacy are treated with care. Second, if the soap audience remains primarily female, if the audience is aging as we type this, if older women want to see older women on-screen, and if elders' buying power is on the rise, shouldn't we be seeing *more* older faces rather than *fewer* on our shows—assuming the industry wants to keep the viewers it now has and will have (given demographic trends) in the future? We emphasize that these studies referenced focus on TV viewers in general, not those of specific genres. We turn below to the long-term viewers interviewed for this project. With twenty or more years of soap watching behind them, what does this devoted group hope to see of age and aging on-screen?

REPRESENTATIONS OF AGE AND AGING: VIEWERS' PERSPECTIVES

The current challenges facing the soap opera industry are not lost on the viewers we surveyed. Most participants are aware of the disconnect between the industry's drive to attract a younger audience with youth-oriented storylines, and the reality of who is actually watching the genre and holds economic power as consumers—namely older adults. Viewer Ten (age fifty-two) articulates this incongruence when she says, "I find I no longer trust that the writers care what I think or what people of my age in general want. Ironic because I have a lot more money to spend than do my daughters, whose viewing the networks covet." Most viewers also refer to the pattern of marginalization documented by Cassata and Irwin (1997) and experienced directly by veteran actors: Older characters are put on the backburner, become props for younger characters, and are then seen less and less until they disappear altogether from the canvas. Viewer Twenty-One (age forty-seven) comments that, "[W]hen an actor/actress reaches a certain age, it's like they don't have a life anymore." Viewer Twenty-Four (age sixty) articulates this disconnect, as well as the negative treatment of older characters, when she says that, if the current focus on the youth demographic continues:

> [W]e will see the veteran characters placed more and more in the background.... I think this is a mistake [because of] the increasing number of baby boomers who will be retiring and who will have both the time to watch the programs and the money to buy the advertised products, the daily programs need to cater to this audience also.

When asked what the future holds for veteran actors and mature characters on soaps, few viewers express optimism. With the exception of one viewer who actively *wants* to see less of older characters, overwhelmingly viewers' predictions for veteran characters are described as "not good or promising," "grim," "bleak," "dim," and "death."

Despite the pessimistic outlook of these viewers, they all describe potentially compelling storylines for older characters and have ideas about how to bring them back into the center of soap narratives. Common themes that emerged from the interviews include relationships (romantic and mentoring) and work or careers. When asked what types of storylines veteran characters do not get to play, viewers of all ages point to the genre's core narrative focus: romance or love stories. However, when asked what types of storylines they would like to see older actors play, very few viewers in the eighteen to forty-nine age category mention wanting to see more romantic narratives for the vets, while many viewers over age fifty want precisely those stories. For example, Viewer Twenty-Six (age fifty), who watches *ATWT*, states:

> Romantic storylines! I would love to see Lucinda Walsh, Emma Snyder or Susan Stewart get a love interest. One of the best storylines in *ATWT* history was when Nancy Hughes, a widow, became the love interest of Police Captain Dan McCloskey. The romance ended in marriage, and it was done so lovely and sweet, and proved that life, love and happiness does not end at age forty.

However, some viewers distinguish between romantic storylines and seeing physical lovemaking depicted on-screen (Harwood 1997). Viewer Ten (age fifty-two) states, "I'm not looking to see explicit love scenes of older actors (or young ones for that matter) but I would like to see them in relationships, both romantic and platonic." Viewer Nine (age fifty-six) stresses the importance of how a romantic storyline is told, explaining that she wants to see a "full-blown love story with all the possible complications.... I'd really like to see a former love come into the life of a vet." Viewer Fifteen (age twenty-eight) notes the distinctiveness of later-life romances when he states that romantic pairings or triangles among veteran characters should be featured because they provide "a different type of dynamic than 'young love.'"

Romantic relationships between older characters are not the only type of ties that longtime viewers want to see. The desire for more storylines with veteran characters mentoring younger characters is mentioned by viewers of all ages. Viewer Thirty-One (age forty) elaborates:

> Veteran characters should be more involved in their children's and grandchildren's lives. It used to be that older characters were the repository of wisdom

on soaps.... I'd like to see them do more of that. And if they can reference their life experiences as they were actually shown on the screen, this is a bonus for long-term fans.

While mentoring is one way to bring older characters into the folds of a soap, this subject is part of a larger theme articulated by many viewers: the desire for soaps to return to what they deem truly multi-generational storytelling with actors of all ages in vital and meaningful roles. Viewer Sixteen (age fifty), who watches *GH*, advises the network to:

> ... rethink their obsession with youth and start to write family drama, using the veteran actors in their roles as core family matriarchs, patriarchs and in their positions within the working communities [and] begin addressing age-related social issues.... I believe that [this strategy] could gain back the core audience they have chased away with their fixation on violence, catastrophe and youth.

The perceived lack of family drama or multi-generational storytelling can be rectified, according to Viewer Seven (age fifty-nine), by new writers with talent for family-based narratives, which "will make it fashionable again for characters of every generation to be featured in storylines."

In addition to multi-generational storylines, some viewers reference the relative absence of work or careers (and, thus, retirement) for most veteran characters. Viewer Sixteen (age fifty) notes that older characters "are not present in places where it would make sense to see them, such as at their jobs." Many viewers want to see more of the work worlds of veteran characters and witness their navigation through retirement. Viewer Twenty-Six (age fifty) articulates some of the realistic challenges facing retirees and offers this storyline idea for *ATWT*:

> Although now that Craig has stolen Lucinda Walsh's company, it might be a good idea for a story on "forced" retirement, and what she might do next to fill her days. I know many retired people who are at loose ends, and are so defined by the job that when it's not there anymore they have a hard time adjusting.

Some viewers point to ways this stage of life may be experienced differently by men than by women. Viewer Twenty-Two (age seventy) observes regarding *ATWT*: "Retirement is difficult for many men.... [T]his would be a good storyline for Tom Hughes."[9] Finally, in accordance with the desire to see more multigenerational storylines, Viewer Thirty-Two (age forty-nine) responds,

"[R]etirement would be another good [storyline for veteran characters]. It would be interesting to show how that transition affects other family members, as the retiree tries to find something else to do with their time."

Head writer Kay Alden's earlier comment about how viewers in the target demographic may not want to see end-of-life issues, such as taking care of a parent, is somewhat contradicted by the viewers we spoke with. When asked what type of age-related storylines they would like to see, viewers eighteen to forty-nine are more likely to mention storylines considered stereotypical for older characters, like dealing with Alzheimer's, menopause, caregiving, and widowhood. For example, a twenty-four year-old viewer of *ATWT* (Viewer Twenty-Seven) suggests:

> For a character who can only appear infrequently these days (such as Helen Wagner), a caregiver storyline could be interesting [with] discussions among Nancy Hughes' son, grandson, and great grandson about what to do with her, having her at a rest home or retirement home . . . . These issues are sensitive, since the sensitivity some of these actors may have about their own age comes into play, but it would be great if soaps could tackle these issues more often.

Older viewers (fifty plus) are more likely to express *both* an aversion to seeing more storylines about later life (many times because they are dealing with these issues in their own lives) *and* a desire to see such storylines told within the context of the family. For example, when asked whether viewers would like more or less airtime devoted to aging issues (e.g., menopause, widowhood, long-term care), Viewer Eight (in her fifties) responded, "I absolutely detest storylines of this type. I see all that I want in real life." In contrast, Viewer Nine (age fifty-six) said, "I think any or all of these issues could and should be integrated into multi-generational storytelling."

In short, while viewers in different age brackets disagree on some aspects of age-based narratives, such as depictions of later life romantic intimacy, they agree on the need to maintain (or return to) key aspects of the genre—a focus on multigenerational relationships, richly drawn characters across the life span, and compelling narratives unfolding in the context of family dynamics. Indeed, writer Tom Casiello suggests that current trends in daytime have helped spur this preference:

> I think that with the declining quality in the soaps now, that [viewers] are trying to latch on to the things that made them fall in love with their soap in the first place, and hope that the show will go on.

An underappreciated feature of the soap genre is viewers' ability to grow up and grow older alongside the fictional characters and communities depicted on-screen. Due to their potential longevity, soaps become a touchstone or through-line to viewers' personal lives, where their own milestones and memories are referenced according to story events of the time and generations of family members share the pleasures of a commonly consumed text. No other form of popular entertainment offers quite this possibility of engagement over the life course.[10] The bedrocks of soap storytelling thus remain salient to long-term viewers in our study, regardless of their own age or that of the actors/characters on-screen.

CONCLUSION

As discussed earlier, industry efforts to construct audiences are ultimately efforts to identify which people in society matter, and, by implication, which do not (Turow 1997). As such, the longstanding marginalization of older adults/characters throughout the entertainment industry is "merely" a reflection of larger societal age-based stereotyping and discrimination. In the context of television, media marketers have eschewed older audiences due to assumptions about low spending power and entrenched brand loyalty, despite growing evidence to the contrary. In the context of media studies, heavy television consumption among older adults is presumed to indicate passivity, loneliness, and vulnerability, despite the paucity of research exploring the meanings that older viewers themselves make of television. In sum, older adults seem to be neither the "right" kind of consumers nor the "right" kind of viewers.

Trade reports suggest that network television producers are beginning to question their "relentless" focus on the eighteen to forty-nine demographic (Collins 2009), and we find recent soap narratives encouraging in their emphasis on veteran actor/characters and inter-generational storytelling (such as the re-centralization of the characters Erica and Adam on *AMC*). As noted, soaps have a long tradition of appealing to viewers across the lifespan through compelling narratives featuring richly drawn characters whose stories unfold across that same lifespan. Our study finds that veteran soap actors, other industry experts, and long-term viewers generally agree on the potential power of narratives that reflect the realities of aging and the vitality of later life, and that more fully explore multi-generational relationships in the context of family structures. Given this consensus, we should be seeing more (and better) representations of age and aging on-screen, not fewer. As long-term viewers ourselves, we believe this is both a moral *and* an economic argument. In a book written for media marketers, Wolfe (2003) calls for the abandonment of age-based marketing practices in favor of "ageless marketing" based on "values and desires" that span traditional generational divides. This is where the money is, and soap operas are uniquely positioned to answer this call. If the genre fails to attract new (younger) viewers by employing slash-and-burn techniques (documented throughout this volume) that drive extant (older) viewers away, who will be left to watch?

NOTES

Acknowledgements: Thanks to Melissa Scardaville, Sam Ford, Kay Alden, and Lynn Liccardo for their assistance in identifying interview participants. Thanks to Andrea Parks for transcription services.

1. Noting this omission, Scodari (2004) takes an innovative approach to issues of gender, age, and romantic fantasy in her extended analysis of a May–December romance on NBC's now-defunct *Another World* (*AW*). Her empirically grounded study focuses on the implications of age/gender intersections in character portrayal and storyline development in the context of soaps' target demographics. See Scodari in this volume for a discussion of gender narrowcasting and the *AW* storyline.

2. See Scardaville's essay in this collection for more on this concept.

3. The following actors were interviewed (years playing the role refers to years at time of interview): Bryan Dattilo (fifteen years as Lucas Roberts on *DOOL*); Linda Dano (seventeen years as Felicia Gallant on *AW*); John McCook (twenty years as Eric Forrester on *B&B*); Peggy McCay (twenty-one years as Caroline Brady on *DOOL*); Elizabeth Hubbard (twenty-three years as Lucinda Walsh on *ATWT*, and eighteen years as Althea Davis on *The Doctors*); Steve Schnetzer (twenty-five years as Cass Winthrop on *AW*, *ATWT*, and *Guiding Light* [*GL*]); Kate Linder (twenty-five years as Esther Valentine on *Y&R*); Robert Newman (twenty-six years as Josh Lewis on *GL*); Jacklyn Zeman (thirty years as Bobbie Spencer on *GH*); Suzanne Rogers (thirty-four years as Maggie Horton on *DOOL*); and Don Hastings (forty-seven years as Bob Hughes on *ATWT*).

4. The following outside experts were interviewed: Mimi Torchin, founding editor of *Soap Opera Weekly* (*SOW*); Kay Alden, former head writer of *Y&R* and current head writer of *B&B* (See Alden's piece in this collection.); Tom Casiello, former associate head writer at *DOOL*, *Y&R*, and *One Life to Live*, and current writer at *Y&R* (See Casiello's essay in this volume.); Lynn Liccardo, former contributor to *SOW* and well known online soap personage (See Liccardo's essay in this collection.); and Michael Logan, soap opera critic for *TV Guide* and regular columnist for *Soap Opera Digest*. Please note that the interview with Casiello occurred outside of the formal data collection period.

5. We did not ask actors their chronological age; instead, age was obtained through online sources and confirmed when possible during the interviews. Given uncertainty about the veracity of online sources, the age range and median age we have indicated here should be considered suggestive rather than definitive.

6. For more information on data collection procedures and characteristics of our sample, see Harrington and Brothers (2010).

7. Some material in this section was also published in different form in Harrington and Brothers (2010).

8. For more on these cuts, see Bibel's essay in this book.

9. For more on the aging of the Tom Hughes character, see Ford's essay in this collection.

10. Again, see Ford's essay in this book and Harringon and Brothers (2010).

REFERENCES

ABC. 2008. Fringe ratings report: ABC daytime programming week of June 2, 2008. (June 12), http://www.abcmedianet.com/web/dnr/dispDNR.aspx?id=061208_04 (accessed July 16, 2009).

ABC Soaps In Depth. 2007. Port Charles' two worlds collide!, December 4, 4.

Adalian, Josef. 2007. Emmys channel daytime's traumas. *Variety* (May 7), http://www.variety.com/index.asp?layout=awardcentral&jump=features&id=daytimeem&articleid=VR1117964403 (accessed July16, 2009).

Adalian, Josef. 2009. CBS will turn off "Light" in September. *Television Week* (April 1), http://www.tvweek.com/news/2009/04/cbs_will_turn_off_light_in_sep.php (accessed July 16, 2009).

Adler, Renata. 1972. The air (on television)—afternoon television: Unhappiness enough, and time. *New Yorker*, February 12, 74.

Ahrens, Frank. 2004. Networks debate age groups' value to advertisers. *Washington Post*, May 21, E1.

Alarr, Albert. 2009. Interview by Erick Green. 20 February 2009.

Alba, Ernest. 2008a. The effect of the Youtube phenomenon on the soap opera text. *MITOpenCourseWare*, http://ocw.mit.edu/NR/rdonlyres/Comparative-Media-Studies/CMS-603Spring-2008/8423CE88-DF71-47B1-A85F-2D91BE276896/0/alba.pdf (accessed July 21, 2009).

Alba, Ernest. 2008b. Examining camera technique: Part two. *MIT CMS: The American Soap Opera* (March 31), http://mitsoaps.wordpress.com/2008/03/31/examining-camera-technique-part-two/ (accessed July 20, 2009).

Albiniak, Paige. 2008. NAPTE '08: NBCU CEO Zucker calls for change. *Broadcasting & Cable* (January 29), http://www.broadcastingcable.com/article/112217-NATPE_08_NBCU_CEO_Zucker_Calls_for_Change.php (accessed July 15, 2009).

Allen, Robert C. 1983. On reading soaps: A semiotic primer. In *Regarding television: Critical approaches - An anthology*, ed. E. Ann Kaplan, 97–108. Frederick, MD: University Publications of America.

Allen, Robert C. 1985. *Speaking of soap operas*. Chapel Hill, NC: University of North Carolina Press.

Allen, Robert C., ed. 1995. *To be continued . . . Soap operas around the world*. London: Routledge.

Allen, Robert C. 2004. Making sense of soaps. In *The television studies reader*, ed. Robert C. Allen and Annette Hill, 242–57. London: Routledge.

Ang, Ien. 1985. *Watching Dallas: Soap opera and the melodramatic imagination*. London: Methuen.

Arlen, Michael. 1981. *Thirty seconds*. Markham, Canada: Penguin Books.

Aspenson, Carolyn. 2003. GH women. *Eye on Soaps* (May), http://www.eyeonsoaps.com/spotlightghwomen.htm (accessed July 16, 2009).

Associated Press. 2009. Daytime Emmys find new home at CW; set for Aug. 30 (April 8). Available at http://www.accesshollywood.com/emmys/daytime-emmys-find-new-home-at-the-cw_article_16500 (accessed July 16, 2009).

Bauder, David. 2001. Do soaps have more lives to live? *Cincinnati Enquirer*, August 2, E3.

Bauder, David. 2008. Unexpected protest at a soap. Associated Press (March 2). Available at http://www.welovesoaps.net/2008/03/associated-press-unexpected-protest-at.html (accessed July 16, 2009).

Baym, Nancy K. 2000. *Tune in, log on: Soaps, fandom, and the online community*. Thousand Oaks, CA: Sage.

Beale, Lewis. 2001. A cable channel seeks new life beyond its niche. *New York Times* (August 26), http://www.nytimes.com/2001/08/26/arts/television-radio-a-cable-channel-seeks-new-life-beyond-its-niche.html (accessed July 16, 2009).

Becker, Howard S. 1982. *Art worlds*. Berkeley, CA: University of California Press.

Bibel, Sara. 2008. Deep soap: Low self-esteem. *Fancast* (May 27), http://www.fancast.com/blogs/deep-soap/deep-soap-low-self-esteem/ (accessed July 16, 2009).

Bielby, Denise D., and William T. Bielby. 1994. "All hits are flukes": Institutionalization decision-making and the rhetoric of network prime-time program development. *American Journal of Sociology* 99: 1287–313.

Bielby, Denise D., and William T. Bielby. 2004. Audience aesthetics and popular culture. In *Matters of culture: Cultural sociology in practice*, ed. Roger Friedland and John Mohr, 295–317. Cambridge: Cambridge University Press.

Bielby, Denise D., C. Lee Harrington, and William T. Bielby. 1999. Whose stories are they? Fans' engagement with soap opera narratives in three sites of fan activity. *Journal of Broadcasting and Electronic Media* 43(1): 35–51.

Bielby, Denise D., Molly Moloney, and Bob Ngo. 2005. Aesthetics of television criticism: Mapping critics' reviews in an era of industry transformation. In *Research in the sociology organizations: Transformations in cultural industries* (23), ed. Candace Jones and Patricia Thornton, 1–43. Greenwich, CT, and London: JAI Press.

Blair, Elizabeth. 2009. Auto ads decline in step with industry woes. *All Things Considered*, National Public Radio (April 28), http://www.npr.org/templates/ story/story.php?storyId=103578652 (accessed July 16, 2009).

Blumenthal, Dannielle. 1997. *Women and soap opera: A cultural feminist perspective*. Westport, CT: Praeger.

Bordwell, David. 1985. *Narration in the fiction film*. London: Routledge.

Bourdon, Jérôme. 2000. Live television is still alive: On television as an unfulfilled promise. *Media, Culture & Society* 22(5): 531–56.

Boutin, Paul. 2006. A video history of YouTube. *Slate* (October 18), http://www.slate.com/id/2151744/ (accessed July 16, 2009).

Branco, Nelson. 2008. So who wants to clone Sri Rao? *TV Guide Canada* (October 10), http://tvguide.sympatico.msn.ca/So+who+wants+to+clone+Sri+Rao/Soaps/Features/Articles/081015_Sri_Rao_NB.htm?isfa=1 (accessed July 16, 2009).

Branco, Nelson. 2009. The soapgeist. *TV Guide Canada* (September 28), http://tvguide.ca/Soaps/Nelson_Ratings/Articles/090928_soapgeist_NB (accessed November 24, 2009).

Brinckerhoff, Peter. 2009. Interview by Erick Green. 14 March 2009.

Britton, Connie. 2008. Comments on Connie Britton. *Charlie Rose*, PBS, June 19. Available at http://www.charlierose.com/view/interview/9140 (accessed July 16, 2009).

Brodesser-Akner, Claude. 2008. How to save the soap opera? Buy a car. *Advertising Age*, 79(44): 3, 28.

Brooks, Peter. 1995. *The melodramatic imagination*. New Haven, CT: Yale University Press.

Brown, Mary Ellen. 1987. The politics of soaps: Pleasure and feminine empowerment. *Australian Journal of Cultural Studies* 4(2): 1–25.

Brown, Mary Ellen. 1994. *Soap opera and women's talk: The pleasure of resistance*. Newbury Park, CA: Sage.

Brown, Scott. 2008. 'As the World Turns' fans demand more gay kissing. *Entertainment Weekly*, March 14, 10.

Brunsdon, Charlotte. 1997. *Screen tastes: Soap opera to satellite dishes*. London: Routledge.

Brunsdon, Charlotte. 2000. *The feminist, the housewife, and the soap opera*. Oxford and New York: Clarendon, Oxford University Press.

Bury, Rhiannon. 2005. *Cyberspaces of their own: Female fandoms online*. New York: Peter Lang Publishing.

Butler, Jeremy. 1986. Notes on the soap opera apparatus: Television style and *As the World Turns*. *Cinema Journal* 25(3): 53–70.

Buxton, Rodney A. 1997. Sexual orientation and television. In *Encyclopedia of television*, ed. Horace Newcomb, 1477–82. Chicago: Fitzroy Dearborn.

Cagle, Jess. 1990. Who shot J. R.? *Entertainment Weekly* (November 16), http://www.ew.com/ew/article/0,,318587,00.html (accessed July 16, 2009).

Caldwell, John Thornton. 2000. Slipping and programming televisual liveness. *Aura* 6(1): 44–61.

Cantor, Muriel G. 1971. *The Hollywood TV producer: His work and his audience*. New Brunswick, NJ: Transaction Books.

Cantor, Muriel, and Joel M. Cantor. 1992. *Prime-time television: Content and control*, second edition. Newbury Park, CA: Sage.

Cantor, Muriel, and Suzanne Pingree. 1983. *The soap opera*. Beverly Hills, CA: Sage.

Carter, Bill. 2006a. *Desperate networks*. New York: Doubleday.

Carter, Bill. 2006b. Sizzling a year ago, but now pfffft!" *New York Times* (December 25), http://www.nytimes.com/2006/12/25/business/media/25telenovela.html (accessed July 16, 2009).

Carter, Bill, and Brian Stelter. 2009. CBS cancels 'As the World Turns,' Procter & Gamble's last soap opera. *New York Times* (December 8), http://www.nytimes.com/2009/12/09/arts/television/09soap.html (accessed January 4, 2010).

Casiello, Tom. 2008. Daytime's shift into the night—and where it could go. *Damn the Man! Save the Empire!* (October 15), http://casiello.blogspot.com/2008/12/daytimes-shift-into-night-and-where-it.html (accessed July 16, 2009).

Casiello, Tom. 2009a. Great Marland's ghost! - Part one. *Damn the Man! Save the Empire!* (January 5), http://casiello.blogspot.com/2009/01/great-marlands-ghost-part-one.html (accessed July 16, 2009).

Casiello, Tom. 2009b. Great Marland's ghost! - Part two (The Clem years). *Damn the Man! Save the Empire!* (January 5), http://casiello.blogspot.com/2009/01/great-marlands-ghost-part-two-clem.html (accessed July 16, 2009).

Casiello, Tom. 2009c. Great Marland's ghost! - Part three (Barbara to the dark side). *Damn the Man! Save the Empire!* (January 7), http://casiello.blogspot.com/2009/01/great-marlands-ghost-part-two-barbara.html (accessed July 16, 2009).

Casiello, Tom. 2009d. Great Marland's ghost! - Part four (Final thoughts). *Damn the Man! Save the Empire!* (January 8), http://casiello.blogspot.com/2009/01/great-marlands-ghost-part-four-final.html (accessed July 16, 2009).

Cassata, Mary, and Barbara Irwin. 1996. The Young and the Restless: *Most memorable moments*. Toronto: Stoddart.

Cassata, Mary, and Barbara Irwin. 1997. Young by day: The older person on daytime serial drama. In *Cross-cultural communication and aging in the United States*, ed. Hana S. Noor Al-Deen, 215–30. Mahwah, NJ: Lawrence Erlbaum Associates.

Cassata, Mary, and Barbara Irwin. 1998. The Young and the Restless: *Special silver anniversary collector's edition*. Toronto: Stoddart.

Cassata, Mary, and Thomas Skill, eds. 1983. *Life on daytime television: Tuning-in American serial drama*. Norwood, NJ: Ablex.

CBS Soaps In Depth. 2009. Fans to show: "Slow things down." January 5, 6.

Chura, Hillary, Jon Fine, and Wayne Friedman. 2002. Ripe old age. *Advertising Age* 73(19): 16.

Cole, Tim. 1999. *Selling the Holocaust: From Auschwitz to Schindler, how history is bought, packaged, and sold.* London: Routledge.

Colker, David. 2009. More pulling TV cord, yet staying plugged in via computer. *Chicago Tribune* (November 2), http://www.chicagotribune.com/business/chi-mon-no-tv-1102nov02,0,7815992.story (accessed November 14, 2009).

Collins, Scott. 2009. Television is starting to look beyond the 18-to-49-year-old demographic. *Los Angeles Times*, January 11, A1.

Collins, Sue. 2008. Making the most out of 15 minutes: Reality TV's dispensable celebrity. *Television and New Media* 9(2): 87–110.

Cook, Terry. 2006. Remembering the future: Appraisal of records and the role of archives in constructing social memory. In *Archives, documentation and institutions of social memory: Essays from the Sawyer Seminar*, ed. Francis Blouin Jr. and William G. Rosenberg, 169–81. Ann Arbor, MI: University of Michigan Press.

Coppa, Francesca. 2008. Women, Star Trek, and the early development of fannish vidding. *Transformative Works and Cultures* 1, http://journal.transformativeworks.org/index.php/twc/article/view/44/64 (accessed July 22, 2009).

Cortez, Carl. 2008. Review: *Guiding Light*—New production format. *IFMagazine* (February 29), http://www.ifmagazine.com/review.asp?article=2343 (accessed July 22, 2009).

Coward, Rosalind. 1986. Come back Miss Ellie: On character and narrative in soap operas. *Critical Quarterly* 28(1–2): 171–8.

Crane, Diana. 1992. *The production of culture: Media and the urban arts.* Newbury Park, CA: Sage.

Craven, Ian Peter. 2008. Southern stars and secret lives: International exchange in Australian television. *Continuum* 22(1): 51–67.

Crew, Adrienne. 2003. Can "Farscape" fans reinvent TV? *Salon* (March 13), http://dir.salon.com/story/ent/tv/feature/2003/03/13/farscape/ (accessed July 22, 2009).

Cridlin, Jay. 2006. A new kind of prime time. *St. Petersburg Times* (September 2), http://www.sptimes.com/2006/09/02/Floridian/A_new_kind_of_prime_t.shtml (accessed July 22, 2009).

Crofts, Stephen. 1995. Global *Neighbours*? In *To be continued . . . Soap operas around the World*, ed. Robert. C. Allen, 98–121. London and New York: Routledge.

Da Ros, Giada. 2004. When, where, and how much is Buffy a soap opera? *Slayage: The Online International Journal of Buffy Studies* 13–14, http://www.slayageonline.com/essays/slayage13_14/Da_Ros.htm (accessed July 30, 2009).

Da Ros, Giada. 2008. L'uso del linguaggio gnomico nel *Queer As Folk* britannico. Il suo uso e non uso politico. *Ol3Media* 1(3): http://host.uniroma3.it/riviste/Ol3Media/Ol3Media/Archivio_files/Ol3Media%2003.pdf (accessed July 30, 2009).

Da Ros, Giada. 2009a. Lez girls: Un ilancio finale dopo sei stagioni di *The L Word*. *Ol3Media* 2(2): http://host.uniroma3.it/riviste/Ol3Media/Ol3Media/DaRos.html (accessed July 30, 2009).

Da Ros, Giada. 2009b. *Lost* come metafora della vita: Sottrazioni, addizioni, teorie, fra tensione metafisica erecherche. *Ol3Media* 2(2), http://host.uniroma3.it/riviste/Ol3Media/Ol3Media/Archivio_files/Ol3Media%2004.pdf (accessed July 30, 2009).

Da Ros, Giada. Forthcoming. "Dramedy" and the double-sided "liturgy" of *Gilmore Girls*. In *Screwball television: Critical perspectives on* Gilmore Girls, ed. David Scott Diffrient and David Lavery. Syracuse, NY: Syracuse University Press.

De Lacroix, Marlena. 2007a. *General Hospital Night Shift*. Good night and good riddance. *Marlena De Lacroix* (October 11), http://marlenadelacroix.com/?p=70 (accessed July 22, 2009).

De Lacroix, Marlena. 2007b. Apocalypse now: *General Hospital*'s murder of Georgie. *Marlena De Lacroix* (December 17), http://marlenadelacroix.com/?p=28 (accessed July 22, 2009).

De Lacroix, Marlena. 2008. *General Hospital Night Shift 2*: If they could bottle this show, they'd save soaps. *Marlena De Lacroix* (October 22), http://marlenadelacroix.com/?p=152 (accessed July 22, 2008).

de Moraes, Lisa. 2008. The TV column: Old perceptions die hard for flourishing CBS. *Washington Post* (December 29), http://www.washingtonpost.com/wp-dyn/content/article/2008/12/28/AR2008122801876.html (accessed July 22, 2009).

Denis, Paul. 1985. *Inside the soaps*. Secaucus, NJ: Citadel Press.

Diliberto, Joe. 2008. The bosses of *Night Shift*: Lisa De Cazotte and Sri Rao. *Soap Opera Weekly*, July 29, 4–5.

DiMaggio, Paul, and Walter Powell. 1983. The iron cage revisited: Institutional isomorphism and collective rationality in organizational fields. *American Journal of Sociology* 48: 147–60.

Dominguez, Robert. 2006. Make way for the supersoaps. *New York Daily News* (September 4), http://www.nydailynews.com/archives/entertainment/2006/09/04/2006-09-04_make_way_for_the_supersoaps_.html (accessed July 22, 2009).

Dugas, Christine. 1986. TV networks have time on their hands. *BusinessWeek*, July 14, 30.

Edmondson, Madeleine, and David Rounds. 1976. *From Mary Noble to Mary Hartman: The complete soap opera book*. New York: Stein and Day.

Elliott, Stuart. 2006. To bolster audience, soaps turn to the web. *New York Times* (March 2) http://query.nytimes.com/gst/fullpage.html?res=9904E3DB1731F931A35750C0A9609C8B63 (accessed July 22, 2009).

Erwin, Patrick. 2009. Blake's take: Liz Keifer on Coop, Phillip and the new Springfield. *A Thousand Other Worlds* (February 17), http://1000worlds.wordpress.com/2009/02/17/blakes-take-liz-keifer-on-coop-phillip-and-the-new-springfield (accessed July 22, 2009).

Fadner, Ross. 2004. Broadband penetration shows steady growth. *Online Media Daily* (March 9), http://www.mediapost.com/publications/index.cfm?fa=Articles.showArticle&art_aid=1549 (accessed July 22, 2009).

Federal Communications Commission. 2005. Historical periods in television technology: Golden age, 1930's through 1950's. (November 21), http://www.fcc.gov/omd/history/tv/1930-1959.html (accessed July 22, 2009).

Fernandez, Maria Elena. 2006. Back into the labyrinth of Luke and Laura. *Los Angeles Times* (November 4), http://articles.latimes.com/2006/nov/04/entertainment/et-soapnet4 (accessed July 22, 2009).

Feuer, Jane. 1984. MTM style. In *MTM: 'Quality television'*, ed. Jan Feuer, Paul Kerr, and Tise Vahimagi, 52–6. London: BFI Publishing.

Fiske, John. 1987. *Television culture*. London and New York: Methuen.

Fitzgerald, Toni. 2007. Adding a shift to Port Charles Hospital. *Media Life Magazine* (July 12), http://www.medialifemagazine.com/artman2/publish/Dayparts_update_51/Adding_a_shift_a_t_Port_Charles_hospital.asp (accessed July 22, 2008).

Fitzgerald, Toni. 2008. 'As the World Turns' turning on a kiss. *Media Life Magazine* (April 24), http://www.medialifemagazine.com/artman2/publish/Dayparts_update_51/As_the_World_Turns_turning_on_a_kiss.asp (accessed July 22, 2009).

Flint, Joe. 1999. NBC Internet to sell jewelry inspired by soap opera. *Wall Street Journal*, December 8, B6.

Flynn, Lauren. 2008. Smart money. *Soap Opera Digest*, July 1, 69–71.

Ford, Sam. 2007. *As the World Turns* in a convergence culture. MS thesis, Massachusetts Institute of Technology, http://cms.mit.edu/research/theses/SamFord2007.pdf (accessed July 22, 2009).

Ford, Sam. 2008a. Soap operas and the history of fan discussion. *Transformative Works and Cultures* 1, http://journal.transformativeworks.org/index.php/twc/article/ view/42/50 (accessed July 22, 2009).

Ford, Sam. 2008b. More on cultural bias and soaps. *MIT Convergence Culture Consortium* (June 13), http://www.convergenceculture.org/weblog/2008/06/more_on_cultural_biases_and_so.php, (accessed July 22, 2009).

Freeman, Michael. 2002. Upcoming shows getting makeovers. *Electronic Media*, June 17, 3.

Frentz, Suzanne, ed. 1992. *Staying tuned: Contemporary soap opera criticism*. Bowling Green, OH: Bowling Green University Popular Press.

Fulton, Eileen, with Desmond Atholl and Michael Cherkinian. 1995. *As my world still turns: The uncensored memoirs of America's soap opera queen*. New York: Birch Lane Press.

Fulton, Eileen. 2006. Comments at tribute to 50 years on television: *As the World Turns*. Museum of Television and Radio event. March 28.

Gans, Herbert. 1979. *Deciding what's news*. New York: Random House.

Geraghty, Christine. 1991. *Women and soap opera*. Cambridge, England: Polity.

Geraghty, Christine. 2005. The study of soap opera. In *A companion to television*, ed. Janet Wasko, 308–23. Malden, MA: Blackwell.

Gibbons, Douglas. 1997. Museum of Television and Radio submission. In *Library of Congress television and video preservation 1997: A study of the current state of American television and video preservation, V.5: Submissions*, 223–72. Washington, D.C.: U.S. Government Printing Office.

Gibson, William. 1986. *Count zero*. New York: Arbor House.

Gilbert, Matthew. 2002. The rise of the McDramas. *Boston Globe*, December 15, N1–2.

Giles, David. 2003. *Media psychology*. Mahwah, NJ: Lawrence Erlbaum.

Gillespie, Marie. 1995. *Television, ethnicity and cultural change*. New York: Routledge.

Glaser, Barney G., and Anselm L. Strauss. 1967. *The discovery of grounded theory*. New York: Aldine de Gruyter.

Glynn, Mary Ann. 2000. When cymbals become symbols: Conflict over organizational identity within a symphony orchestra. *Organization Science* 11(3): 285–98.

Goodwin, Paige E., Robert C. Intrieri, and Dennis R. Papini. 2005. Older adults affect while watching television. *Activities, Adaptation & Aging* 29(2): 55–72.

Gray, Jonathan. 2005. Antifandom and the moral text: Television without pity and textual dislike. *American Behavioral Scientist* 48(7): 840–58.

Greppi, Michele. 2006. Challenge is to sell audiences on novelas: WMYO-TV. *Television Week* (August 14), http://www.tvweek.com/news/2006/08/Wmyotv_challenge_is_to_sell_au.php (accessed July 22, 2009).

Guinness World Records. 2006. Longest time in the same role in a TV series. 308. New York: Bantam.

Hagman, Larry. 2008. Comments on Larry Hagman. *Tavis Smiley*, PBS, July 3. Available at http://www.pbs.org/kcet/tavissmiley/archive/200807/20080703_hagman.html (accessed July 22, 2009).

Hall, Stuart. 1980. Encoding/decoding. In *Culture, media and language: Working papers in cultural studies 1972–79*, ed. Stuart Hall, Dorothy Hobson, Andrew Lowe, and Paul Willis. London: Hutchinson.

Hanes, Rosemary. 2006. Interview with Mary Jeanne Wilson. February.

Harding, Mark. 2008a. Another ratings reality check (or don't blame Ellen Wheeler). *MarkH's Soap Musings* (November 30), http://markhsoap.blogspot.com/2008/11/another-ratings-reality-check-or-dont.html (accessed July 22, 2009).

Harding, Mark. 2008b. E-mail correspondence with Lynn Liccardo. December 9.

Harding, Mark. 2009. Optimism/Another graph/GL's fate is not Ellen Wheeler's fault. *MarkH's Soap Musings* (April 4), http://markhsoap.blogspot.com/2009/04/optimismanother-graphgls-fate-is-not.html (accessed July 22, 2009).

Harrington, C. Lee. 2003. Homosexuality on *All My Children*: Transforming the daytime landscape. *Journal of Broadcasting & Electronic Media* 47: 216–35.

Harrington, C. Lee, and Denise D. Bielby. 1995. *Soap fans: Pursuing pleasure and making meaning in everyday life*. Philadelphia, PA: Temple University Press.

Harrington, C. Lee, and Denise Brothers. 2010. A life course built for two: Acting, aging, and soap operas. *Journal of Aging Studies* 24(1): 20–9.

Harwood, Jake. 1997. Viewing age: Lifespan identity and television viewing choices. *Journal of Broadcasting & Electronic Media* 41: 203–13.

Hauck, Charlie. 2006. My plan to save network television. *New York Times* (September 16), http://www.nytimes.com/2006/09/16/opinion/16hauck.html (accessed July 22, 2009).

Havrilesky, Heather. 2003. Murphy's *Law and Order*. *Salon* (March 19), http://dir.salon.com/story/ent/tv/feature/2003/03/19/law_order/ (accessed July 22, 2009).

Hayward, Jennifer. 1997. *Consuming pleasures: Active audiences and serial fictions from Dickens to soap opera*. Lexington, KY: University Press of Kentucky.

Healey, Joe. 2003. Soaps online may appeal to young but not restless. *Los Angeles Times*, February 26, Proquest.

Herbert, James. 2005. Soap opera sees the (cyber) light. *San Diego Tribune*, September 18, Proquest.

Hibberd, James. 2007. MyNetTV changes production script; Limp ratings prompt shift from telenovelas to reality. *Television Week*, March 5, 6.

Hibberd, James. 2009. "Guiding Light" ending after 72 years. *Hollywood Reporter* (April 1), http://www.hollywoodreporter.com/hr/content_display/television/news/e3i3d34378f2d844a000aafbd413178864d (accessed July 22, 2009.)

Hilton, Perez. 2008. It was real sex, not just porn? *Perez Hilton* (September 4), http://www.perezhilton.com/2008-09-04-it-was-real-sex-not-just-porn (accessed July 22, 2009).

Hinckley, David. 2008. Gay kiss is missed on soap. *New York Daily News* (February 22), http://www.nydailynews.com/entertainment/tv/2008/02/22/2008-02-22_gay_kiss_is_missed_on_soap.html (accessed June 30, 2009).

Hinsey, Carolyn. 2008. Carolyn's corner. *Soap Opera Weekly*, August 19, 13.

Hiss, Anthony. 1975. The talk of the town: Daytime drama. *New Yorker*, May 5, 31.

Hobson, Dorothy. 1982. *Crossroads: The drama of a soap*. London: Methuen.

Holmes, Su, and Sean Redmond. 2006. Introduction: Understanding celebrity culture. In *Framing celebrity: New directions in celebrity culture*, ed. Su Holmes and Sean Redmond, 1–16. New York: Routledge.

Hurst, Jill Lorie. 2008. Interview by Patrick Erwin. December 9, 2008.

Hyatt, Wesley. 1997. *The encyclopedia of daytime television*. New York: Billboard Books.

Ibarra, Sergio. 2008. Nielsen: Older audiences getting bigger. *TVBizWire* (August 29), http://www.tvweek.com/blogs/tvbizwire/2008/08/nielsen_older_audiences_gettin.php (accessed July 22, 2009).

Intintoli, Michael James. 1984. *Taking soaps seriously: The world of Guiding Light*. New York: Praeger.

Irwin, Barbara. 1990. An oral history of a piece of Americana: The soap opera experience. PhD dissertation, University at Buffalo, the State University of New York.

Iser, Wolfgang. 1978. *The act of reading: A theory of aesthetic response*. London and Henley-on-Thames, England: Routledge and Kegan Paul.

Jackall, Robert. 1988. *Moral mazes: The world of corporate managers.* New York: Oxford University Press.

Jacobs, Karre. 2005. A soapy slide in the ratings. *Broadcasting & Cable* (February 13), http://www.broadcastingcable.com/article/156159-A_Soapy_Slide_in_the_Ratings.php (accessed July 22, 2009).

James, Meg. 2009. Lights go out for CBS' "Guiding Light." *Los Angeles Times Company Town* (April 1), http://latimesblogs.latimes.com/entertainmentnewsbuzz/ 2009/04/lights-go-out-for-cbs-guiding-light.html (accessed July 22, 2009).

Janofsky, Michael. 2007. NBC's Zucker sees "beginning of end" for network soap operas. *Bloomberg Online* (January 28), http://www.bloomberg.com/apps/news?pid=newsarchive&sid=aYjWjFMq.oMU (accessed July 22, 2009).

Jenkins, Henry. 1988. Star Trek: rerun, reread, rewritten. *Critical Studies in Mass Communication* 5: 85–107.

Jenkins, Henry. 1992. *Textual poachers: Television fans and participatory cultures.* London: Routledge.

Jenkins, Henry. 2001. Convergence? I diverge. *Technology Review* (June), http://www.technologyreview.com/business/12434/ (accessed July 22, 2009).

Jenkins, Henry. 2006a. *Convergence culture.* New York: New York University Press.

Jenkins, Henry. 2006b. The magic of the back-story: Further reflections on the mainstreaming of fan culture. *Confessions of an Aca-Fan: The Official Weblog of Henry Jenkins* (December 5), http://henryjenkins.org/2006/12/the_magic_of_back_story_furthe.html (accessed July 22, 2009).

Jenrette, Jerra, Sherrie McIntosh, and Suzanne Winterberger. 1999. "Carlotta!": Changing images of Hispanic-American women in daytime soap operas. *Journal of Popular Culture* 3(2): 37–48.

Jensen, Michael. 2008. And you thought the "mistletoe" moment was bad. *AfterElton* (February 15), http://www.afterelton.com/bgwe/2-15-08?page=0%2C1 (accessed July 22, 2009).

Johnson, Derek. 2007. Fan-tagonism: Factions, institutions, and constitutive hegemonies of fandom. In *Fandom: Identities and communities in a mediated world,* ed. Jonathan Gray, Cornell Sandvoss, and C. Lee Harrington, 285–300. New York: New York University Press.

Johnson, Richard. 1986–1987. What is cultural studies anyway? *Social Text* 16: 38–80.

Johnson, Sherry. 1982. How soaps whitewash blacks. *American Film* 7(5): 36–7.

Jossip. 2008. *Soap Opera Weekly* editor Carolyn Hinsey fired. (August 7), http://www.jossip.com/soap-opera-weekly-editor-carolyn-hinsey-fired-20080807/ (accessed July 22, 2009).

Kaufman, Debra. 2001. The Young and the Restless: Bringing HD to daytime television. *HighDef* 3(4): 6–10, http://www.highdef.com/magazine/archive/HighDef_Aug-Dec_2001H.pdf (accessed July 22, 2009).

Kaylin, Lucy. 2005. American Idol. *GQ* (June), http://men.style.com/gq/features/landing?id=content_2008 (accessed July 22, 2009).

Kemper, David. 2006. *Farscape* wrap speech. *Farscape* Starburst Edition, Volume 4 3. DVD. USA: ADV Films.

Kennedy, Dana. 1993. Soaps on the ropes. *Entertainment Weekly* (October 29), http://www.ew.com/ew/article/0,,308587,00.html (accessed July 15, 2009).

Kiesewetter, John. 2009. P&G's 'As the World Turns' canceled. *Cincinnati Enquirer* (December 8), http://news.cincinnati.com/article/20091208/ENT11/912090360/ (accessed January 4, 2010).

Kincaid, Jamaica. 1978. The talk of the town: Soap. *New Yorker,* September 18, 31.

Kompare, Derek. 2005. *Rerun nation: How repeats invented American television.* New York: Routledge.

Kroll, Dan J. 2008. Deidre Hall, Drake Hogestyn fired as DAYS trims its budget. *Soap Central* (November 16), http://www.soapcentral.com/days/news/2008/1117-hall_hogestyn.php (accessed July 23, 2009).

Kubicek, John. 2007. Exclusive interview with Minae Noji from *General Hospital*. *buddyTV.com* (July 12), http://www.buddytv.com/articles/general-hospital/exclusive-inertview-minae-noji-8185.aspx (accessed July 23, 2009).

Kung, Michelle. 2009. James Franco's upcoming 'General Hospital' appearance explained. *Wall Street Journal Speakeasy* (November 6), http://blogs.wsj.com/speakeasy/2009/11/06/james-francos-upcoming-general-hospital-appearance-explained/ (accessed November 25, 2009).

Kustritz, Anne. 2007. Ownership and desire: Fans' and producers' polymorphous triangulations. *Flow* (December 19), http://flowtv.org/?p=1034 (accessed July 23, 2009).

Lafayette, Jon. 2007. Originals clean up SOAPnet summer: Three new series bolster channel's upfront numbers. *TelevisionWeek* (August 12), http://www.tvweek.com/news/2007/08/originals_clean_up_soapnets_su.php (accessed July 23, 2009).

LaGuardia, Robert. 1974. *The wonderful world of soaps*. New York: Ballantine Books.

LaGuardia, Robert. 1983. *Soap world*. New York: Arbor House.

Lang, Kurt. 1958. Mass, class, and the reviewer. *Social Problems* 6: 11–21.

Larmonth, Michael. 2006. MyNetwork turns up heat on spicy telenovelas. *Variety* (July 30), http://www.variety.com/article/VR1117947630.html (accessed July 23, 2009).

Lavigne, Carlen. 1998. Space opera, melodrama, feminism, and the women of *Farscape*. *Femspec* 6(2): 54–64.

Lawson, Mark. 2006. The beauty myth. *Guardian* (December 19), http://www.guardian.co.uk/media/2006/dec/19/broadcasting.tvandradio2 (accessed July 23, 2009).

Lee, Richard A. 1997. The youth bias in advertising. *American Demographics* 19(1): 47–50.

Lenhart, Jennifer. 2008a. Is daytime still worth the investment? *Soap Opera Digest*, April 8, 62–5.

Lenhart, Jennifer. 2008b. Interview with Christopher Goutman. *Soap Opera Digest*, May 27, 42–5.

Levin, Gary. 2008. Network audiences are showing their age. *USA Today* (May 13), http://www.usatoday.com/life/television/news/2007-06-26-median-TV-age_N.htm?loc=interstitialskip (accessed July 23, 2009).

Levin, Gary. 2009. DVR viewers give a big boost to ratings for some shows. *USA Today* (June 19), http://www.usatoday.com/life/television/news/2009-06-17-dvr-ratings_N.htm (accessed July 23, 2009).

Levine, Elana. 2001. Toward a paradigm for media production research: Behind the scenes at *General Hospital*. *Critical Studies in Media Communication* 18(1): 66–82.

Levine, Elana. 2006. What's happening on the soaps? And why should we care? *Flow* (May 26), http://flowtv.org/?p=168 (accessed July 23, 2009).

Levine, Elana. 2007. *Wallowing in sex: The new sexual culture of 1970s American television*. Durham, NC: Duke University Press.

Levine, Elana. 2009a. Doing soap opera history: Challenges and triumphs. In *Convergence media history*, eds. Janet Staiger and Sabine Hake, 173–81. New York: Routledge.

Levine, Elana. 2009b. Like sands through the hourglass: The changing fortunes of the daytime television soap opera. In *Beyond prime time: Television programming in the post-network era*, ed. Amanda D. Lotz, 36–54. New York: Routledge.

Levinsky, Mara. 2008. Reflections on 2008: *Digest* administers daytime's 12-month checkup. *Soap Opera Digest*, December 30, 48–51.

Levinsky, Mara. 2009. The end of "Zen"? *Soap Opera Digest*, April 28, 70.

Lewis, Errol. 2008. Daytime ratings: Week of April 28 edition. *Soap Opera Network* (April 28), http://www.soapoperanetwork.com/news/ratings/29-report/283-daytimeratings042808 (accessed July 23, 2009).

Lewis, Lisa A. 1992. *The adoring audience: Fan culture and popular media*. London: Routledge.

Library of Congress. 1997. *Library of Congress television and video preservation 1997: A study of the current state of American television and video preservation.* Washington, D.C.: U.S. Government Printing Office.

Library of Congress. Undated. Television in the Library of Congress, http://www.loc.gov/rr/mopic/tvcoll.html (accessed July 23, 2009).

Liccardo, Lynn. 1996. Who really watches the daytime soaps? *Soap Opera Weekly*, April 30, 36–8.

Liccardo, Lynn. 2008. A Critfan yearns for the world as it was. *Confessions of an Aca-Fan* (April 9), http://henryjenkins.org/2008/04/a_critfan_yearns_for_the_world.html (accessed July 23, 2009).

Lidz, Franz. 2009. Slippery soaps. *Portfolio* (January 29), http://www.portfolio.com/news-markets/top-5/2009/01/09/The-Decline-of-Soap-Operas (accessed July 23, 2009).

Lisotta, Christopher. 2006a. Breaking new programming ground with telenovelas; MyNetworkTV. *TelevisionWeek* (May 15), www.tvweek.com/pdf/tw20p22.pdf (accessed August 11, 2009).

Lisotta, Christopher. 2006b. Soaps shine as lab for ventures. *TelevisionWeek*, July 17, 5.

Littlejohn, David. 1976. Thoughts on television criticism. In *Television as a cultural force*, ed. Richard Adler and Douglass Cater, 147–73. New York: Praeger.

Logan, Michael. 2008. Soaps: Lesbians unite! *TV Guide*, November 3, 77.

Lotz, Amanda. 2006. Rethinking meaning making: Watching serial TV on DVD. *Flow* (September 22), http://flowtv.org/?p=13 (accessed November 25, 2009).

Lotz, Amanda D. 2007. *The television will be revolutionized*. New York: New York University Press.

Lucas, Michael. 1999. Farscape blends love, adventure—and puppets. *Los Angeles Times* (July 20), http://articles.latimes.com/1999/jul/20/entertainment/ca-57650 (accessed July 23, 2009).

Marc, David. 1984. *Demographic vistas: Television in American culture.* Philadelphia, PA: University of Pennsylvania Press.

Mares, Marie-Louise, and Emory H. Woodard IV. 2006. In search of the older audience: Adult age differences in television viewing. *Journal of Broadcasting & Electronic Media* 50(4): 595–614.

Martin, Denise. 2007. "Battlestar Galactica": Say it ain't frakkin so! *Los Angeles Times* (May 31), http://latimesblogs.latimes.com/showtracker/2007/05/say_it_aint_fra.html (accessed July 23, 2009).

Martin, Ed. 2005. ABC's *General Hospital* is a lightweight *Sopranos* in disguise. *Jack Myers MediaVillage* (April 21). No longer available online.

Martin, Ed. 2008. *General Hospital: Night Shift* is a sexy summer surprise. *Jack Myers MediaBizBloggers* (August 19), http://www.jackmyers.com/commentary/ed-martin-watercooler/27127704.html (accessed July 23, 2009).

Matelski, Marilyn J. *Soap operas worldwide: Cultural and serial realities.* Jefferson, NC: McFarland.

McClure, David. 2008. Deployment of broadband to rural America. *US Internet Industry Association* (March 4), http://www.usiia.org/pubs/Rural.pdf (accessed July 23, 2009).

McNeil, Alex. 1996. *Total television*, fourth edition. New York: Penguin Books.

Miller, Laura. 2007. The man behind "Battlestar Galactica." *Salon* (March 24), http://www.salon.com/ent/feature/2007/03/24/battlestar/ (accessed July 23, 2009).

Mittell, Jason. 2004. *Genre and television: From cop shows to cartoons in American culture.* New York: Routledge.

Mittell, Jason. 2006. Narrative complexity in contemporary American television. *Velvet Light Trap* 58: 29–40.

Mittell, Jason. 2009. *Television and American culture.* New York: Oxford University Press.

Modleski, Tania. 1979. The search for tomorrow in today's soap operas: Notes on a feminine narrative form. *Film Quarterly* 33(1): 12–21.

Modleski, Tania. 1982. *Loving with a vengeance: Mass-produced fantasies for women*. Hamden, CT: Archon.

Moran, Jonathon. 2004. A new day for Nine. *The Age* (August 26), http://www.theage.com.au/articles/2004/08/26/1093456731202.html (accessed July 23, 2009).

Mumford, Laura Stempel. 1994. Stripping on the girl channel: Lifetime, *thirtysomething*, and television form. *Camera Obscura* 11–12(33–34): 166–91.

Mumford, Laura Stempel. 1995. *Love and ideology in the afternoon: Soap opera, women, and television genre*. Bloomington, IN: Indiana University Press.

Mundorf, Norbert, and Winifred Brownell. 1990. Media preferences of younger and older adults. *The Gerontologist* 30(5): 685–91.

Museum of Television and Radio. 1997. *Worlds without end: The art and history of the soap opera*. New York: H. N. Abrams.

Nasser, Jaime. 2008. Exporting tears and fantasies of (under)development: Popular television genres and nationalism in Mexico after World War II. PhD dissertation, University of Southern California.

Newcomb, Horace. 1974. *TV: The most popular art*. Garden City, NY: Anchor.

Newcomb, Horace. 1985. Magnum: The champagne of TV. *Channels of Communication*, May/June, 23–6.

Newcomb, Horace, and Robert Alley. 1983. *The producer's medium*. New York: Oxford University Press.

Newcomb, Roger. 2009a. TeleNext: Exhausted Every Option for Guiding Light. *We Love Soaps* (July 24), http://www.welovesoaps.net/2009/07/telenext-exhausted-every-option-for.html (accessed July 28, 2009).

Newcomb, Roger. 2009b. ABC.com adding AMC and OLTL starting today. *We Love Soaps* (December 15), http://www.welovesoaps.net/2009/12/abccom-adding-amc-and-oltl-starting.html (accessed December 28, 2009).

Newcomb, Roger. 2009c. Soap opera ratings for Dec. 7–13, 2009. *We Love Soaps* (December 17), http://www.welovesoaps.net/2009/12/soap-opera-tv-ratings-for-dec-7-13-2009.html (accessed December 28, 2009).

Ng, Eve. 2008. Reading the romance of fan cultural production: Music videos of a television lesbian couple. *Popular Communication* 6(2): 102–21.

Nielsen. 2009. "Lost," "Saturday Night Live" and "Grey's Anatomy" most popular entertainment TV programs streamed from tagged broadcast network Web sites in December. (February 12), http://www.nielsen-online.com/pr/pr_090212_2.pdf (accessed July 23, 2009).

Nochimson, Martha. 1992. *No end to her: Soap opera and the female subject*. Berkeley, CA: University of California Press.

O'Meara, Radha. 2004. *The Dorm*: Taking the reality out of reality TV. *Metro* 140, 110–11.

O'Meara, Radha. Forthcoming. "I will try harder to merge the worlds": Expanding narrative and navigating spaces in Gilmore Girls. In *Screwball television: Critical perspectives on Gilmore Girls*, ed. David Scott Diffrient and David Lavery. Syracuse, NY: Syracuse University Press.

Ouellette, Laurie. 2002. *Viewers like you? How public TV failed the people*. New York: Columbia University Press.

Owen, Rob. 2003. Tuned in: "Farscape"—The blame goes around. *Pittsburgh Post Gazette* (January 10), http://www.post-gazette.com/tv/20030110owen3.asp (accessed July 23, 2009).

Passanante, Jean. 2006. Interview by Sam Ford. March 29.

Poll, Julie. 1996. As the World Turns *scrapbook: Special 40th anniversary edition*. Los Angeles: General Publishing.

Pritchard, William H. 1986. Why the professor must have his soap along with his Henry James. *TV Guide*, May 10, 31.

Radway, Janice A. 1984. *Reading the romance: Women, patriarchy, and popular literature*. Chapel Hill, NC: University of North Carolina Press.

Reardon, Marguerite. 2009. Free TV for cell phones and mobile devices. *CNET News* (April 20), http://news.cnet.com/8301-1035_3-10223478-94.html (accessed November 14, 2009).

Regev, Motti. 1994. Producing artistic value: The case of rock music. *Sociological Quarterly* 35: 85–102.

Reynolds, Mike. 2008. SoapNet swings into second "Night Shift." *Multichannel News* (May 24), http://www.multichannel.com/article/133402-SoapNet_Swings_Into_Second_Night_Shift_.php (accessed July 23, 2009).

Rice, Lynette. 2009a. Eric Braeden talks about controversial exit as Victor Newman on "Young and the Restless." *Entertainment Weekly Hollywood Insider Blog* (October 7), http://hollywoodinsider.ew.com/2009/10/07/exclusive-eric-braeden/ (accessed November 24, 2009).

Rice, Lynette. 2009b. Eric Braeden stays put on "The Young and the Restless." *Entertainment Weekly Hollywood Insider Blog* (October 23), http://news-briefs.ew.com/2009/10/23/eric-braeden-stays-put-on-the-young-and-the-restless/ (accessed November 24, 2009).

Riggs, Karen E. 1998. *Mature audiences: Television in the life of elders*. New Brunswick, NJ: Rutgers University Press.

Robertson, Roland. 1995. Glocalization: Time-space and homogeneity-heterogeneity. In *Global Modernities*, ed. Mike Featherstone, Scott Lash, and Roland Robertson, 25–44. London: Sage.

Robinson, Julie Clark. 2007. Interview with GH's Minae Noji, Part II. *Soaps.com* (August 6), http://www.soaps.com/generalhospital/news/924/Interview_with_GHs_Minae_Noji__Part_II (accessed July 23, 2009).

Rohter, Larry. 2007 How "Ugly Betty" changed on the flight from Bogotá. *New York Times* (January 7), http://www.nytimes.com/2007/01/07/weekinreview/07rohter.html (accessed July 23, 2009).

Rosen, Christine. 2007. Virtual friendship and the new narcissism. *The New Atlantis* 17 (Summer), http://www.thenewatlantis.com/publications/virtual-friendship-and-the-new-narcissism (accessed July 23, 2009).

Rosen, Robert. 1997. UCLA Film and Television Archive submission. In *Library of Congress television and video preservation 1997: A study of the current state of American television and video preservation, V.5: Submissions*, 450–60. Washington, D.C.: U.S. Government Printing Office.

Rosenberg, Howard. 1987. Requiem for best TV "Blues." *Los Angeles Times*, April 15, Calendar Section 1.

Ross, Sharon Marie. 2008. *Beyond the box: Television and the Internet*. Malden, MA: Blackwell Publishing.

Roush, Matt. 2004. Dispatches. *TV Guide* (October 15). Available at http://www.farscapeworld.com/news/shownews.php?id=1066 (accessed July 23, 2009).

Roush, Matt. 2007. Comments on a discussion about the fall television season. *Charlie Rose*, PBS, November 15. Available at http://www.charlierose.com/view/interview/8782 (accessed July 23, 2009).

Rowe, Kathleen. 1995. *The unruly woman: Gender and the genres of laughter*. Austin: University of Texas Press.

Russell, Cheryl. 1997. The ungraying of America. *American Demographics* 19(7): 12–5.

Samuels, David. 2008. Shooting Britney. *The Atlantic* (April), http://www.theatlantic.com/doc/200804/britney-spears (accessed July 23, 2009).

Scardaville, Melissa C. 2005. Accidental activists: Fan activism in the soap opera community. *American Behavioral Scientist* 48(7): 881–901.

Schemering, Christopher. 1988. *The soap opera encyclopedia*. New York, NY: Ballantine Books.

Schiller, Abbie. 2006. E-mail correspondence with Elana Levine. August 18.

Schilling, Mary Kay. 1999. Secrets and lies. *Entertainment Weekly* (February 5), http://www.ew.com/ew/article/0,,20036782_20037403_274347_4,00.html (accessed July 23, 2009).

Schlossberg, Suzanne. 2002. The end of a seasonal affair. *AlterNet* (April 11), http://www.alternet.org/mediaculture/12846/ (accessed July 23, 2009).

Schoenherr, Steven E. 2007. Golden age of radio 1935–50. *History Department at the University of San Diego* (March 20), http://history.sandiego.edu/gen/recording/radio2.html (accessed July 23, 2009).

Scodari, Christine. 2004. *Serial monogamy: Soap opera, lifespan, and the gendered politics of fantasy*. Cresskill, NJ: Hampton Press.

Scodari, Christine, and Jenna L. Felder. 2000. Creating a pocket universe: "Shippers," fan fiction, and *The X-Files* online. *Communication Studies* 51: 238–57.

Sconce, Jeffrey. 2004. What if?: Charting television's new textual boundaries. In *Television after TV*, ed. Lynn Spigel and Jan Olsson, 93–112. Durham, NC, and London: Duke University Press.

Seiter, Ellen. 1982. Promise and contradiction: The daytime television serial. *Film Reader* 5, 150–63.

Seiter, Ellen, Hans Borchers, Gabriele Kreutzner, and Eva-Maria Warth. 1989. "Don't treat us like we're stupid and naïve": Towards an ethnography of soap opera viewers. In *Remote control: Television, audiences, and cultural power*, ed. Ellen Seiter, Hans Borchers, Gabriele Kreutzner, and Eva-Maria Warth, 223–47. London and New York: Routledge.

Seiter, Ellen, and Mary Jeanne Wilson. 2005. Soap opera survival tactics. In *Thinking outside the box: A contemporary television genre reader*, ed. Gary R. Edgerton and Brian G. Rose, 136–55. Lexington, KY: University Press of Kentucky.

SFX Magazine. 2003. Farscape unplugged. April, 65–76.

Shanks, Bob. 1976. *The cool fire*. New York: Vintage Books.

Simon, Ron. 2002. Interview by Mary Jeanne Wilson. December.

Smith, Lynn. 2008. Soap fans claim bias against gay characters. *Los Angeles Times* (February 22), http://articles.latimes.com/2008/feb/22/entertainment/et-atwtweb21 (accessed July 23, 2009).

Snark Weighs In. 2005. GL podcast review. (September 25), http://snarkweighsin.blog-city.com/glpodreview.htm (accessed July 23, 2009).

Snark Weighs In. 2008. Kids, guns, and Marlena: A rebuttal, I. (February 7), http://snarkweighsin.blog-city.com/rebuttalone.htm (accessed July 23, 2009).

SOAPnet Medianet. 2007a. Upcoming SOAPnet original series to get multiplatform play. (July 10). No longer available online.

SOAPnet Medianet. 2007b. July 07 Stands as SOAPnet's most watched July of all time in the Net's sales prime and total day in total viewers and women 18–49 (April 8). Available at http://www.cableu.tv/soapnet-press-releases-2007/ (accessed August 13, 2009).

SOAPnet Medianet. 2008a. Antonio Sabato Jr. and Tristan Rogers join cast of SOAPnet original series *General Hospital: Night Shift*. (June 30), http://www.soapnetmedianet.com/DNR/2008/doc/june/SabatoRogersEntireCast.doc (accessed July 23, 2009).

SOAPnet Medianet. 2008b. SOAPnet surpasses 70 million households. (September 15), http://soapnetmedianet.com/DNR/2008/doc/september/SN_70million_release_update.doc (accessed July 23, 2009).

Soap Opera Digest. 1995. A matter of degrees. March 24, 76–8.

Soap Opera Digest. 2007a. Editor's choice: The difference between night and day, *GH/GH:NS*. August 14, 33.

Soap Opera Digest. 2007b. Guza previews *GH: Night Shift*. May 22, 6.

Soap Opera Digest. 2007c. *Night Shift*: Executive suite with Jill Faren Phelps. July 17, 39.

Soap Opera Digest. 2008a. Scorpio returns. August 12, 19.
Soap Opera Digest. 2008b. Editor's choice: It happened one night. November 25, 33.
Soap Opera Digest. 2009a. Zach and Kendall divorce on *AMC*. April 21, 7.
Soap Opera Digest. 2009b. Braun and Rigel exit *AMC*. April 28, 4.
Soap Opera Digest. 2009c. May sweeps preview!. April 28, 40.
Soap Opera Weekly. 2007. The doctor is in. October 16, 34.
Spehr, Paul. 1992. Selection and rejection at the Library of Congress. In *Documents that move and speak: Audiovisual archives in the new information age: National Archives of Canada, Ottawa, Canada, April 30, 1990–May 3, 1990: proceedings of a symposium*. ed. Naugler, Harold, 49–57. Munchen, Germany: K. G. Saur.
Spence, Louise. 2005. *Watching daytime soap operas: The power of pleasure*. Middletown, CT: Wesleyan University Press.
Stein, Joshua David. 2009. James Franco explains what he's doing on *General Hospital*, sort of. *New York Magazine* (November 9), http://nymag.com/daily/entertainment/ 2009/11/james_franco_on_30_rock_genera.html (accessed November 25, 2009).
Steinberg, Brian. 2008. Soaps: They're not your mother's daytime dramas anymore. *Boston Globe* (April 27), http://www.boston.com/business/articles/2008/04/27/soaps/ (accessed July 23, 2009).
Streeter, Leslie Gray. 2008. *General Hospital* finally admits that mobsters kill people! *The Pop Shop* (April 15), http://www.pbpulse.com/tv/all-shows/2008/04/25/general-hospital-finally-admits-that-mobsters-kill-people/ (accessed August 13, 2009).
Suratt, Samuel. 1995. One hundred years: Television heritage, the mirror of our society. *Moving Image Review* (Winter), http://www.archive.org/stream/movimarevi19882007nortrich/movimarevi19882007nortrich_djvu.txt (accessed June 22, 2009).
Tedeschi, Bob. 2006. Older consumers flex their muscle (and money) online. *New York Times* (June 12), http://www.nytimes.com/2006/06/12/technology/12ecom.html (accessed July 23, 2009).
The Futon Critic. 2007. SOAPnet announces first original scripted drama *General Hospital: Night Shift*, a weekly continuation of the Emmy winning, top rated daytime drama *General Hospital* (February 12), http://www.thefutoncritic.com/news.aspx?id=20070212soapnet01 (accessed July 23, 2009).
Thomas, Becca. 2007a. Nominee for worst character in daytime: Sonny Corinthos. *Serial Drama* (February 22), http://serialdrama.typepad.com/serial_drama/2007/02/nominee_for_wor_1.html (accessed July 23, 2009).
Thomas, Becca. 2007b. The people in the little box aren't real. *Serial Drama* (December 19), http://serialdrama.typepad.com/serial_drama/2007/12/the-people-in-t.html (accessed July 23, 2009).
Thompson, Robert J. 1996. *Television's second golden age: From* Hill Street Blues *to* ER. New York: Continuum.
Thornton, Patricia H., and William Ocasio. 1999. Institutional logics and the historical contingency of power in organizations: Executive succession in the higher education publishing industry, 1958–1990. *American Journal of Sociology* 105: 801–43.
Thurber, James. 1948. *The beast in me and other animals*. New York: Harcourt, Brace, and Co.
Timberg, Bernard. 1981. The rhetoric of the camera in television soap opera. In *Television: The critical view*, third edition, ed. Horace Newcomb, 132–47. New York: Oxford University Press.
Timberg, Bernard. 2002. *Television talk: A history of the TV talk show*. Austin, Texas: University of Texas Press.
Time. 1976. Sex and suffering in the afternoon. (January 12), http://www.time.com/time/magazine/article/0,9171,913850,00.html (accessed July 23, 2009).

Time. 2008. The kiss campaign (March 17), http://www.time.com/time/specials/2007/article/0,28804,1638826_1720383_1720379,00.html (accessed July 23, 2009).

Torchin, Mimi. 2009. Speaking her mind—Exclusive interview: Mimi Torchin catches up with GL's Maureen Garrett (Holly). *TV Guide Canada* (July 27), http://tvguide.sympatico.msn.ca/The+Soapgeist+July+27+2009/Soaps/Nelson_Ratings/Articles/090727_soapgeist_NB.htm?isfa=1 (accessed July 28, 2009).

Tuchman, Gaye. 1979. *Making news*. New York: Free Press.

Tulloch, John. 1989. Approaching the audience: The elderly. In *Remote control: Television, audiences and cultural power*, ed. Ellen Seiter, Hans Borchers, Gabriele Kreutzner, and Eva-Maria Warth, 180–202. London and New York: Routledge.

Tulloch, John. 1990. *Television drama: Agency, audience, and myth*. London: Routledge.

Turow, Joseph. 1997. *Breaking up America: Advertisers and the new media world*. Chicago and London: University of Chicago Press.

Turow, Joseph. 2005. Audience construction and culture production: Marketing surveillance in the digital age. *Annals of the American Academy of Political and Social Science* 597: 103–21.

UCLA Film & Television Archive. 2003. UCLA film and television archive collection policy: Television. Unpublished document. March 17.

UCLA Film & Television Archive. Undated a. The ABC collection. http://www.cinema.ucla.edu/collections/Profiles/abc.html (accessed July 23, 2009).

UCLA Film & Television Archive. Undated b. Television collections. http://www.cinema.ucla.edu/collections/television.html (accessed July 23, 2009).

Umstead, R. Thomas. 2007. SOAPnet aims younger: "Night Shift," off-net buys part of push to lower viewer age. *Multichannel News* (July 9), http://www.multichannel.com/article/106252-SoapNet_Aims_Younger.php (accessed July 23, 2009).

United States Copyright Office. Undated. Copyright Basics. http://www.copyright.gov/circs/circ01.pdf (accessed July 23, 2009).

USA Today. 1996. Truth behind "X-Files" move. May 22, 3D.

Vara, Vauhini. 2005. New *Guiding Light* podcasts aim for a younger audience. *Wall Street Journal* (October 18), http://online.wsj.com/article/SB112956084439070707.html (accessed June 15, 2009).

Waldman, Allison J. 2008. TV Land focuses on baby boomers. *TelevisionWeek* (February 4), http://www.tvweek.com/news/2008/02/tv_land_focuses_on_baby_boomer.php (accessed July 23, 2009).

Weiss, Joanna. 2008. Their soap smooch made history. Fans ask: Will it happen again? *Boston Globe* (March 1), http://www.boston.com/ae/tv/articles/2008/03/01/their_soap_smooch_made_history_fans_ask_will_it_happen_again/ (accessed July 23, 2009).

Wesch, Michael. 2008. YouTube statistics. *Digital Ethnography* (March 18), http://mediatedcultures.net/ksudigg/?p=163 (accessed July 23, 2009).

West, Abby. 2008. Sabato Jr.: "Night" to day? *Entertainment Weekly* (July 18), http://www.ew.com/ew/article/0,,20213473,00.html (accessed July 23, 2009).

Wheeler, Ellen. 2008. Interview by Patrick Erwin. December 9, 2008.

White, Mimi. 1994. Women, memory and serial melodrama. *Screen* 35(4): 336–53.

Williams, Carol Traynor. 1992. *"It's time for my story": Soap opera sources, structure, and response*. Westport, CT: Praeger.

Williams, Garry. 2004. A daze of their lives. *Sunday Herald Sun* (September 12). No longer available online.

Wilson, Benji. 2007. How Ugly Betty turned into a swan. *Telegraph* (September 29), http://www.telegraph.co.uk/culture/tvandradio/3668243/How-Ugly-Betty-turned-into-a-swan.html (accessed July 23, 2009).

Wittebols, James H. 2004. *The soap opera paradigm: Television programming and corporate priorities*. Lanthan, MD: Rowman & Littlefield.

Wolfe, David B., with Robert E. Snyder. 2003. *Ageless marketing*. Chicago: Dearborn Trade Publishing.

Wood, Michael. 2004. Taking reality buy surprise. *New York Review of Books* (November 4), http://www.nybooks.com/articles/17522 (accessed July 23, 2009).

Wynter, Karen. 2008. "World" of controversy. *CNN ShowbizTonight*, CNN, March 3.

INDEX

ABC (American Broadcasting Company), 4–5, 9, 11, 14–15, 23–24, 28–29, 31, 34, 36, 38, 40–41, 43, 44, 49–53, 56, 61, 64, 68, 70–71, 104–5, 107–9, 112–14, 119, 122–24, 126, 128, 129n10, 130, 133–35, 140–41, 144, 147, 153n14, 155–56, 163–64, 181, 186–87, 190n1, 192–93, 196, 199n1, 204–5, 209–10, 214, 216n2, 217n29, 218n32, 219–20, 223–24, 229–30, 233, 256, 261–62, 264n6, 265–67, 269, 272, 275, 294, 300, 301, 305, 308; Daytime, 40, 101, 108, 147, 191–93, 202, 205, 207, 230n3, 266
ABC Soaps in Depth, 195
Abe Carver (character), 288
"Accidental Activists," 58
ACP/PBL Media, 292n5
Adalian, Josef, 21n1, 198
Adam Chandler (character), 123, 313
Adam Drake (character), 29
Adam Hughes (character). *See* Adam Munson
Adam Munson (Hughes) (character), 95–97
addiction, 9, 92, 105, 130, 233–34, 240
Adler, Renata, 252
advertising, 3, 19, 43, 61–62, 68, 71, 85, 112, 131, 156, 221–22, 226, 238, 246, 283, 287, 295, 301
Aeryn Sun (character), 110–11
Affleck, Ben, 235
age, 15, 18, 20, 24, 26–27, 36, 85, 89, 91–92, 100n14, 300–14; artificial youth, 176; Soap Opera Rapid Aging Syndrome (SORAS), 91–92, 96, 98, 294, 299n1
Ahrens, Frank, 301, 303
Alan Quartermaine (character), 227
Alarr, Albert, 188–89
Alba, Ernest, 18, 100, 186, 200
Albiniak, Paige, 35
Alden, Kay, 17, 101–3, 155, 306, 312, 314, 314n4
Aleksander, Grant, 186
Alexis Davis (character), 212

Alias, 133
Alice Horton (character), 6
Alice in Wonderland, 94
Alinsky, Saul, 46
Alison Stewart (character), 205
All My Children, 4, 6, 15, 21n3, 36, 38–39, 41, 44–48, 61, 64, 74, 106, 114–15, 123, 153n14, 164, 181, 190n1, 204–5, 219, 223, 265, 269, 294, 296, 308, 313
Allen, Robert, 13–14, 17, 60, 81, 83–85, 141, 175–76, 178–79, 191, 195, 197, 199n3, 200n10, 208, 286, 288, 290
Althea Davis (character), 314n3
American Behavioral Scientist, 58
American Dreams, 124
American Idol, 112, 114
America's Most Wanted, 146
America's Next Top Model, 10
amnesia, 104, 285, 291–92
AMPTP (Association of Motion Picture and Television Producers), 155, 158
Anderson, Gillian, 236, 245
Ang, Ien, 57n3, 84
Angel, 111, 133
Angela Moroni (character), 286–87
Angie Costello (character), 165, 167–69, 171–72, 173n4
Aniston, Jennifer, 237–38, 241–42, 245
Anna Devane (character), 194, 227
Anna DiMera (character). *See* Anna Fredericks
Anna Fredericks (DiMera)(character), 276
Another World, 15, 28, 40, 44, 58, 81, 106, 113, 116, 120–21, 216n4, 275, 293, 300, 314n1
AOL, 193, 216n4
Apprentice, The, 10
archiving, 3, 140–41, 150, 152, 153n17, 230
Argus, The, 96
Arlen, Michael, 164
As the World Turns (*ATWT*), 4–6, 15–16, 23,

331

29, 39, 43, 61, 64, 75, 86–99, 101, 107, 115–16, 120–21, 123, 126–28, 154–55, 158, 175–80, 203, 265, 276, 293–99, 305
"*As the World Turns* in a Convergence Culture," 86
Aspenson, Carolyn, 217n21
Associated Press, 4, 297
Aubrey Cross (character), 216n9
audience, 3, 6, 8, 11, 13, 19, 24–25, 36, 43, 45, 53, 64–65, 67, 84–89, 106–13, 115–17, 134, 141, 198, 202, 233–48, 250–64, 275–78, 282, 285, 300–13; daytime, 43, 67, 102–3, 113, 134; interpretation, 241; segmentation/fragmentation, 107, 112–13, 117, 301; surplus, 19–20, 295. *See also* fandom
Austin Reed (character), 288

Baby Boomers, 18, 36, 85, 302–3, 305, 309
Bachelor, The, 112, 114
Banned Club, The, 269
Bannon, Natalie, 93
Barbara Ryan (character), 88, 93–94, 96
Bash Board, 269
Battlestar Galactica, 111
Bauder, David, 297, 303
Bay City, 238
Baym, Nancy, 14, 17, 104–5, 191, 201, 222–23, 225, 227
BBC (British Broadcasting Corporation), 54, 116
Beale, Lewis, 110
Becker, Howard S., 264n1
Bell, Bradley, 32
Bell, Maria Arena, 157–59
Bell, William J., 26, 32, 154–55, 157–58
Bell, William J., Jr., 153n13, 157
Bell-Phillip Television Productions, 32, 153n13, 154, 157
Betty Suarez (character), 52–54
Beverly Hills 90210, 24
Bianca Montgomery (character), 45–47, 223, 294, 296
Bibel, Sara, 15, 120, 122, 125, 128n4, 129n17, 159n1, 179n1, 186n1, 190n2, 216n7, 314n8
Big Brother, 10, 103
Bielby, Denise, 14, 19, 61, 69, 191–92, 195, 222, 233, 248, 264n3, 281, 300
Bielby, William T., 69, 222, 253
Billy Douglas (character), 104, 294
Black, Claudia, 111
Black Entertainment Television (BET), 153n7
Blackwell, Deborah, 193
Blair, Elizabeth, 4
Blake Marler (character), 184
blogs, 4, 25, 38, 137, 180, 202, 204–15, 235, 238, 240–45, 277, 294
Bloom, Barbara, 297
Blumenthal, Dannielle, 13
Bo Brady (character), 291
Bob Hughes (character), 6, 16–17, 29, 88–89, 91–92, 96, 100n3, 121, 314n3
Bobbie Jones (character). *See* Bobbie Spencer
Bobbie Spencer (character), 294, 314n3
Bold and the Beautiful, The (*B&B*), 5, 31–32, 101–2, 145, 153, 157, 186, 204, 279, 306, 314n3, 314n4
Bordwell, David, 286–87, 289–90
Boston Globe, The, 297
Bourdon, Jérôme, 240
Boutin, Paul, 221
Braeden, Eric, 4, 21n2, 41
Branco, Nelson, 5, 198, 200n14, 214
Brandon Walker (character), 286–87
Braun, Tamara, 45
Bravo, 10
Breaking Up America, 112
Brinckerhoff, Peter, 165, 173n3, 188–89
Britton, Connie, 125
Brodesser-Akner, Claude, 176
Brooke Logan (character), 102
Brooks, Peter, 81
Brot Monroe (character), 46
Brothers, Denise, 20, 100n11, 100n14, 118n9
Brown, Chris, 243
Brown, Mary Ellen, 9, 106
Brown, Scott, 297
Brownell, Winifred, 309
Browning, Robert, 94
browsing, 82, 221, 238
Brunsdon, Charlotte, 7, 21n6, 54, 141
Buchanan family (characters), 123
Buffy the Vampire Slayer, 31, 111, 114, 133
Bury, Rhiannon, 222
Butler, Jeremy, 13
Buxton, Rodney A., 301

cable television. *See* television
Cagle, Jess, 122
Cagney & Lacey, 23, 108
Caldwell, John Thornton, 283
cancellation, 4–6, 23, 30, 34, 39–40, 88–89, 112, 114, 119, 121, 126, 128, 138, 153n7, 154, 158, 172, 186, 222
Cantor, Joel, 164
Cantor, Muriel, 13, 16, 60
Carlivati, Ron, 127
Carly Corinthos Jacks (character), 213, 266, 273
Carly Tenney (character), 88
Carol Deming (character), 93
Caroline Brady (character), 314n3
Carter, Bill, 5, 52, 221
Carter, Chris, 110
Casey Hughes (character), 95–97
Casiello, Tom, 19, 100n11, 129n16, 198, 216n7, 312, 314n4
Cass Winthrop (character), 314n3
Cassata, Mary, 13, 15, 22–28, 304, 309
CBS (Columbia Broadcasting System), 4–6, 8, 10, 15, 18, 23–24, 28–32, 36, 38–40, 43–44, 61, 64, 70, 81, 87–88, 97, 100n13, 101, 103, 107–8, 112–14, 119–22, 124, 126, 128, 129n10, 134, 141, 147, 153n13, 154–55, 158, 164, 167–68, 173n2, 175–76, 180, 186–89, 190n3, 192, 199n1, 202–4, 222, 229, 233, 240, 256, 264n6, 265, 267, 275–77, 292n1, 293, 295, 297, 299, 301, 305–6
CBS Evening News, 173n2
CBS Soaps in Depth, 126
Cedric Daniels (character), 136
celebrity. *See* stardom
Charles Tyler (character), 46–47
Charlie (character), 52
Cheers, 113, 134
Chicago Sun-Times, The, 56
Childress, Clayton, 264
Chloe Lane (character), 286
Chris Hughes (character), 93, 121
Chura, Hillary, 303
Cinderella, 54–55
City, The, 14, 144, 153n8, 181
Clay Davis (character), 136
CNN, 297
Cole, Tim, 131
Colin, Margaret, 93, 100n10

Colker, David, 189
Collins, Scott, 302, 313
Collins, Sue, 10
Colombino, Terri, 88
Colson, David, 89, 93
commercialism, 13, 51, 84–85, 107–9, 113, 116, 140, 143, 152, 153n10, 251, 253, 281; and commercials, 68, 114–15, 135, 144, 176, 177, 187, 197, 201, 225–26, 239, 286
complexity, 8, 16, 84, 121, 126, 133, 195; narrative, 133–35, 194–95, 196–97, 199
Compton, Forrest, 30
Conboy, John, 32
consecutive viewing, 7, 10–12, 144–45
consumers, 62, 76–77, 107, 212, 238, 247–48, 250, 254, 301–3, 309, 313. *See also* production: consumer and producer conflict
Cook, Terry, 143
Cooper, Jeanne, 158, 307
Coppa, Francesca, 220
Cops, 146
copyright, 142, 145–48, 151–52, 153n11, 230
Corday, Ken, 39, 71, 147
Corrington, Bill, 31
Corrington, Joyce, 31
Cortez, Carl, 115
Court TV, 23
Coward, Rosalind, 291
Craig Montgomery (character), 176, 311
Crane, Dagne, 29
Crane, Diana, 257
Cranecouture.com, 203
Craven, Ian Peter, 284
Crew, Adrienne, 111
Cridlin, Jay, 57
Crofts, Stephen, 284
Cronkite, Walter, 173n2
Cruise, Tom, 235, 237–38
CSI: Crime Scene Investigation, 101, 112, 114
cultural significance, 4, 6–8, 50–51, 54–56, 81, 84–86, 92–93, 143, 202, 222, 239–41, 250–54, 263–64, 302
CW (television network), 4, 9–10, 31, 41, 49, 56, 103

Da Ros, Giada, 15
Dad's Army, 54
daily stripping, 10–11

INDEX   333

Dallas, 23, 58, 81, 122–26, 129n10, 129n11, 134
Damian Grimaldi (character), 88
Damian Spinelli (character), 200n8
Damn the Man! Save the Empire, 275
Damon, Stuart, 308
Dan McCloskey (character), 310
Dana Scully (character), 109–10
Daniel Hughes (character), 96–97
Dano, Linda, 307, 314n3
Dark Shadows, 114
Dattilo, Bryan, 314n3
Dawson, Trent, 88
Dawson's Creek, 108
Days of Our Lives (DOOL), 4, 6, 19, 24, 39, 40–41, 43–44, 61, 70–71, 76, 104, 106, 114, 147, 181, 188–89, 202–4, 216n3, 256, 275–76, 279–92, 292n1, 292n2, 292n3, 300, 307–8, 314n3, 314n4
Daytime Dish, 224, 230n3
Daytime Dollars, 204, 216n6
Daytime Emmy Awards. *See* Emmy Awards/Daytime Emmy Awards
Daytime Television: Tuning in American Serial Drama, 22
Daytime TV, 252, 282
De Kosnik, Abigail, 19
De Lacroix, Marlena, 128n3, 180, 193, 196–98, 216n16, 217n21, 264n8
de Moraes, Lisa, 126
Deas, Justin, 89–90, 93–94, 100n10
decline, 3–5, 15–17, 20, 23, 39, 64, 66–67, 77, 87, 120, 125–27, 152, 156, 187, 201, 222, 262–63, 275
Deep Soap, 38
Defamer, 235, 239, 248n2
demographics, 6, 13, 18–19, 35–36, 47, 61, 64, 76, 85, 108, 111, 119, 176, 182, 196, 264, 291, 300–14
Denis, Paul, 252, 264n2
Desire, 50, 57n2
Desperate Housewives, 9, 109, 130
Devil Wears Prada, The, 54
Dibbern, Tad, 122, 129n11
Dickens, Charles, 7, 69
Diliberto, Joe, 200n12
DiMaggio, Paul, 69
DirecTV, 39–40, 126, 147, 182, 205–6, 216n5
Dirty Sexy Money (DSM), 126, 135
disengagement, 8, 302

Disney, 24, 40–41, 115, 262
Dobson, Bridget, 32
Dobson, Jerry, 32
Doctors, The, 314n3
docusoap, 10, 130
"Doing Soap Opera History," 201
Dolan, Ellen, 88, 95
Dominguez, Robert, 49, 55
Duchovny, David, 110, 236, 245
Dugas, Christine, 221
Dusty Donovan (character), 176–78
DVR, 5, 24, 33, 68, 115, 225–26
Dwight Schrute (character), 216n2
Dyer, Richard, 242
Dynasty, 23, 122–24, 126, 129n13, 134, 136

EastEnders, 116
Edge of Night, The, 15, 29–30, 87, 122, 250, 264n6, 267
Edmondson, Madeleine, 60
Edwards, John, 243
"Effect of the Youtube Phenomenon on the Soap Opera Text, The," 163
Efron, Zac, 238
Elfman, Doug, 56
Eliot Gerard (character), 176–78
Elizabeth Webber (character), 170, 220
Elliott, Stuart, 216n6
Emerald City Bar, 216n2
Emergency!, 105
Emily Quartermaine (character), 170–71, 212–13
Emily Stewart (character), 88, 96, 176–78
Emma Snyder (character), 88, 310
Emmy Awards/Daytime Emmy Awards, 4, 25, 30–31, 40, 44, 46, 89, 101, 107, 145, 165, 180, 188, 254
Entertainment Tonight, 40
Entertainment Weekly, 257, 297
ephemerality, 140, 178
episodic structure, 62, 81, 85, 87, 106, 108, 110, 112, 114, 117, 126–27, 175, 194, 197, 247
ER, 108, 124, 138
Eric Brady (character), 286
Eric Forrester (character), 102, 314n3
Erica Kane (character), 6, 44–45, 267, 294, 313
Erwin, Patrick, 18, 40, 118n8, 120–21, 128n3, 159n2, 179n1, 190n3

Esther Valentine (character), 314n3
Evans, Mary Beth, 39, 44
excess, 13, 131, 153n1
experimentation, 3, 16, 18, 144, 198; formal, 15, 18, 56, 133, 158, 165, 176
"Exporting Tears and Fantasies of (under) Development," 49

Fadner, Ross, 220
Falcon Crest, 23, 122, 134
Fan Fussin' Forums, 269
Fancast, 38, 216n4
Fancy Crane (character), 203, 205
fandom, 8, 9, 17, 19, 59–63, 66, 67, 69, 70, 74–77, 104, 219–32, 233–49, 267, 273, 276, 293, 295, 297, 298–99; creations, 234, 275–77; feedback, 19, 25–26, 74–76, 297
Farscape, 109–12, 117, 118n7
Fashion House, 50, 57n2
Fawlty Towers, 54
Federal Communications Commission, 266
Felder, Jenna L., 110
Felicia Gallant (character), 314n3
Fernandez, Maria Elena, 222
Ferrera, America, 52, 54
Feuer, Jane, 153n15
Fickett, Mary, 46
Fine, Jon, 303
Firefly, 109, 111–12, 117
Fiske, John, 9, 21n5, 106–8, 110, 116, 286
Fitzgerald, Toni, 192, 293, 295–96
Flannery, Susan, 102, 186
Flint, Joe, 202
Flood, Ann, 30
Flow TV, 201
Flynn, Lauren, 308
Ford, Sam, 21n4, 122, 129n8, 203, 206, 216n1, 216n9, 216n10, 240, 295, 299n1, 314, 314n9, 314n10
Fox (Fox Broadcasting Company), 24, 49, 51, 57n2, 108–12, 114, 117n6, 130, 133, 135, 138, 146, 236, 301, 303
Fox Crane (character), 205
Fox Mulder (character), 109–10
Franco, James, 16
Franklin, Hugh, 46
Frannie Hughes (character), 17
Freeman, Michael, 112

Frentz, Suzanne, 14
Friday Night Lights (*FNL*), 39, 125–26, 135
Friedman, Wayne, 303
Friends, 10
From Hill Street Blues to ER, 112
Frons, Brian, 192, 194, 200, 262
Fry, Ed, 88
Fulton, Eileen, 6, 88–89, 306
Futon Critic, The, 192
FX, 9, 111, 130

Galman, Peter, 89–90, 92
game shows, 5, 40, 85, 144
Gans, Herbert, 164
Garner, Jennifer, 235
Garrett, Maureen, 185
Gawker, 235, 239, 248n2
gay: characters, 19, 104, 293–99; themes, 293–99
Geary, Anthony, 119–21, 126, 128n2
gender, 8, 9, 106–11, 113, 115–17, 167, 209, 211, 213, 300, 304, 309, 314n1
General Hospital (*GH*), 15–16, 18–19, 31, 34, 37–38, 40–41, 44, 61, 64, 104–6, 108, 114, 119–21, 133, 140–41, 144–47, 163–74, 173n5, 174n8, 187, 188, 189–90, 199n1, 199n5, 199–200n7, 200n8, 200n10, 200n11, 200n12, 200n15, 204–13, 216n15, 217n28, 217n29, 217–18n32, 230n4, 233, 250, 256, 265–69, 272–74, 294, 300, 308, 311, 314n3
General Hospital: Night Shift, 18, 34–37, 40, 188, 191–200, 199n4, 199–200n7, 200n8, 200n9, 200n12, 217–18n32
"*General Hospital*: Twist of Fate," 199n1
Generations, 14, 144, 153n7, 153n8
Genoa City, 123, 205
Genre and Television, 133
George Foster Peabody Awards, 58
Geraghty, Christine, 21n6, 141
Gibbons, Douglas, 148
Gibson, William, 233–34, 240
Gilbert, Matthew, 111–12
Giles, David, 87
Gillespie, Marie, 236–37
Gilmore Girls, 31, 279
Glaser, Barney G., 305
Glynn, Mary Ann, 60
Gonzales, Racquel, 18, 129n22, 190n4, 217–18n32

Goodwin, Paige E., 302
Gossip Girl, 9, 103
Goutman, Chris, 75
Gray, Jonathan, 222
Green, Erick Yates, 18, 173n1, 173n3, 179n1, 200n15
Greenlee Smythe (character), 47
Greppi, Michele, 51
Grey's Anatomy, 38, 108, 117n4, 193, 196, 198, 199n4, 216n2
Griffith, Josh, 157–58
Guardian, The, 54
Guiding Light (GL), 4–5, 15, 18, 23, 31, 36, 39–44, 58, 61, 64, 88, 101, 114–16, 119, 123, 141, 145, 153n13, 154, 158, 180–88, 190n3, 202–6, 216, 256–57, 266, 314n3
Guinness World Records, 88
Guza, Robert "Bob," 194
Gyllenhall, Jake, 238

Hagman, Larry, 122
Hal Munson (character), 94–97
Hall, Deidre, 4, 39, 44, 308
Hall, Stuart, 163
Hammer, Bonnie, 110
Hanes, Rosemary, 146–47
Hank Elliot (character), 294
Hansis, Van, 205, 293, 295, 298
Harding, Mark, 129n16, 129n19
Harley Cooper (character), 114
Harmon, Angie, 112
Harrington, C. Lee, 61, 100n11, 100n14, 118n9, 191–92, 195, 222, 233, 248, 250–53, 260, 264n3, 281, 296, 314n6, 314n7
Hart to Heart, 205, 216n8
Harwood, Jake, 309–10
Haslem, Wendy, 292
Hastings, Don, 6, 88, 306, 314n3
Hattie Adams (character), 288
Hauck, Charlie, 117
Havrilesky, Heather, 108
Hayman, Connie Passalaqua, 254
Hays, Kathryn, 88
Hayward, Jennifer, 14
HBO, 9, 31, 114, 130, 133, 136, 138
Healey, Joe, 203
Hennessy, Jill, 112
Henry Coleman (character), 88, 216n9

Henry Grubstick (character), 52
Hensley, Jon, 88
Herbert, James, 203
Herlie, Eileen, 47
Heroes, 109, 117n5
Herrera, Anthony, 88
Hibberd, James, 4, 56
High School Musical, 115
high-budget shows, 52–53, 56
Higley, Dena, 39
Hill Street Blues, 9, 23, 81, 117n3, 124–25, 134
Hills, The, 36, 115
Hilton, Paris, 235, 239–40, 242
Hilton, Perez, 235–36, 244, 248n2, 248n3
Hinckley, David, 297
Hinsey, Carolyn, 184, 244
Hiss, Anthony, 252
Hobson, Dorothy, 84, 141
Hogestyn, Drake, 4, 39, 41, 44
Holden Snyder (character), 88
Holly Norris (character), 185
Holly Scorpio (character), 227
Holmes, Katie, 235, 237–38
Holmes, Scott, 88–90, 92–97
Holmes, Su, 242
Holocaust, 131
Homer, 69
Hope Brady (character). *See* Hope Williams
Hope Williams (character), 286, 291
Horta, Silvio, 55
Horton family (characters), 241
How I Met Your Mother, 10
Hubbard, Elizabeth, 88, 314n3
Hudgens, Vanessa, 238
Hulu, 216n4
Hurst, Jill Lorie, 181–82
Hyatt, Wesley, 141
hybridity, 8–10, 55–56, 108–9, 126, 134, 197

I Want to Believe, 110, 236
Ibarra, Sergio, 303
IMDb (Internet Movie Database), 122, 129n11, 129n13
immersion, 8, 12, 16–17, 139, 240, 244
Internet communities, 35, 102, 201–15, 219–30, 233–48, 266–68, 272–75. *See also* blogs
intimacy, 3, 11–12, 72, 107, 159, 171, 173, 174n11, 245, 269, 309, 312

Intintoli, Michael James, 13–14, 164
Intrieri, Robert C., 302
Irwin, Barbara, 14–15, 22–28, 304, 309
Iser, Wolfgang, 286
"It's Time for My Story": Soap Opera Sources, Structure, and Response, 44
iTunes, 43, 193, 201, 203–4, 216n5

J. R. Ewing (character), 122
Jack & Bobby, 124–25
Jack Deveraux (character), 104
Jack Snyder (character), 88
Jackall, Robert, 59
Jacobs, Karre, 222
James, Meg, 21n1
James Stenbeck (character), 88, 93, 123
Janofsky, Michael, 191
Jaramillo, Deborah, 18
Jason Morgan (character), 193, 210–13, 217n24, 220
Jay-Z, 238
Jeff Baker (character), 93, 100n8
Jenkins, Henry, 109, 212, 216n1, 222–23
Jennifer Horton Deveraux (character), 104
Jenrette, Jerra, 14
Jensen, Michael, 296
Jessica Buchanan (character), 205
Jessie Brewer (character), 167–68
Jill Abbott (character), 154
Jill Foster (character). *See* Jill Abbott
Joan of Arcadia, 124
Joann Curtis (character), 293
Joe Martin (character), 21n3, 46
Johanssen, Scarlett, 242
John Abbott (character), 123
John Black (character), 39, 44, 285–86, 288–89, 291
John Crichton (character), 110–11
Johnny Ryan (character), 123
Johnson, Derek, 112–13
Johnson, Richard, 16
Johnson, Sherry, 21n7
Jolie, Angelina, 235, 237–38, 242, 245
Jonah Lockwood (character), 30
Jonathan Randall (character), 185, 216n9
Jonathan's Story, 216n9
Josh Lewis (character), 314n3
Jossip, 244

Judge Judy, 33
Julie Jamison (character), 30
Just Jared, 235, 242, 248n2
Just TV, 133

Kate Martin (character), 47
Kate Roberts (character), 288, 291
Katherine Chancellor (character), 154, 158, 293
Katie Peretti (character), 88, 205–6, 214
Katie Peretti Kasnoff (character). *See* Katie Peretti
Kaufman, Debra, 189
Kayla Brady (character), 44
Kaylin, Lucy, 241
Keifer, Elizabeth, 184
Keith Whitney (character), 30
Kelly Lee (character), 199–200n7
Kemper, David, 111
Kendall Hart (character), 45, 47, 205
Kennedy, Dana, 113
Kennedy family, 30
Kiesewetter, John, 5, 87
Kim Sullivan Hughes (character), 17, 88, 97
Kimberly Brady (character), 104
Kincaid, Jamaica, 252
Kincaid, Jason, 89, 94
Knots Landing, 23, 134, 155
Knowles, Beyoncé, 238
Kompare, Derek, 142, 148, 153n5
Kroll, Dan J., 4
Kubicek, John, 194
Kung, Michelle, 16
Kustritz, Anne, 109

L Word, The, 31
L.A. Diaries, 205, 216n10
L.A. Law, 124, 134
Labine, Claire, 123, 149
Lafayette, Jon, 196
LaGuardia, Robert, 119–20, 252
Lamb, Bill, 51
Lang, Katherine Kelly, 102
Lang, Kurt, 251, 264n5
Larmonth, Michael, 50
Larry McDermott (character), 88
Larry Wolek (character), 123
Latham, Lynn Marie, 155–56, 158

Laura Webber Spencer (character), 119–24, 129n19, 133, 170, 187, 222, 227
Lavigne, Carlen, 110
Law & Order, 112
Lawson, Mark, 54–55
Lee, Richard A., 301–2
Lena Kundera (character), 223
Lenhart, Jennifer, 75, 308
Leno, Jay, 126
Levin, Gary, 115, 303
Levine, Elana, 19, 129n22, 163, 191, 193, 200n11, 230n5
Levinsky, Mara, 45, 183
Lewis, Errol, 199
Lewis, Lisa A., 223
Lewis family (characters), 123
Library of Congress Motion Picture, Broadcasting and Recorded Sound division (MBRS) 142, 145–48, 150
Liccardo, Lynn, 17, 77n1, 77n2, 100n2, 100n9, 139n1, 159n2, 308, 314, 314n4
Lidz, Franz, 41, 129n21, 303
Lien Hughes (character), 95–96
Lifetime Television, 10–11
"Like Sands through the Hourglass," 201
Linder, Kate, 314n3
Lisa Grimaldi (character). *See* Lisa Miller
Lisa Miller (Grimaldi) (character), 6, 88–89, 91–92, 100n3, 121, 306–7
Lisotta, Christopher, 49, 57, 205
Littlejohn, David, 257
LiveJournal, 245, 249n6
Llanview, 123, 239
Logan, Michael, 46, 252, 264n3, 308, 314n4
Lohan, Lindsay, 235, 238, 242
Lopez, Jennifer, 240
Lord of the Rings, 246
Los Angeles, 4, 21, 42, 238
Los Angeles Times, 254, 297
Lost, 31, 105, 109, 117, 130, 135, 138
Lotz, Amanda, 10, 216n11
Love of Life, 130, 250
Loving, 14, 31, 44, 144
low-budget shows, 51–53, 198
Lowell, Carey, 112
Lucas, Michael, 110
Lucas Jones (character), 294
Lucas Roberts (character), 276, 287–88

Lucci, Susan, 6, 41, 44, 308
Lucinda Walsh (character), 88, 123, 176, 224, 310–11, 314n4
Luke Snyder (character), 19, 205, 293–99
Luke Spencer (character), 119–24, 129n19, 133, 187, 222, 227
LukeVanFan, 294, 295, 299n2
Lulu Spencer (character), 205
Lumet, Sidney, 166
Lynn Carson (character), 293–94

MacDonnell, Ray, 21n3
Maeve Ryan (character), 123
Maggie Horton (character), 290, 314n3
Man from Oakdale, The, 216n9
Manhattanites, 293
Marc, David, 11
Mares, Marie-Louise, 302
Margo Hughes (character). *See* Margo Montgomery
Margo Montgomery (Hughes) (character), 88, 93–97, 121
Mariah Maximilliana "Maxie" Jones (character), 200n8
Marland, Douglas, 31–32, 72–73, 94–96, 100n11, 100n12, 121, 127, 129n19, 276–77, 306
Marlena Evans (character), 39, 44, 276, 286, 288–89, 291
Martin, Denise, 111
Martin, Ed, 198, 217n20, 217n21
Martinez, J. R., 46
Marx, Gregg, 89, 94
Mary Hartman, Mary Hartman, 134
Masters, Marie, 88
Matelski, Marilyn, 14
Mayer, John, 245
Mayer, Paul Alva, 123
McCay, Peggy, 307, 314n3
McClure, David, 220
McCook, John, 102, 306–7, 314n3
McCullough, Kimberly, 199n5
McIntosh, Sherrie, 14
McNeil, Alex, 144
meaning, 13–15, 18, 86, 113, 116, 131, 139n1, 172, 176, 194, 208–9, 212–13, 235, 241, 246–47, 250, 280, 285–87, 290–91, 313
Media Domain, 127, 129n18, 264n8
Meg Snyder (character), 176–78

MEGAUPLOAD, 220, 230n2
Melrose Place, 108
Menighan Hensley, Kelly, 88
Meet the Press, 4
Metro, 279
Metzler, J. A., 18, 43n1, 129n17, 186n3
Mickey Horton (character), 291
Mike Karr (character), 30
Miller, Laura, 111
Minshaw, Alicia, 45
MIT CMS: The American Soap Opera, 163
MIT OpenCourseWare, 163
Mittell, Jason, 8–9, 18, 133–38
mobile devices, 25, 35, 193
Modleski, Tania, 13, 54, 84, 141
Moloney, Molly, 251
Mona Kane (character), 47
Monica Quartermaine (character), 105
Monjardim, Jayme, 55
Montel Williams Show, The, 33
Monty, Gloria, 34, 119–21, 126, 128n2, 149
Moonlighting, 113
Moonves, Leslie, 4
Moore, Julianne, 17
Moran, Jonathon, 282
Moss, Ronn, 102
Motion Picture, Broadcasting and Recorded Sound (MBRS). *See* Library of Congress
Mr. Big (character), 94, 121
MSN, 268, 271n2
MTV, 10, 36, 115
multigenerationality, 6, 17, 19, 88, 98–99, 125, 311–12
Mumford, Laura Stempel, 10–11, 13, 106, 113, 247
Mundorf, Norbert, 309
Museum of Television and Radio. *See* Paley Center for Media
MyNetworkTV (MNTV), 49–51, 56, 57n2
Myrtle Fargate (character), 47
MySpace, 205, 276

Nancy Hughes (character), 6, 88, 100n3, 121, 310, 312
Nancy Pollock (character), 30
narrative: cumulative, 23, 117n2; discontinuities, 214, 279–91. *See* storytelling
Nasser, Jaime, 15, 153n3

Natalia Rivera (character), 186
National Academy of Television Arts and Sciences (NATAS), 25, 254
National Football League (NFL), 117n6
NBC (National Broadcasting Company), 4, 8–10, 14–15, 23–25, 28, 31–32, 34, 39–41, 43–44, 58, 61, 64, 68, 70, 81, 104–5, 108–9, 112–14, 117n3, 121, 124–26, 133–35, 138, 144, 147, 152, 181–82, 187–88, 192, 203–5, 216n4, 216n5, 229, 233, 256, 275–76, 279, 281, 283, 293, 300–1, 305, 307–8, 314n1
Neighbours, 279
New York City, 4, 42, 52, 142, 173n2, 181–82, 235, 238, 254
New York Daily News, 297
New York Times, 4, 7, 52–53
New York Post, 239, 252
New Yorker, 252
Newcomb, Horace, 13, 17, 81–82, 107, 163–64, 174n11
Newcomb, Roger, 19, 186n1, 192, 204, 222, 230
Newman, Robert, 314n3
News Corporation, 49, 57
Ng, Eve, 223, 230n4
Ngo, Bob, 251
Nichols, Stephen, 39, 44
Nicole Walker (character), 286
Nielsen, 4, 43, 50, 76, 117n4, 226, 303
Nikolas Cassadine (character), 170–72
Nine Network, 19, 279–91
Nixon, Agnes, 15, 26, 44–48, 72, 123, 149, 153n14, 264n6, 276
Noah Drake (character), 227
Noah Mayer (character), 19, 293–99
Nochimson, Martha, 7, 106
Noji, Minae, 199–200n7
Northern Exposure, 113
nostalgia, 77, 138, 159, 196–97, 265
Nuke Fancast, The, 299n7, 299n8
NYPD Blue, 108, 124

Oakdale, 88–89, 91–98, 121, 123, 178, 205–6, 265
Oakdale Confidential, 205–6, 216n9
O.C., The, 57n3, 135
Ocasio, William, 59–60
Office, The, 54, 216n2
Olivia Spencer (character), 186
O'Meara, Radha, 19

INDEX 339

One Life to Live (OLTL), 5, 15, 28, 44, 61, 64, 104, 106, 114–15, 123, 127–28, 153n14, 157, 186, 190n1, 204–5, 219, 256, 275, 294, 314n4
Online Fandom, 104
Only Fools & Horses, 54
"Oral History of a Piece of Americana, An," 22
Ouellette, Laurie, 116
Our Private World, 107, 199n1
Owen, Rob, 111

Paley Center for Media, 142, 147–48, 153n12, 153n8, 153n13, 173n2
Palin, Sarah, 243
Palmer Courtland (character), 123
Papini, Dennis R., 302
Park, Michael, 88
Passanante, Jean, 91, 126, 294
Passions, 15, 25, 29, 39–40, 114, 116, 147, 152, 182, 203–5, 216n5, 279
Patrick Drake (character), 193–94, 199n5, 200n8, 207–10, 213–14, 216n15, 217n28, 217–18n32
Pawnbroker, The, 166
PBS (Public Broadcasting Service), 116
Peapack, New Jersey, 182, 184
Penghlis, Thaao, 39
Penn, Sean, 242
Penny Hughes (character), 93
People, 235, 248n1, 257
Peyton Place, 107, 129n10, 134
Phelps, Jill Farren, 16, 194
Philip Kiriakis (character), 286
Phillip Brent (character), 46
Phillip Spaulding (character), 186
Phillips, Irna, 26, 44, 87, 100n12, 149, 306
Phoebe Tyler (character), 46–47
Pine Valley, 123, 238, 270
Pingree, Susan, 13, 60
Pitt, Brad, 235, 237–38, 241, 245
play, 135, 176, 241, 248; active spectatorship, 302; viewing games, 81–82; playful spectatorship, 248; puzzle narrative, 8–9, 109
Playhouse 90, 168
pleasure, 8, 10, 17, 19, 56, 122, 131, 136–37, 139, 223, 234, 247–48, 281, 283–84, 286–87, 292; in blurring reality/fantasy, 234; in complexity, 195; in watching, 81–82, 242–43

podcasts, 25, 204, 206
Poll, Julie, 90
popular culture, 4, 6, 8, 14, 22, 29, 84, 131, 250–51, 253, 260, 263, 264
Popular Culture Association, 14, 22
Port Charles, 120, 165, 174n8, 211, 241
Port Charles, 14, 114, 199
Port Charles Online, 219, 266
Porter, Julie, 19
Portfolio, 41
Portia Faces Life, 44
Powell, Walter, 69
Pratt, Charles, 44–47
Pritchard, William H., 122
Procter & Gamble, 4, 5, 15, 23, 39, 68, 87, 148, 152, 153n13, 180, 203–4, 216n4, 216n6, 293, 295
production: consumer and producer conflict, 6, 77, 280, 283; new models, 25, 180–81, 183–85, 187–90; text/paratext division, 246
Project Daytime, 14–15, 22, 26–28
Project Holiday Spirit, 298, 299n5
Project Runway, 10
Puff Daddy, 239–40

Quaid, Dennis, 240
quality, in art, 15–17, 43, 52–53, 62–63, 66–67, 77, 87, 124, 129n16, 141
Quartermaine family (characters), 241
QueenEve, 19, 248, 272–74
Queer as Folk, 31

radio, 4, 6, 30, 44, 51, 56, 87, 130, 141–42, 147–49, 180–86, 192, 199n3, 206, 251, 266, 293; Golden Age, 266
Radway, Janice, 84
Rao, Sri, 196–98, 200n12
Real World, The, 10
Reardon, Marguerite, 189
Rebecca Shaw (character), 170
Red Room, 119
Redmond, Sean, 242
Reese Williams (character), 45–47, 294
Regev, Motti, 60
Reid, Frances, 6
Reilly, James E., 39, 76, 283
Requiem for a Heavyweight, 168
rerun, 5, 10, 24, 41, 108, 117n5, 140, 152, 153n2, 204

Rerun Nation, 142
Restlessstyle.com, 205
Reva Shane (character), 114
Reynolds, Mike, 200n12
"Rhetoric of the Camera in Television Soap Opera, The," 163
Rhianna, 243
Rice, Lynette, 4
Rick Bauer (character), 114
Ridge Forrester (character), 102
Riegel, Eden, 45–46
Riggs, Karen E., 301–2
Riley Morgan (character), 97
Ritchie, Nicole, 235, 242
ritual, 5, 11–12, 24, 137, 151
Robert Scorpio (character), 34, 197, 227
Robertson, Roland, 284
Robin's Daily Dose, 202, 205–7, 213–14, 216n2, 217n28
Robinson, Julie Clark, 199
Rogers, Suzanne, 306, 314n3
Rogers, Tristan, 15, 34–37
Rohter, Larry, 53, 55
Roman Brady (character), 288, 291
Rosa Salvaje, 49
Rosen, Christine, 270
Rosen, Robert, 150
Rosenberg, Howard, 124
Rosenbloom, Ralph, 166
Ross, Sharon Marie, 7–9
Ross Marler (character), 180
Rounds, David, 60
Rourke, Mickey, 242
Roush, Matt, 111, 125
Rowe, Kathleen, 153
Russell, Cheryl, 301–2
Ruth Martin (character), 46
Ryan, Meg, 239–40
Ryan Lavery (character), 45
Ryan's Hope, 44, 123, 156

Salem, 241, 276, 280–81, 283, 285, 287, 289–90
Salon, 111
Sam McCall (character), 212
Samantha Evans (character), 288
Samantha "Sami" Brady (character), 276, 287–88
Samuels, David, 245, 252

Santa Barbara, 14, 31–32, 40, 42, 191
Scardaville, Melissa, 16, 100n2, 100n11, 254, 264n4, 314, 314n2
Scare Tactics, 111
Schemering, Christopher, 192, 199n2
Schiller, Abbie, 206–7
Schilling, Mary Kay, 110
Schlossberg, Suzanne, 112
Schnetzer, Steve, 307, 314n3
Schoenherr, Steven E., 266
Schrute-Space, 216n2
Sci Fi Channel (SyFy), 109–11
Scodari, Christine, 14, 17, 77n2, 100n2, 309, 314n1
Sconce, Jeffrey, 117
Scorpio, Robin, 19, 105, 193–95, 197, 199n5, 200n8, 200n11, 202, 205–15, 216n15, 216n17, 217n24, 217n28, 217n32, 227
Scott, Ed, 39
Scott, Melody Thomas, 41
Scott Baldwin (character), 227
Screwball Television: Gilmore Girls, 279
Search for Tomorrow, 8, 31, 44, 130
Seganti, Paolo, 88
Seiter, Ellen, 13, 54, 57n3, 140–41, 193, 285
Serial Monogamy, 106
seriality, 3–4, 6–9, 12, 14, 17–19, 81, 84–86, 103, 106–12, 116, 126, 131, 133–39
SFX Magazine, 111
Shane Donovan (character), 104
Shanks, Bob, 164
Shetter, Hogan, 39
Shield, The, 9, 130
Showtime, 31
Sidler, Lauren, 282
Silbermann, Jake, 293, 298
Silverman, Ben, 52
Simon, Ron, 148
Simpson, O. J., murder trial, 17, 23, 67, 104–5, 113, 125, 129n19
Simpsons, The, 117n6, 138
Six Feet Under, 133
Skill, Thomas, 13, 22
Slesar, Henry, 30, 264n6
Slezak, Erika, 44
Smith, Hillary Bailey, 94
Smith, Jack, 155
Smith, Jackie, 32

INDEX 341

Smith, Lynn, 297
Snark Weighs In, 125, 206
Soap, 134
Soap Central, 100n10, 264n8
Soap Dispenser, The, 264n8
Soap Fans, 250, 300
Soap Opera Digest, 45–47, 58–59, 75, 113, 136, 185, 194, 196–97, 244, 252, 254, 314n4
Soap Opera Encyclopedia, The, 199n2
Soap Opera Network, 243, 249n5
Soap Opera Now, 254
Soap Opera Rapid Aging Syndrome (SORAS). *See* aging
"Soap Opera Survival Tactics," 140
Soap Opera Weekly, 119, 184–85, 194, 244, 252, 254, 261–62, 314n4
SoapCity, 202–3, 216n3
Soapdom, 264n8
SOAPnet, 5, 18, 24, 35, 40–41, 43, 115, 128, 152, 153n2, 188, 192–93, 196–98, 204, 214, 217n28, 217n32, 261–62
Soapnetic, 193
Soaps of Our Lives, 265
Soaps.com, 199–200n7, 264
SoapZone, 219–21, 224–25, 230, 230n1, 243, 249n5, 264n8, 266, 269, 271n1, 272
Sonny Corinthos (character), 210–11, 213, 266, 273
Sony Pictures Television, 153n13
Sopranos, The, 9, 114, 130, 138
Speaking of Soap Operas, 13, 84–85
Spears, Britney, 235, 238
Spehr, Paul, 145–46
Spence, Louise, 7, 14, 17, 130–32, 199n6, 243, 281
Split Reflections, 205, 216n8
Springfield, 114, 123, 182, 239
St. Elsewhere, 124, 134, 136
Star Trek, 8, 246
Star Wars, 246
stardom, 256, 308; gossip, 235–48
Stargate SG-1, 111
Stefano DiMera (character), 276, 286
Stein, Joshua David, 16
Steinberg, Brian, 192
Stelter, Brian, 5
Stephanie Forrester (character), 102, 186
Stephen "Patch" Johnson (character), 44, 291
Sterling's Magazines, 264n2

Steve Hardy (character), 165, 167–72, 173n4
storytelling, 3, 6–9, 12, 15–17, 20, 24–25, 62–67, 77, 83, 85, 109, 114, 131, 136, 182–83; narrative complexity, 133–35, 194–95, 196–97, 199; formal aspects, 176–79; immersive worlds, 240; structure, 175; subtext, 32; transmedia, 15, 205
Strauss, Anselm L., 305
Streeter, Leslie Gray, 216–17
Suarez family (characters), 52
Sunset Beach, 14, 40
Suratt, Samuel, 150
Survivor, 10, 103, 112
Susan Stewart (character), 88, 177, 310
Susman, Linda, 254
syndication, 108, 140

Tabloid Truth, 205
Tad Martin (character), 46–47
talk shows, 23, 70, 85, 131, 144, 163, 224
talk!talk!, 265
Tammi, Tom, 89, 93
Taylor, Elizabeth, 119
Taylor Thompson (character), 46
technology, 11, 25, 33, 164, 172, 186, 189, 192, 220, 226, 284; digital cameras, 180–81; high definition broadcast, 187; screen size, 189
Tedeschi, Bob, 301–2
Telegraph, The, 53–54
TeleNext, 5, 39, 43, 87–88, 148, 153n13, 180, 182–83, 186, 295, 297
telenovela, 15–16, 49–52, 55–56, 57n1, 85, 114, 163
television, 85; daytime, 23–24, 67, 69–70, 85, 103, 113, 119–28, 133–39, 191–200; live, 29; primetime, 23–24, 69, 81, 86, 103, 107–8, 112, 119–28, 133–39, 191–200; reality, 23, 35; streaming, 11–12, 201
Television and American Culture, 133
Television Critics Association, 264n5
Television Without Pity, 8
Television Week, 50–51, 56
Tess Buchanan (character), 205
Texas, 31
thirtysomething, 11, 134
Thomas, Becca, 217
Thompson, Jason, 194, 199n5

Thompson, Robert, 108, 112, 117n3, 123–24, 194, 199n5
Thornton, Patricia, 59–60
Thousand Other Worlds, A, 180
Thurber, James, 251–52
TIIC (The Idiots in Charge), 58, 201–18, 227
Timberg, Bernard, 13, 18, 100, 174n6, 186, 200
Time, 253, 297
To Be Continued: Soap Operas around the World, 83, 85
Tom Hughes (character), 17, 86–100, 121, 129n8, 311, 314n9
Tomlin, Gary, 39
Tonight Show, The, 126
Tony DiMera (character), 276, 286, 288–89
Top Chef, 10
Torchin, Mimi, 185, 254, 305, 308, 314n4
"Toward a Paradigm for Media Production Research," 201
TPTB (The Powers That Be), 36, 120, 125, 127, 219, 296–98
Transformative Works and Cultures, 86
Tuchman, Gaye, 164
Tulloch, John, 11, 302
Tune In, Log On, 104
Turow, Joseph, 108, 112, 116, 301, 303–4, 313
TV: The Most Popular Art, 81
TV Guide, 111, 125, 252, 254, 314n4
TV Week, 282, 289, 292n5
Twentieth Television, 57n2
24, 109, 114, 130, 133
Twilight Zone, The, 167–68, 172
Twin Peaks, 134

UCLA Film & TV Archive, 142–48, 150, 153n4, 153n6, 153n9
Ugly Betty, 49–57
Umstead, R. Thomas, 193
unions, 38, 42–43, 47, 68, 155–57, 212–13
United States Copyright Office, 142, 145
UPN (United Paramount Network), 31, 49, 57n2, 111, 133
Us Weekly, 235, 248n1
USA Today, 117n6
Usenet Newsgroup, 110

Vara, Vauhini, 203
Vaughn, Vince, 245

Variety, 50
verDorn, Jerry, 180
Verizon, 193
Victor Kiriakis (character), 286
Victor Newman (character), 123, 326
Victoria "Viki" Lord (character), 44
video, 3, 8, 19, 85, 220. *See also* television
Vietnam War, 46, 91–92, 95, 98
Viewers Like You?, 116
Vogue, 54–55

Wagner, Helen, 6, 88, 121, 312
Waldman, Allison J., 303
Wallowing in Sex, 201
Walton, Jess, 41
Walton, Kip, 165
Warrick, Ruth, 46
Watching Daytime Soap Operas, 130–31
WB (Warner Brothers) Television Network, 31, 49, 57, 108, 111, 124, 133
We Love Soaps, 293, 296, 299n4
Webisodes, 25, 35, 202, 205
Weiss, Joanna, 296–97
Wesch, Michael, 221
West, Abby, 196
West, Maura, 88
West Wing, The, 133
Wheeler, Ellen, 181–83, 186, 188
White, Mimi, 290
"Who Really Watches Soap Operas," 119
Widdoes, Kathleen, 88
William "Will" Horton (character), 288
Williams, Carol Traynor, 14–15, 153n14
Williams, Garry, 285
Wilson, Benji, 53, 57n5
Wilson, Mary Jeanne, 18, 53, 57n3, 100n6, 193
Winfrey, Oprah, 243
Winterberger, Suzanne, 14
Wire, The, 130, 136, 138
Witherspoon, Reese, 238
Wittebols, James H., 107–8, 112, 116, 303
WMYO (television station), 51
Wolfe, David B., 301, 313
Wood, Michael, 132
Woodard, Emory H., IV, 302
wrestling, 12, 108
Writers Guild of America (WGA), 18, 31,

38–39, 43n1, 44, 62, 67, 101, 126, 129n17, 155–59, 184, 186, 207, 216n7, 276–78
Wynter, Karen, 297

X-Files, 109, 114, 117

Yo Soy Betty La Fea, 50, 52–53, 55
Young and the Restless, The (*Y&R*), 4–5, 18, 21–22, 24, 38, 41–43, 101, 120, 123, 129n16, 153n13, 154–59, 175–76, 189, 202–5, 256, 275, 277, 292n1, 293, 303, 314n3, 314n4
YouSendIt, 220, 230n2
YouTube, 19, 25, 90, 100n5, 152, 153n17, 163, 173n5, 174n7, 174n8, 219–30, 277, 294–95, 299n2

Zeman, Jacklyn, 314n3
Zenk-Pinter, Colleen, 88
Zucker, Jeff, 34, 39

www.ingramcontent.com/pod-product-compliance
Lightning Source LLC
Chambersburg PA
CBHW050332230426
43663CB00010B/1825